Best Explanations

Best Explanations

New Essays on Inference to the Best Explanation

EDITED BY
Kevin McCain
and Ted Poston

OXFORD
UNIVERSITY PRESS

OXFORD
UNIVERSITY PRESS

Great Clarendon Street, Oxford, OX2 6DP,
United Kingdom

Oxford University Press is a department of the University of Oxford.
It furthers the University's objective of excellence in research, scholarship,
and education by publishing worldwide. Oxford is a registered trade mark of
Oxford University Press in the UK and in certain other countries

The moral rights of the authors have been asserted

First Edition published in 2017
Impression: 2

Published in the United States of America by Oxford University Press
198 Madison Avenue, New York, NY 10016, United States of America

British Library Cataloguing in Publication Data
Data available

Library of Congress Control Number: 2017944922

ISBN 978-0-19-874690-4

Printed and bound by
CPI Group (UK) Ltd, Croydon, CR0 4YY

For our better halves, Molly and Christine

Contents

Part IV. Inference to the Best Explanation and Skepticism

Part V. Applications of Inference to the Best Explanation

List of Figures and Table

Figures:

Table:

List of Contributors

JAMES R. BEEBE, State University of New York at Buffalo

ALEXANDER BIRD, University of Bristol

J. ADAM CARTER, University of Glasgow

IGOR DOUVEN, French National Centre for Scientific Research, Institut des Sciences humaines et sociales

ELIZABETH FRICKER, University of Oxford

RICHARD FUMERTON, University of Iowa

ALI HASAN, University of Iowa

LEAH HENDERSON, University of Groningen

KAREEM KHALIFA, Middlebury College

KEVIN McCAIN, University of Alabama at Birmingham

TIMOTHY McGREW, Western Michigan University

JARED MILLSON, Agnes Scott College

CHERYL MISAK, University of Toronto

TED POSTON, University of South Alabama

DUNCAN PRITCHARD, University of Edinburgh

SUSANNA RINARD, Harvard University

MARK RISJORD, Emory University

WILLIAM ROCHE, Texas Christian University

JONAH N. SCHUPBACH, University of Utah

RUTH WEINTRAUB, Tel Aviv University

1

Best Explanations

An Introduction

Kevin McCain and Ted Poston

Explanatory reasoning is quite common. Not only are rigorous inferences to the best explanation (IBE) used pervasively in the sciences, explanatory reasoning is virtually ubiquitous in everyday life. It is not a stretch to say that we implement explanatory reasoning in a way that is "so routine and automatic that it easily goes unnoticed" (Douven, 2011). IBE is used to help increase crop yields by producing accurate agricultural models (Gauch, 1992, 2012). It is widely used by health professionals—some even claim that explanatory reasoning is the primary method of medical diagnosis (Josephson & Josephson, 1994). Many philosophers argue explanatory reasoning plays a key role in the epistemology of testimony because it is by way of such reasoning that we decide whether or not to trust what we are told (Adler, 1994; Fricker, 1994; Harman, 1965; Lipton, 1998). Some maintain that explanatory reasoning is at the heart of all epistemic justification (Lycan, 1988; McCain, 2014, 2016; Poston, 2014). It may even be that we cannot comprehend language without employing IBE (Dascal, 1979; Hobbs, 2004).

Despite its widespread use in science and everyday life, the nature and use of IBE are hotly contested. The chapters collected in this volume represent the cutting edge of research on a variety of issues concerning IBE. Here we have the most recent work of prominent epistemologists and philosophers of science incorporating both traditional and formal approaches to a number of issues. The chapters in this volume run the gamut from utilizing IBE to respond to classic philosophical problems to addressing particular issues concerning IBE itself to exploring new applications of and problems for IBE. Let us take a brief look at the contents to follow.

The three chapters in Part I are focused on understanding the nature of IBE. Igor Douven gets things started by offering an explication of how we should understand IBE (Chapter 2). After making the nature of IBE more precise Douven argues that statistical experiments show that IBE performs quite well when compared to other probabilistic rules for updating credences—notably, Bayesian updating. Next, Cheryl Misak offers insights into what two great pragmatists, Peirce and Ramsey, thought about IBE (Chapter 3). In the final chapter of this section, Jonah Schupbach (Chapter 4) proposes to "clean up" IBE by construing what it takes to be the best explanation in terms of explanatory power and giving a formal explication of explanatory power.

The question of whether IBE is a fundamental form of reasoning is the focus of the chapters in Part II of this volume. IBE's claim to being a fundamental source of rational belief is challenged in different ways in the first three chapters of this part. Richard Fumerton (Chapter 5) begins the attack on IBE's fundamentality by arguing that IBE cannot be the ultimate ground of rational belief because, insofar as it is successful, IBE is parasitic on other grounds for rational belief such as direct acquaintance and induction. Kareem Khalifa, Jared Millson, and Mark Risjord take a different approach to attacking IBE's fundamentality in their chapter (Chapter 6). They argue that explanatory pluralism is widespread in philosophy of science, and pluralism of this variety undermines the claim that IBE is a fundamental source of justification. While Alexander Bird's discussion is perhaps less critical of IBE than the previous two, he still contends that IBE is not a fundamental method of reasoning (Chapter 7). He argues that IBE is useful and produces rational beliefs, but only because it is a good heuristic for appropriate Bayesian reasoning. In the final chapter of this part of the volume Kevin McCain and Ted Poston argue that explanatory reasoning is evidentially relevant in and of itself (Chapter 8). In other words McCain and Poston argue in favor of understanding IBE as a fundamental source for rational beliefs.

Part III is composed of two chapters attempting to justify IBE. J. Adam Carter and Duncan Pritchard consider a variety of ways that an argument might be rule-circular (Chapter 9). They argue that while IBE can be defended in several rule-circular fashions, some rule-circular defenses of IBE are better than others. In the other chapter in this part of the volume Ali Hasan offers a rationalist defense of IBE (Chapter 10). By arguing that we can have a priori access to probabilities, Hasan defends the claim that IBE can be justified a priori by way of rational reflection.

Skepticism is the theme for the chapters in Part IV. This part of the volume begins with James Beebe's case for thinking that skeptical arguments presuppose explanationism, i.e. that explanatory reasoning is justifying (Chapter 11). As a result, Beebe maintains that explanationist principles provide the best way to understand skeptical arguments, and one cannot at the same time doubt explanationism while granting the force of skeptical challenges. In the subsequent chapter Ruth Weintraub compares skepticism about IBE to the more familiar inductive skepticism (Chapter 12). Although there are many parallels, Weintraub argues that skepticism about IBE and skepticism about induction are distinct in important ways. In light of these differences, Weintraub claims that IBE seems to be more threatened by skeptical challenges. In the final chapter of this part of the book Susanna Rinard attacks the viability of IBE responses to skepticism (Chapter 13). Ultimately, Rinard not only argues that such responses are problematic, but also that explanatory virtues are not linked to truth.

Finally, the chapters in Part V explore applications of IBE. In the first chapter of this part of the volume William Roche considers how appealing to explanatory reasoning may help address Hempel's paradox (Chapter 14). Roche argues that in the end explanatory reasoning does not offer a solution to this paradox, rather the way to solve this paradox is to appeal to a more straightforwardly Bayesian notion of screening off. In the chapter that follows Timothy McGrew seeks to put IBE to work in resolving a problem that Bayesianism has with a plausible constraint on setting prior probabilities (Chapter 15). McGrew argues that IBE offers a reasonable way to support this constraint, Cromwell's Rule, which is compatible with Bayesianism. Completing the essays focused on IBE and Bayesian themes, Leah Henderson argues that biologist George C. Williams' attack on group selection is best understood in terms of an emergent compatibilist account of IBE and Bayesianism (Chapter 16). On Henderson's model considerations operative in IBE emerge from an independently plausible system of Bayesian reasoning. The final chapter of this part of the volume, and of the volume as a whole, concerns the epistemology of testimony (Chapter 17). In this chapter Elizabeth Fricker argues against the idea that testimonial justification is fundamental. Instead, Fricker argues that testimonial justification is grounded in explanatory reasoning.

The chapters in this volume represent the latest thinking on IBE and related themes. Hopefully, they will not only contribute to the resurgence of interest in explanationism in general and IBE in particular directly, but they will also spur additional research in this fruitful domain.[1]

References

Adler, J. (1994). Testimony, trust, knowing. *Journal of Philosophy*, 91, 264–75.

Dascal, M. (1979). Conversational relevance. In A. Margalit (Ed.), *Meaning and use* (pp. 153–74). Dordrecht: Kluwer.

Douven, I. (2011). Abduction. In E.N. Zalta (Ed.), *The Stanford encyclopedia of philosophy* (Spring 2011 Edition), URL: <http://plato.stanford.edu/archives/spr2011/entries/abduction/>.

Fricker, E. (1994). Against gullibility. In B.K. Matilal & A. Chakrabarti (Eds), *Knowing from words* (pp. 125–61). Dordrecht: Kluwer.

Gauch, Jr., H.G. (1992). *Statistical analysis of regional yield trials: AMMI analysis of factorial designs.* New York: Elsevier.

Gauch, Jr., H.G. (2012). *Scientific method in brief.* Cambridge: Cambridge University Press.

Harman, G. (1965). The inference to the best explanation. *Philosophical Review*, 74, 88–95.

Hobbs, J.R. (2004). Abduction in natural language understanding. In L. Horn & G. Ward (Eds), *The Handbook of Pragmatics* (pp. 724–41). Oxford: Blackwell.

Josephson, J.R. & Josephson, G.S. (Eds) (1994). *Abductive inference.* Cambridge: Cambridge University Press.

Lipton, P. (1998). The epistemology of testimony. *Studies in History and Philosophy of Science*, 29, 1–31.

Lycan, W. (1988). *Judgement and justification.* Cambridge: Cambridge University Press.

McCain, K. (2014). *Evidentialism and epistemic justification.* New York: Routledge.

McCain, K. (2016). *The nature of scientific knowledge: An explanatory approach.* Cham, Switzerland: Springer.

Poston, T. (2014). *Reason and explanation: A defense of explanatory coherentism.* New York: Palgrave-Macmillan.

[1] Thanks to Olaf Dammann for helpful comments on the proofs for this volume.

PART I

Inference to the Best Explanation

2

Inference to the Best Explanation
What Is It? And Why Should We Care?

Igor Douven

It is almost an understatement to say that Bayesianism dominates current theorizing about how evidence ought to impact the confidence we invest in our hypotheses. Indeed, the overwhelming popularity of Bayesianism makes it easy to forget that not so long ago the so-called Inference to the Best Explanation (IBE) was much more central to our thinking about the relationship between evidence and hypotheses than Bayes' rule, the eponymous rule at the heart of Bayesianism. According to IBE, explanatory considerations have confirmation-theoretic import, meaning that hypotheses are to be assessed at least partly on the basis of their explanatory virtues. As Bayesians and others have pointed out, however, in spite of its erstwhile prominence, it is difficult to find a statement of IBE that is more specific than the slogan-like characterization of it in the previous sentence. And that alone may be a reason to prefer Bayes' rule over IBE: rather than a slogan, Bayes' rule is a mathematically precise formula.

It would thus seem that those of us who are still attracted to the idea that explanation has a role to play in confirmation theory—a role that cannot be completely captured by Bayesian means—will have to provide a formulation of IBE that is at least nearly as precise as the formulation of Bayes' rule. But even supposing we are able to answer to everyone's satisfaction the question of what IBE is, Bayesians will probably still not care, for lack of precision appears to be the least of their misgivings about IBE. In their view, IBE—in whichever precise version that does not boil down to Bayes' rule—can lead us to change our confidence in hypotheses in ways that are to be condemned as irrational. So, why bother about IBE in the first place?

One answer is: because IBE might be descriptively correct—people might, as a matter of fact, respond to the receipt of new evidence as prescribed by (some version of) IBE—even if they are wrong to do so. That would be a valid reason to study IBE (even if perhaps more for psychologists than for philosophers). A second answer is that the

Bayesian arguments aiming to show that IBE, however exactly spelled out, is irrational may not be quite as airtight as Bayesians like to believe.

I start by considering the question of what IBE is, suggesting that it is best thought of as a slogan after all, but that this is not necessarily a drawback. Next, I discuss some recent evidence supporting the claim that explanatory considerations do come into play when people assess a hypothesis or set of hypotheses, and that those considerations come into play in an essentially non-Bayesian way. This discussion will then help to show what is wrong with Bayesian arguments against IBE as a normative principle, and will also help to drive home the point that we should think of IBE as a slogan.

1. What Is Inference to the Best Explanation?

Here is a more or less random sampling of purported statements of IBE that one finds in the literature:

In making this inference [i.e., in inferring to the best explanation] one infers, from the fact that a certain hypothesis would explain the evidence, to the truth of that hypothesis. In general, there will be several hypotheses which might explain the evidence, so one must be able to reject all such alternative hypotheses before one is warranted in making the inference. Thus one infers, from the premise that a given hypothesis would provide a "better" explanation for the evidence than would any other hypothesis, to the conclusion that the given hypothesis is true.   (Harman [1965:89])

It is reasonable to accept a satisfactory explanation of any fact, which is the best available explanation of that fact, as true.   (Musgrave [1988:239]; italics omitted)

Inference to the best explanation is the procedure of choosing the hypothesis or theory that best explains the available data.   (Vogel [1998])

IBE authorises the acceptance of a hypothesis H, on the basis that it is the best explanation of the evidence.   (Psillos [2004:83]; italics omitted)

[W]hen one is choosing between competing candidates for belief A and B, one has good reason to accept A rather than B if A provides a better explanation of a relevant body of facts than B does.   (Vogel [2005:74f])

As other commentators have remarked, these and other formulations of IBE raise several questions of clarification, most notably, the question of how to determine which of a number of competing explanations is best. And should we really infer to the best explanation no matter how poorly it might explain the relevant facts, as some of the formulations seem to imply? This apparent problem is avoided by Musgrave's version, which requires that the explanation be satisfactory to begin with (see in the same vein Lipton [1993]), but then Musgrave is silent on what it takes for an explanation to qualify as satisfactory.

Authors have also noted a peculiar asymmetry in most of these and other formulations of IBE, in that they license an inference to an *absolute* verdict—that a given hypothesis is *true*—from what will typically only be a *relative* judgment, namely, that the hypothesis is the best explanation *among those on the table*.[1] What justification could there be for that? The asymmetry problem does not arise for Vogel's 2005 formulation, in which IBE involves the act of accepting one hypothesis rather than another, and not the act of accepting a hypothesis tout court, and Kuipers [1992] proffers a version of IBE that circumvents this problem by licensing acceptance of the best explanation not as true but merely as being closer to the truth than any of the other hypotheses under consideration. This version of IBE might be deemed somewhat uninformative, however, because it gives no indication of how close to the truth the best explanation is. Similarly, Vogel's 2005 version says nothing about when it is permissible to accept a hypothesis tout court on the grounds of its explanatory goodness.

As already intimated, an even bigger problem facing IBE may be that it has none of the precision—not even remotely—of its current main contender, to wit, Bayes' rule. Bayesians assume that people assign credences, or degrees of belief, to the propositions expressible in their language. Bayesians further hold that for a person to be rational at a given point in time, his or her credences must obey the probability axioms at that time. They further hold that a new piece of information E is to impact one's credences via Bayes' rule, which it does precisely if, for all propositions H expressible in one's language,

$$\mathrm{Pr}_E(H) = \mathrm{Pr}(H|E) = \frac{\mathrm{Pr}(H)\mathrm{Pr}(E|H)}{\mathrm{Pr}(E)}.$$

Here, $\mathrm{Pr}(\cdot)$ is the credence function at the time of receipt of E and $\mathrm{Pr}_E(\cdot)$ the credence function right after the learning of E (and no stronger proposition); $\mathrm{Pr}(E)$ and $\mathrm{Pr}(H)$ are the prior degrees of belief in E and H, respectively, and $\mathrm{Pr}(H|E)$ is often referred to as "the likelihood of E on H" (and sometimes also, mistakenly, as "the likelihood of H on E"). We see that Bayes' rule appears as a crystal clear, simple mathematical formula, which is in stark contrast with the formulations of IBE given previously.

However, as van Fraassen [1989, Ch. 6] has shown in the context of his critique of IBE, it is not so difficult to provide a formulation of IBE that is as precise as Bayes' rule. In his formulation, IBE is a probabilistic update rule that has Bayes' rule as a limiting case. To state the rule, let $\{H_i\}_{i \leq n}$ be a set of self-consistent, mutually exclusive, and jointly

---

[1] To be sure, Harman's formulation requires that one be able to dismiss all possible alternative explanations, but if the use of IBE is as pervasive as he believes, both in science and in everyday life, then "all" will have to be read as being somehow limited in scope, most probably as ranging only over the conceived alternatives.

exhaustive hypotheses, and let $\Pr(\cdot)$ again be one's credence function prior to learning E. One's new credence in H_i immediately after learning E (and nothing stronger)—which we shall write as $\Pr[E](H_i)$—then comes from $\Pr(\cdot)$ via van Fraassen's rule precisely if

$$\Pr[E](H_i) = \frac{\Pr(H_i)\Pr(E|H_i) + f(H_i,E)}{\sum_{j=1}^{n}(\Pr(H_j)\Pr(E|H_j) + f(H_j,E))}.$$

In van Fraassen's version, the function f assigns a bonus point to the hypothesis that explains E best (given the background knowledge) and nothing to the other hypotheses. If f is the constant 0 function—all hypotheses receive an explanatory bonus of 0—then this rule amounts to Bayes' rule.[2]

Note that the above probabilistic version of IBE is not just mathematically precise, but also avoids the asymmetry problem mentioned previously (though not the problem of defining explanatory bestness). However, it seems to face a very different problem, for van Fraassen uses it to illustrate Lewis' dynamic Dutch book argument for Bayes' rule, which aims to show that updating one's credences via any non-Bayesian rule can lead to irrational betting behavior, in particular, to engaging in series of bets that guarantee a negative net payoff. I can be brief on this argument here because Dutch book arguments have lost much of their appeal even to the Bayesian community, mainly because—most think now—these arguments are concerned with practical rationality (betting strategies) rather than with epistemic rationality,[3] and also because I have elsewhere given my reasons for believing that specifically the dynamic Dutch book argument fails (Douven [1999], [2013:431–5]). So, *pace* van Fraassen, I believe that aficionados of IBE in search of a precise formulation of the rule could well consider his version of IBE as a useful starting point, especially because the proposal is really a schema that offers room for variation.

Obviously, one can consider different values for the explanatory bonus. More interestingly, in van Fraassen's version the winner takes all as far as explanatory bonuses go: the best explanation gets an explanatory bonus, and all other hypotheses get nothing, regardless of whether, perhaps, some of them are very close in explanatory goodness to the best explanation. That is a design decision one can make, and one that is inessential to van Fraassen's use of the rule. But given that, as stated, advocates of IBE may be interested in van Fraassen's proposal for reasons different from his, they may want to

---

[2] A number of authors (e.g., Lipton [2004, Ch. 7], Poston [2014, Ch. 7]) have argued that IBE is best thought of as a rule placing constraints on prior probabilities, and is thus compatible with Bayes' rule (or even complements it). I will not go into this possible explication of IBE here, except to note that, if the arguments to be given in the present chapter hold water, we are not compelled to choose between explanatory considerations as constraining prior probabilities and explanatory considerations as reasons for giving bonus points (or as playing a more complicated role—see below) in updating.

[3] See, e.g., Joyce [1998]. However, I am not so sure that the two kinds of rationality can be separated as neatly as Joyce's and others' critiques of the Dutch book arguments presuppose; see Douven [2013:430f], but see also Elqayam [2016].

explore other ways of crediting hypotheses for their explanatory goodness. An obvious class of alternatives consists of rules that give each hypothesis what it deserves in light of its specific explanatory qualities.

The literature on explanation contains a number of different probabilistic measures of explanatory goodness that allow one to formalize such alternatives. There is no consensus on which of these measures is best. There may not even be one best measure: different measures may quantify different aspects of what we value in good explanations.[4] However this may be, later on it will be shown how we can obtain a great diversity of versions of IBE by combining, in different ways, the various measures of explanatory goodness with van Fraassen's above schema.

Naturally, advocates of IBE wishing to state a mathematically precise version of that rule are by no means committed to van Fraassen's schema, so there may be many more precise versions of IBE still. But is that not a problem, for how will we pick the *right* one? The answer is that, no, it is not necessarily a problem, because, just as there may not be a unique best measure of explanatory goodness, there may not be a unique best mathematically precise version of IBE. The suggestion I want to make is even more radical, namely, that IBE is best thought of as a slogan—the one we saw above: explanatory power is of confirmation-theoretic significance—which in each situation has to be fleshed out in what is the best possible way for that situation. Finding the best elaboration for a given situation may be a matter of experience or talent: some people may be better at this than others, which may make them better scientists, or generally more successful in life. Finding the elaboration of the slogan that is best for a given situation may sometimes even be a matter of luck.

To clarify this suggestion, I offer an analogy: psychologists studying problem solving have identified a number of general strategies that people use to solve problems. Among them is a strategy often called "divide and conquer," which comes down to dividing up a problem into more manageable sub-problems, solving the sub-problems, and obtaining a solution to the initial problem by collating the solutions to the sub-problems. This strategy will appear familiar to anyone, simply because we all frequently and routinely rely on it. It is a strategy that parents recommend to their children and teachers to their students. For instance, mathematics teachers tell their students that there are a number of ways to calculate the area of a pentagon, but that the simplest one is to cut the pentagon up into three triangles, calculate the areas of those triangles, and then sum the outcomes. Similarly, but at a more advanced level, they instruct students to integrate a function over a complicated region by trying to divide the region into smaller regions such that one obtains nice bounds on one's integrals, and then summing the integrals over those smaller regions. Or, to give a very different example, piano teachers recommend that their students study difficult passages by grouping the

---

[4] And philosophers of science may have been mistaken in holding that there is exactly one concept of explanation; see Colombo [2017].

notes in certain ways, practicing the groups separately, and after each group has been mastered, playing the passage in its entirety.

But note that the divide and conquer strategy needs filling in. For what are the sub-problems? And how is one to collate the solutions to the sub-problems to obtain the solution to the overarching problem? This is difficult, if not impossible, to say in general. Discovering the best way to divide a larger problem, and the best way to put the partial solutions together, is sometimes far from straightforward, and some people may be better at these things than others. In piano practice, it is known that one way of grouping the notes in a passage can be much more helpful than others in reducing the difficulty of that passage, and some piano teachers are famous for having a special knack for finding very efficient groupings for tricky passages. Successful mathematical modeling is also often a matter of identifying the most expedient way to split a problem up into smaller ones.

My claim is that it is no different with IBE: different situations may call for different ways of letting explanatory considerations guide one's belief changes. I will try to support this claim in Section 3. First, however, I want to summarize some recent findings from the psychology of reasoning, which show that people's belief changes are, as a matter of fact, governed by explanatory considerations, at least in some contexts. In Section 3, these findings will serve as a source of inspiration for formulating some elaborations of IBE that are hitherto unknown in the literature. These will then be used to address the normative issues pertaining to IBE.

2. IBE in the Psychology of Reasoning

By now, psychologists have gathered a wealth of evidence showing that explanation plays a key role in various aspects of cognition. For instance, Chi et al. [1994] found that asking people to explain certain facts facilitated both learning those facts and generalizing them. And Williams and Lombrozo [2013] report experiments showing that explanation helps people activate beliefs they already hold in trying to learn criteria for category membership. More relevant to the present topic, there is much experimental support for the claim that explanation influences the attribution of prior probabilities. For instance, in Lombrozo's [2007] studies, people tended to assign higher probabilities to simpler explanations unless they had unambiguous probabilistic information to the contrary.[5]

It is to be noted that most empirical work on the role of explanation in reasoning that has so far been conducted bears only indirectly on the question of whether IBE could at least be descriptively adequate, in the sense that people are inclined to take explanatory considerations into account when they update their credences on incoming evidence. Most of the extant studies on the relationship between explanation and credences basically show that the perceived goodness of explanations can influence the assignment of prior probabilities. Whether explanatory goodness is

---

[5] For similar results, see Read and Marcus-Newhall [1993] and Sloman [1994], [1997].

also a factor that influences the *change* of those probabilities when new evidence is obtained is a different question, and one which, until very recently, has received little to no attention.

Douven and Schupbach [2015a], [2015b] have focused on exactly this question. A main part of their research consisted of a re-analysis of data gathered earlier in the context of Schupbach's [2011] work on the material adequacy of various formal measures of explanatory goodness.[6] To find out which of these measures (if any) accorded with people's judgments of the goodness of explanations, Schupbach had conducted an experiment in which participants were asked questions *both* about the probabilities they assigned to particular hypotheses after they had been given new evidence relevant to those hypotheses *and* about the explanatory goodness of those same hypotheses. More specifically, participants were first shown two urns—called "urn A" and "urn B"—containing the same number of balls, some of which were black, the others white, but with different ratios of black and white balls. After the participants were told about the contents of the urns, one urn was chosen on the basis of a flip with a fair coin, and from that urn, ten balls were drawn, without replacement. The participants could not see which urn had been selected and were asked, after each drawing of a ball, how likely it was, in their opinion, that urn A had been selected, and they were also asked to rate the explanatory goodness both of the hypothesis that urn A had been selected (H_A) and of the hypothesis that urn B had been selected (H_B). On this basis, Schupbach could determine which measure of explanatory goodness did best in predicting participants' judgments of explanatory goodness.

In Douven and Schupbach [2015a], the goal was rather to *use* those judgments of explanatory goodness to *predict* participants' credences in H_A (or, equivalently, H_B). In that paper, the authors first construct a number of different statistical models with participants' credences as the dependent variable and some subset of the following variables as predictors: (i) objective probabilities, specifically, for each draw, the objective probability of H_A given the ball or balls drawn at that point in the series of draws; (ii) judgments of the explanatory goodness of H_A, and (iii) judgments of the explanatory goodness of H_B. Their analysis showed that the "Bayesian" model, with objective probabilities as the only predictor, did fairly well, as one might expect. However, the "explanationist" models they also considered, which had people's assessments of explanatory goodness among the predictors, did better. The model that had the judgments of the explanatory goodness of H_A and the judgments of the explanatory goodness of H_B as predictors, alongside objective probabilities, did particularly well, yielding highly accurate predictions. This model was topped only by a model that Douven and Schupbach constructed at a later stage, to wit, a model that had as predictors objective probabilities as well as the *difference in judged explanatory*

---

[6] The data of Schupbach's original experiment as well as those of two follow-up experiments presented in Douven and Schupbach [2015a] can be downloaded from the website of the journal *Cognition*, at <http://www.sciencedirect.com/science/article/pii/S0010027715000955#MMCvFirst>.

goodness between H_A and H_B. Surprisingly, the model that included *only* the judged explanatory goodness of H_A and the judged explanatory goodness of H_B as predictors, and so excluded objective probabilities, still did significantly better than the Bayesian model, although it did not do nearly as well as the two previously mentioned explanationist models. All in all, the findings reported in Douven and Schupbach [2015a] constitute strong support for the claim that, at least in some contexts, explanationism—the hypothesis that people update via some form of IBE—is superior to Bayesianism, understood as the descriptive claim that people update via Bayes' rule.

Even more pertinent to the current chapter are the results from Douven and Schupbach [2015b]. The question that paper sought to answer is this: Supposing subjective judgments of explanatory goodness are unavailable, could *objective* degrees of explanatory goodness, as determined by one of the measures of explanatory goodness studied in Schupbach [2011], still help us to arrive at more accurate predictions of a person's credences than we would arrive at on the basis of objective probabilities alone (when such objective probabilities are available)? To answer this question, they calculated, for each measure considered by Schupbach [2011], the explanatory goodness of H_A as well as the explanatory goodness of H_B according to the given measure, for each participant and each draw. They then basically repeated the analysis of their [2015a] paper, but now with the objective degrees of explanatory goodness as predictor variables, along with objective probabilities, finding that adding explanatory goodness of H_A and explanatory goodness of H_B as predictors to objective probabilities resulted in a model that was significantly more accurate than the Bayesian model with only objective probabilities as predictor. This was so for *each* measure of explanatory goodness which they considered. Nevertheless, two measures stood out in the analysis in terms of predictive accuracy, to wit, Popper's [1959] measure, according to which the explanatory goodness of a hypothesis H in light of evidence E equals

$$\frac{\Pr(E|H) - \Pr(E)}{\Pr(E|H) + \Pr(E)},$$

and Good's [1960] measure,

$$\ln\left(\frac{\Pr(E|H)}{\Pr(E)}\right),$$

or rather a particular functional rescaling of that measure, called "L_2" in Douven and Schupbach [2015b].[7]

This finding is actually quite surprising. For what it means is that if only objective probabilities are available to help us predict people's credences, we can improve on the

---

[7] The functional rescaling facilitates the comparison of Good's measure with the other measures, given that the range of Good's measure as shown here is $(-\infty, \infty)$, while that of the other measures, as well as that of the rescaling of Good's measure, is $[-1, 1]$. See Douven and Schupbach [2015b, Sect. 3] for details.

Bayesian model by making the objective probabilities do double duty and use them also as input in Popper's or Good's measure of explanatory goodness. At least this is so in the kind of updating context studied in Schupbach [2011]. How broadly this generalizes across updating contexts remains to be seen. For our present purposes, this question is immaterial; as already mentioned, here the results from Douven and Schupbach [2015b] will mainly serve as inspiration for delving deeper into the normative aspects of the debate between Bayesians and advocates of IBE.

3. Fleshing Out the Slogan

Arguably, what we want from a probabilistic update rule is that it be *fast*—fast in making us believe the truth to at least a high degree, that is—and that it be *accurate*, very roughly meaning that in the course of making us believe the *truth* to a high degree, it does not, along the way, make us believe *falsehoods* to a high degree.[8] This is not just to speak roughly but also to speak relatively: in some domains, paucity of evidence or reasons having to do with underdetermination (Douven [2008]) may prevent any reasonable update rule from leading us to believe the truth to a high degree. But if update rules *can* lead us to believe the truth about a domain with a high degree of confidence, then we would hope to be using the one that is fastest and most accurate, or if these two desiderata pull in different directions, the one that strikes the best balance between them (where it may depend on context what counts as "best"; more on this below).

In Douven [2013], computer simulations were used to compare the performance of Bayes' rule and that of van Fraassen's probabilistic version of IBE—which from here on we will refer to as "EXPL"—with respect to the aforementioned desiderata. The simulation setup required a formal explication of these desiderata. Making us believe the truth to a high degree was understood as making us believe the truth to a degree above 0.9, and accuracy was spelled out in terms of a scoring rule, where both the Brier rule and the log rule were used. With $\{H_i\}_{i \le n}$ a set of mutually exclusive and jointly exhaustive hypotheses and $[\![H_i]\!] \in \{0,1\}$ the truth value of H_i, the Brier rule assigns a penalty of $\sum_{i=1}^{n}([\![H_i]\!] - \Pr(H_i))^2$ to an agent whose degrees of belief are given by $\Pr(\cdot)$. And where H_i is the true hypothesis, the log rule assigns a penalty of $-\ln(\Pr(H_i))$ to the agent.

In the simulations in Douven [2013], two agents—a Bayesian and an agent who updates via EXPL—are trying to determine the bias of a coin for turning up heads, where it is given in advance that this bias is a multiple of 1/10; thus, the agents are considering eleven mutually exclusive and jointly exhaustive bias hypotheses. They try to determine which of those eleven hypotheses is true by updating their credences after each toss of the coin, which in total is tossed 1,000 times; initially, they have the same

---

[8] Which an update rule might do even if it is fast. And conversely, an update rule can be accurate in that it makes us believe the truth to a moderately high degree very quickly, but then keeps our degree of belief more or less there, without ever making us believe the truth to a high degree.

credence of 1/11 in each of the bias hypotheses. For each bias that the coin can have, 1,000 such simulated sequences of 1,000 coin tosses were run.

A first result of these simulations was that one will on average be faster—typically *much* faster—in believing the true bias hypothesis to a high degree if one updates via EXPL than if one updates via Bayes' rule. To also compare the two rules with respect to accuracy, for both agents and in each simulation, Brier scores and log scores were calculated after the 100th, 250th, 500th, 750th, and 1,000th updates, and it was found that, for all biases and in the vast majority of simulations, the agent updating via EXPL incurred at *each* reference point a lower penalty than the Bayesian. So, in a clear sense, EXPL is more accurate than Bayes' rule. On the other hand, it was also found that while there is typically no difference between the agents in terms of the *average* penalty— averaged over all 1,000 simulated sequences—incurred at the reference points, *when* there was a difference, it was always a small difference in favor of the Bayesian. So, it could be said that, in another clear sense, Bayes' rule is more accurate than EXPL.

It is worth briefly explaining these mixed findings. One could say that by updating via EXPL, one picks up the lead in the evidence more quickly than by updating via Bayes' rule; that is why EXPL is mostly much faster in bringing one's credence in the true bias hypothesis above the 0.9 threshold. However, this very same fact also makes the EXPL updater more vulnerable to misleading cues in the evidence, which in the simulations sometimes occur in the form of longer subsequences of consecutive tosses in which the relative frequency of heads deviates strongly from the coin's bias for heads. Such subsequences put both the Bayesian and the EXPL updater temporarily on a wrong path, but the latter goes much further on that path than the former, precisely because the EXPL updater is quicker to follow cues in the evidence. The EXPL updater can in fact be led so far astray by such misleading evidence that averaged over all simulations, she incurs a slightly higher Brier and log penalty than the Bayesian, despite the fact that in almost all simulations, she incurs lower Brier and log penalties. That is why EXPL gives more accurate results in virtually all simulations, and why in the very few simulations in which Bayes' rule is more accurate, it tends to be a *lot* more accurate, so much so, that on average Bayes' rule is still *slightly* more accurate than EXPL.

Douven and Wenmackers [2017] compared Bayes' rule with EXPL in a social setting, in which agents update their credences partly based on new evidence they receive and partly based on exchanges with certain of their fellow agents.[9] In that setting, EXPL was found to outperform Bayes' rule on both counts, speed *and* accuracy. But we are not always in a social setting, and one may wonder whether there are versions of IBE that outperform Bayes' rule on the said counts even in settings in which evidence coming from the world is all we have to go on.

---

[9] The agents basically update in the way in which agents update in the Hegselmann–Krause model (Hegselmann and Krause [2002], [2006]), which roughly means that, in updating, an agent averages, on the one hand, the information contained in the new evidence and, on the other hand, the pooled credences of those fellow agents whose credence functions are close enough to his or her own credence function. See Douven and Wenmackers [2017] for technical details.

It was previously stated that EXPL gives all credit for explanatory goodness to the hypothesis that explains the evidence best, and nothing to any other hypothesis. The objective measures of explanatory goodness stated in Section 2 (and others like it) can help to give content to the claim that probabilistic versions of IBE may instead credit each hypothesis separately, in proportion to that hypothesis' explanatory power. The results now to be presented will be seen to indicate that some such versions of IBE may do better than Bayes' rule across the board. More importantly for our present concerns, the results will appear to also be instrumental in driving home the point that IBE may be best thought of as a schema to be filled in differently in different contexts.

EXPL was seen to work very much like Bayes' rule, except that it adds a bonus point to the best explanation and then normalizes to obtain probabilities again. We cannot quite follow this procedure if we want to credit all hypotheses separately, at least not if the crediting is to be done on the basis of an objective measure of explanatory goodness of the type encountered in Section 2. In particular, given evidence E and a set $\{H_i\}_{i \leq n}$ of mutually exclusive and jointly exhaustive hypotheses, we cannot first update the hypotheses on E via Bayes' rule, then add $\mathcal{M}(H_i, E)$ to H_i (with $\mathcal{M}$ being some measure of explanatory goodness), and finally normalize. The measures of objective goodness considered have a range of $[-1, 1]$ (see note 7), with 0 being the neutral point. This means that, actually, these measures can assign bonus points as well as malus points; where a hypothesis is an extremely poor explanation of the evidence, they can even assign a malus point of -1, which, when added to the hypothesis' probability, could result in a negative value, which cannot be "normalized" to a probability.

Therefore, the rules to be considered will first update the hypothesis' probability via Bayes' rule, then calculate the hypothesis' explanatory goodness according to a given objective measure, next add or subtract a percentage of the hypothesis' probability in proportion to its explanatory goodness, and finally normalize. More formally, the rules we will consider are instances of the following schema:

$$\Pr[\![E]\!](H_i) = \frac{\Pr(H_i)\Pr(E|H_i) + c\Pr(H_i)\Pr(E|H_i)\mathcal{M}(H_i, E)}{\sum_{j=1}^{n}(\Pr(H_j)\Pr(E|H_j) + c\Pr(H_j)\Pr(E|H_j)\mathcal{M}(H_j, E))},$$

where $\Pr(\cdot)$ is one's credence function prior to learning E, $\Pr[\![E]\!](H_i)$ is one's new credence in H_i immediately after learning E (and nothing stronger), and $c \in [0, 1]$ a constant determining what percentage of H_i's probability after a Bayesian update on E is added in proportion to this hypothesis' power to explain E. Note, incidentally, that these rules are beset neither by the asymmetry problem mentioned in Section 1 nor by the problem of leaving explanatory bestness undefined: explanatory bestness can be stated straightforwardly in terms of whichever measure of explanatory goodness the rule uses (although the notion of explanatory bestness is no longer called upon here; the measure of explanatory goodness is all we need).

The results to be reported are from simulations that used either Popper's or Good's measure of explanatory goodness (with Good's measure in the version L_2), and that further assumed a value for c of 0.1. The choice of measures is based on the results from Douven and Schupbach [2015b], but the choice of the value for c is largely arbitrary. Also, there are clearly ways of bringing to bear a hypothesis' explanatory goodness on the credence we assign to it other than the one implemented by the above schema. I will not try to motivate the design choices made here, because the point of the following results is not to present the ultimate probabilistic version of IBE, but rather, first, to support the claim that there exist versions of IBE that, at least in certain kinds of updating contexts, do better than Bayes' rule in terms of speed and accuracy, and second, to show that there may be no one best way to explicate the slogan that explanation has a place in confirmation theory.[10]

The updating in the simulations again concerned a coin whose bias is unknown other than that it is a multiple of $1/10$. But now, instead of comparing just a Bayesian agent and an agent updating via EXPL, there were two additional agents who updated via the instance of the above schema with Popper's and with Good's measure of explanatory goodness, respectively, for $\mathcal{M}$. Here, too, the agents all started with an initial credence of $1/11$ in each of the eleven bias hypotheses, and they updated their credences in those hypotheses after each of the 500 tosses with the coin they were shown per simulation. For each possible bias the coin could have, 1,000 simulations were run. The four agents, or better Bayes' rule, EXPL, Popper's rule (as we shall call the rule with Popper's measure of explanatory goodness), and Good's rule, were evaluated and compared with each other both on the count of speed and on that of accuracy.

Again, speed was measured in terms of the number of tosses it took to believe the true bias hypothesis to a degree greater than 0.9. The results of the measurements are shown in Figure 2.1.[11] A first thing to notice is that we find a replication of the results concerning speed from Douven [2013]: EXPL is, for any bias hypothesis, on average faster—and for most bias hypotheses *much* faster—in assigning a high credence to the truth. However, we now see that EXPL is not just faster than Bayes' rule, but that it is also faster than either of the two other rules. A further observation is that Popper's rule is, on average, somewhat faster than Bayes' rule, for all bias hypotheses. It cannot be seen from the figure whether the difference in speed between those rules is significant

---

[10]   Moreover, the code for the simulations, written in R, is made available via my ResearchGate page (doi: 10.13140/RG.2.1.2670.9201), in the hope that this will encourage readers to experiment for themselves with variants of the rules used for the present study.

[11]   Here as well as elsewhere, results are only shown for biases up to 0.5, given that the results are symmetric across that value, because a bias for heads of x is equivalent to a bias for tails of 1 – x. (If this is not clear, suppose a rule were significantly faster in assigning a high probability to the truth if that is the hypothesis that the bias for heads is x than if that is the hypothesis that the bias for heads is 1 – x. Then it would be significantly faster in assigning a high probability to the truth if that is the hypothesis that the bias for heads is x than if that is the hypothesis that the bias for tails is x. That would be inconsistent with the assumption that "heads" and "tails" are arbitrary labels for the two sides of the coin.)

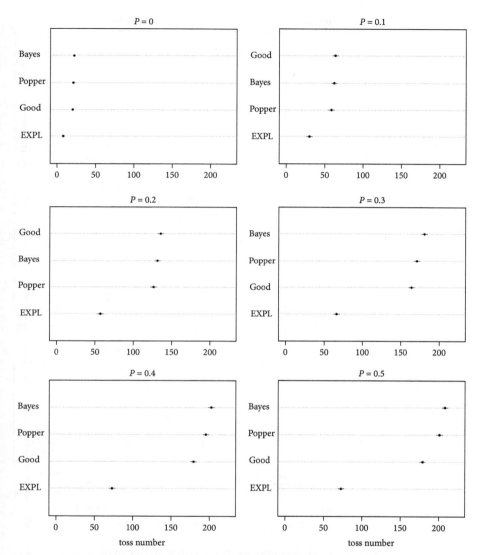

Figure 2.1 Dot charts showing speed of convergence of the various update rules, for six different possible biases (p = x indicates that the objective probability for the coin to come up heads equals x). Dots indicate averages over 1,000 simulations of toss at which the true bias hypothesis receives a probability greater than 0.9; thick lines indicate standard errors. Rules are ordered from top to bottom according to increasing speed.

for all those hypotheses, or even for any, but a series of t-tests confirms that it is, for all hypotheses (highly significantly so in fact: p < 0.0001 for all hypotheses; average Cohen's d effect size: 0.33, SD = 0.05). And while Good's rule is faster (and significantly so) than both Bayes' rule and Popper's rule when the bias of the coin is in the middle range, it is (significantly) *slower* than both for the biases closer to the extremes.

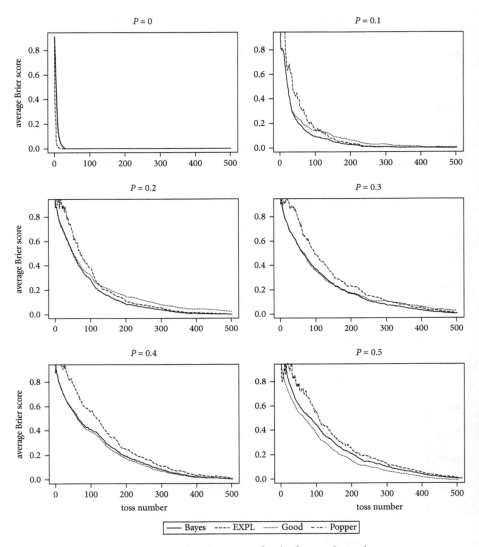

Figure 2.2   Average Brier scores for the four update rules.

Also as in Douven [2013], accuracy of the rules was assessed via both the Brier and the log scoring rule. Figure 2.2 shows the Brier penalties incurred by the updating rules we are considering, after each toss, and averaged over the 1,000 simulations that were run for biases up to 0.5. In further confirmation of the results from Douven [2013], for all but the extreme biases, EXPL does worse than Bayes' rule. On the other hand, for biases in the middle range, Good's rule does better than the other three rules. The results in terms of log penalties are qualitatively the same. These results are not shown here but can be obtained by running the scores.plot function in the accompanying R-script (see note 10).[12]

What about Popper's rule? Where is it in Figure 2.2? It turns out that the results for Popper's rule are not separately visible, because they are extremely close to those for Bayes' rule, so close indeed that their graphs basically coincide. It is worth delving more deeply into this, given that, as we saw, in terms of speed Popper's rule did better than Bayes' rule, for any possible bias. If, in addition, Popper's rule is as accurate as Bayes' rule, then there would be at least one probabilistic version of IBE that outperforms Bayes' rule tout court.

It turns out that the mean difference in average Brier scores over the 500 updates is 0.0001 (SD = 0.001) in favor of Bayes' rule, while the mean difference in average log scores is 0.0004 (SD = 0.003), also in favor of Bayes' rule. Instead of only looking at average penalties, we can also compare the 1,000 Brier/log penalties incurred by one rule after an update with the 1,000 Brier/log penalties incurred by the other rule after the same update. The function bp.scores in the R-script takes as input the name of either the Brier rule or the log rule as well as a bias and performs 500 t-tests, one for each update, on the Brier/log scores for the two rules after the updates. It outputs for each of the rules the proportion of updates after which there was a significant difference in means in favor of that rule (so the proportion of updates after which the rule incurred a significantly lower penalty, on average) as well as the mean effect sizes (plus standard deviations). One verifies that, for the biases closer to the extremes, the proportion of updates after which there is a significant difference in means in favor of Popper's rule exceeds the proportion of updates after which there is a significant difference in means in favor of Bayes' rule, while for the biases closer to 0.5, it is the other way round. Most importantly, however, one sees that in all cases the effect sizes are vanishingly small, with the average value for Cohen's d being always very close to 0 (the standard deviation also being close to 0 in all cases); statisticians universally regard such effect sizes as not worth a mention.[13]

It hence appears that, in the kind of updating context considered here, one is not worse off with Popper's rule than with Bayes' rule if the main concern is accuracy; and we already know that, if the main concern is speed, one is somewhat better off with

[12] The script in fact provides code to produce color plots, which makes the differences in scores between the various rules more visible.

[13] A traditional t-test outputs a p-value (among other things) that indicates whether there is a significant difference between two population means. However, statisticians frequently warn that a non-significant result is not to be interpreted as evidence for the null hypothesis of "no difference." A so-called Bayesian t-test has more to offer in this respect, for such a test can provide evidence both for the hypothesis that there *is* a difference between the means as well as for the null hypothesis, or rather, it provides *relative* evidence for one or the other hypothesis, indicating the extent to which the data support one hypothesis over the other (see Lee and Wagenmakers [2013:104ff] for details). The R-script contains code for running Bayesian t-tests on the penalties as well as a function—pl.func—for plotting the average penalties for the 500 updates together with the so-called Bayes factors after the various updates. The Bayes factors are plotted on a logarithmic scale on the alternative y-axis and give information about whether the averages can indeed be regarded as being the same or whether they differ. I invite readers to run this part of the code to obtain additional information about the difference—or rather lack of difference—in accuracy between Bayes' rule and Popper's rule.

Popper's rule: while the statistical tests we performed on the results concerning speed revealed a small effect of the choice of rule, the effect certainly was not negligible (as it was in the case of accuracy). The further conclusion is that one can often do better still than Popper's rule, just not always. As our results showed, when the bias of the coin is in the middle range, Good's rule is faster than Popper's and also incurs lower, or in any case not higher, Brier or log penalties than Popper's rule. But we also saw that this no longer holds true for biases more toward the extremes.

To see how this can be valuable information, suppose you are a producer of roulette tables. Before the tables leave the factory, they are extensively tested for fairness. This is done by rolling a ball 1,000 times and registering whether it lands on red or on black. You know that the production process is such that if the table has a bias at all, it will be a very slight one. So, one of the update rules that does particularly well in the middle range—Good's, above all—would seem appropriate. By contrast, suppose you have a number of machines that produce widgets. They are all reliable in typically producing good widgets, though they all from time to time make broken widgets. You have enough money to replace one machine, and you want to replace the one that is least reliable. Here the "bias" is going to be close to 1, and so to determine which machine should be replaced, it is recommended that you use a rule that does comparatively well in the extreme ranges, which would be Bayes' or Popper's rule; given the first conclusion, the obvious choice would be the latter.

All of this is assuming that accuracy matters, which in the examples it probably does: selling a roulette table with a bias that you mistakenly believe to be fair might damage the reputation of your brand, and replacing the "wrong" widget machine may be costly. But in some contexts, accuracy might not be a concern. There is a persistent rumor about Mozart's death that tends to come up in conversations among music lovers, namely, that Mozart did not die from natural causes but was poisoned. The rumor is almost certainly false (Zegers, Weigl, and Steptoe [2009]), but who started it? Answer: Mozart did, shortly before his death (Robbins Landon [1999:148ff]). Perhaps you knew, but if not, and if I had not told you, you might, out of sheer curiosity, have wanted to find out about the source of the rumor. And if you were really curious, you might have wanted to find out fast. But if, in the course of your investigations of the matter, you had come to mistakenly hold that it was Mozart's wife who started the rumor, or one of the first newspapers that reported about his death, before you eventually found out the truth, that might have done little to no harm: accuracy will not matter much in this kind of case, in which being mistaken is unlikely to have any dire consequences, or to have consequences at all. For such cases, EXPL may be the rule of choice, as we have seen, for at least among the rules we considered, it is by far the fastest, even if not the most accurate.

4. Conclusion

So how are we to explicate the slogan that we should infer to the best explanation? Supposing that the results reported above generalize across other updating contexts, we could propose Popper's rule as such an explication: it takes explanatory considerations into account while not only having a mathematically precise form, like Bayes' rule, but also being faster than Bayes' rule, and equally accurate. However, we might thereby do ourselves an intellectual disservice: in some contexts, we could be faster in believing the truth to a high degree, or more accurate while on our way to the truth, or even both, by using a different explication of the slogan. Hence my claim that we do best to conceive of IBE as the aforementioned slogan, a slogan that not only *permits* different fleshings-out, but that may even rationally *require* different fleshings-out, depending on what our interests are. And as was noted, the possible precisifications are by no means limited to the ones studied in this chapter; these depended on design decisions that were to some extent arbitrary, the goal here being to make some principled points, not to find the best probabilistic explication or explications of IBE. Indeed, whether the above results do generalize across updating contexts is an open question. But while this question is of interest in itself, it is orthogonal to the points I wanted to make in this chapter: that it is perfectly fine to think of IBE as a slogan and that, even if inferring to the best explanation may expose one to Dutch bookies, doing so may have compensating advantages, given any plausible understanding of "advantage." To the extent that I have been able to support these points, I have answered the questions of what IBE is and of why we should care about it.[14]

References

Chi, M. T. H., de Leeuw, N., Chiu, M.-H., and LaVancher, C. [1994] "Eliciting Selfexplanations Improves Understanding," *Cognitive Science* 18:439–77.

Colombo, M. [2017] "Experimental Philosophy of Explanation Rising: The Case for a Plurality of Concepts of Explanation," *Cognitive Science* 41(22):503–17.

Douven, I. [1999] "Inference to the Best Explanation Made Coherent," *Philosophy of Science* 66:S424–S435.

Douven, I. [2008] "Underdetermination," in S. Psillos and M. Curd (eds), *The Routledge Companion to Philosophy of Science*, London: Routledge, pp. 292–301.

Douven, I. [2013] "Inference to the Best Explanation, Dutch Books, and Inaccuracy Minimisation," *Philosophical Quarterly* 69:428–44.

Douven, I. and Schupbach, J. N. [2015a] "The Role of Explanatory Considerations in Updating," *Cognition* 142:299–311.

Douven, I. and Schupbach, J. N. [2015b] "Probabilistic Alternatives to Bayesianism: The Case of Explanationism," *Frontiers in Psychology* 6:459, doi: 10.3389/fpsyg.2015.00459.

Douven, I. and Wenmackers, S. [2017] "Inference to the Best Explanation versus Bayes' Rule in a Social Setting," *British Journal for the Philosophy of Science* 68(2):535–70.

Elqayam, S. [2016] "Scams and Rationality: Dutch Book Arguments Are Not All They're Cracked Up to Be," in N. Galbraith, D. E. Over, and E. J. Lucas (eds), *The Thinking Mind: A Festschrift for Ken Manktelow*, Hove: Psychology Press.

[14] I am greatly indebted to Kevin McCain, Ted Poston, and Christopher von Bülow for very helpful comments on previous versions of this chapter. Versions of this chapter were presented at the universities of Geneva, Ghent, and Leuven. I am grateful to the audiences on those occasions for stimulating questions and discussion.

Good, I. J. [1960] "Weight of Evidence, Corroboration, Explanatory Power, Information and the Utility of Experiment," *Journal of the Royal Statistical Society* B22:319–31.

Harman, G. [1965] "The Inference to the Best Explanation," *Philosophical Review* 74:88–95.

Hegselmann, R. and Krause, U. [2002] "Opinion Dynamics and Bounded Confidence: Models, Analysis, and Simulations," *Journal of Artificial Societies and Social Simulation* 5, available at <http://jasss.soc.surrey.ac.uk/5/3/2.html>.

Hegselmann, R. and Krause, U. [2006] "Truth and Cognitive Division of Labor: First Steps towards a Computer Aided Social Epistemology," *Journal of Artificial Societies and Social Simulation* 9, available at <http://jasss.soc.surrey.ac.uk/9/3/10.html>.

Joyce, J. [1998] "A Nonpragmatic Vindication of Probabilism," *Philosophy of Science* 65:575–603.

Kuipers, T. A. F. [1992] "Naive and Refined Truth Approximation," *Synthese* 93:299–341.

Lee, M. and Wagenmakers, E.-J. [2013] *Bayesian Cognitive Modeling*, Cambridge: Cambridge University Press.

Lipton, P. [1993] "Is the Best Good Enough?" *Proceedings of the Aristotelian Society* 93:89–104.

Lipton, P. [2004] *Inference to the Best Explanation* (2nd ed.), London: Routledge.

Lombrozo, T. [2007] "Simplicity and Probability in Causal Explanation," *Cognitive Psychology* 55:232–57.

Musgrave, A. [1988] "The Ultimate Argument for Scientific Realism," in R. Nola (ed.), *Relativism and Realism in Science*, Dordrecht: Kluwer, pp. 229–52.

Popper, K. R. [1959] *The Logic of Scientific Discovery*, London: Hutchinson.

Poston, T. [2014] *Reason and Explanation*, Basingstoke: Palgrave Macmillan.

Psillos, S. [2004] "Inference to the Best Explanation and Bayesianism," in F. Stadler (ed.), *Induction and Deduction in the Sciences*, Dordrecht: Kluwer, pp. 83–91.

Read, S. J. and Marcus-Newhall, A. [1993] "Explanatory Coherence in Social Explanations: A Parallel Distributed Processing Account," *Journal of Personality and Social Psychology* 65:429–47.

Robbins Landon, H. C. [1999] *1791: Mozart's Last Year*, New York: Schirmer Books.

Schupbach, J. N. [2011] "Comparing Probabilistic Measures of Explanatory Power," *Philosophy of Science* 78:813–29.

Sloman, S. A. [1994] "When Explanations Compete: The Role of Explanatory Coherence on Judgments of Likelihood," *Cognition* 52:1–21.

Sloman, S. A. [1997] "Explanatory Coherence and the Induction of Properties," *Thinking and Reasoning* 3:81–110.

van Fraassen, B. C. [1989] *Laws and Symmetry*, Oxford: Clarendon Press.

Vogel, J. [1998] "Inference to the Best Explanation," in E. Craig (ed.), *Routledge Encyclopedia of Philosophy*, London: Routledge, available at <https://www.rep.routledge.com/articles/inference-to-the-best-explanation>.

Vogel, J. [2005] "The Refutation of Skepticism," in M. Steup and E. Sosa (eds), *Contemporary Debates in Epistemology*, Malden MA: Blackwell, pp. 72–84.

Williams, J. J. and Lombrozo, T. [2013] "Explanation and Prior Knowledge Interact to Guide Learning," *Cognitive Psychology* 66:55–84.

Zegers, R. H. C., Weigl, A., and Steptoe, A. [2009] "The Death of Wolfgang Amadeus Mozart: An Epidemiological Perspective," *Annals of Internal Medicine* 151:274–8.

3

Peirce and Ramsey

Truth, Pragmatism, and Inference to the Best Explanation

Cheryl Misak

1. Introduction

In the novel *We Are All Completely Beside Ourselves*, the protagonist says: "Scientists have solved the problem of solipsism with a strategy called *inference to the best explanation*. It's a cheap accommodation, and no one is happy about it" (Fowler 2013: 133).[1] As far as philosophical ideas appearing in popular literature go, I've seen worse. To place the argument of this chapter in the context of Fowler's metaphor I would say: how one feels about one's lodgings depends on what one is expecting. If one invokes inference to the best explanation in the expectation of solving the problem of solipsism or refuting the skeptic, the solipsist and the skeptic are going to think the purported solution is a substandard accommodation. The solipsist and skeptic have sought knowledge that gives us certainty, and failing to find it, they conclude that we have no knowledge at all—no knowledge, for instance, that other minds, causes, theoretical entities, and the external world exist. It won't satisfy them to be told that if the best explanation of what we observe is that these things exist, then we know they exist. For they are after a stronger sense of "know" than inference to the best explanation can deliver, and they will deny that inference to the best explanation can yield knowledge. They are after an indubitable foundation for knowledge, perhaps one that seeks a correspondence between true statements and facts. If that's what you want, inference to the best explanation is going to appear to be a cheap and second-best accommodation.

But once one adopts a more realistic expectation—once one adopts a more human account of truth and knowledge—then inference to the best explanation will seem excellent accommodation indeed. It will be an essential part of an alternative position that we can and should adopt in place of the quest for certainty—a quest that leads directly to solipsism and skepticism. That alternative position is a certain kind of pragmatism, first developed by one of the founders of the tradition, C.S. Peirce, in Cambridge,

[1] This chapter builds on Misak (2004) and (2016). It is much better for the excellent comments of Ted Poston and Kevin McCain.

Massachusetts in the late 1880s and then extended by Frank Ramsey in Cambridge, England before he died in 1930 just shy of his twenty-seventh birthday.[2] It is Ramsey who uses the term "realistic" to describe his position, nicely setting it against the kind of "realist" position his friend the early Wittgenstein was promoting in the *Tractatus*.[3] Skepticism and solipsism loomed for Wittgenstein, and inference to the best explanation was not something he thought would suffice to ward it off. But, just like every other project that attempts to conclusively ground our knowledge in experience, the project of the *Tractatus* came unwound. It was Ramsey who pulled at the first loose string. Wittgenstein, depending on your interpretation of him, later became either a Ramseyan pragmatist or a more radical anti-philosophy philosopher, giving up on any attempt to give a second-order justification of how we come to know.

The kind of pragmatist I am interested in here argues that if we start off with appropriate expectations, inference to the best explanation is an important kind of inference in inquiry aimed at truth. Indeed, inference to the best explanation is part and parcel of a sensible pragmatist concept of truth. Peirce and Ramsey offer us a pragmatist—one might even say teleological—justification of inference to the best explanation that is different from standard justifications and interesting in its own light. This justification of inference to the best explanation has been pretty much lost in the contemporary discussion, and my aim here is to bring it back into the light.

As a preliminary, let me set out three key pragmatist theses. The first is about belief: the belief that *p* is in part a disposition to behave or a set of commitments that someone who believes *p* incurs. The second is about knowledge: knowledge grows in a piece-meal and fallible way, when a belief is unsettled by a surprising experience. As Peirce put it, knowledge "is not standing upon the bedrock of fact. It is walking upon a bog, and can only say, this ground seems to hold for the present. Here I will stay till it begins to give way" (*CP* 5. 589; 1898). The third is about truth: were a belief to be indefeasible or as good as it could be, by way of accounting for experience, fitting with our other well-grounded beliefs, and so on, then it is true. These accounts of belief, knowledge, and truth are connected to our human practices or our ways of believing and inquiring. That, if you like, is the fundamental pragmatist insight.

2. Peirce

Peirce argued that logic is a normative science—it is all about evaluating belief. Belief, for Peirce, is in part a habit or disposition to behave:

Every natural or inbred belief manifests itself in natural or inbred ways of acting, which in fact constitute it a belief-habit. (I need not repeat that I do not say that it is the single

---

[2] For the story of how Ramsey got hold of Peirce, who was pretty much unknown in England in the 1920s, and how Ramsey's pragmatism then influenced Wittgenstein, see my *Cambridge Pragmatism: From Peirce and James to Ramsey and Wittgenstein* (2016).

[3] See Methven (2015).

deeds that constitute the habit. It is the single "ways," which are conditional propositions, each general). (*CP* 5. 510; 1905)

I say "in part," because Peirce argued that belief is also a feeling. But he thought that a focus on the subjective "feeling" aspect of belief would only lead us down philosophical blind alleys. So he turned his attention to behavior and observable consequences.

The normative science of logic, along with all other domains of inquiry is about acquiring good belief-habits, habits we can evaluate in terms of whether or not they work out, in what we shall see is a robust sense of "working out." We shall see that the pragmatists varied on what counts as good grounds for belief. But at the heart of all their positions is the idea that belief is a habit or rule for behavior and is to be evaluated as such. Peirce sees that an instinctive reaction might also be taken to be such a habit. Logic, however, is not interested in instinct, but in the pre-meditated, self-controlled, critical evaluation of belief or assertion. It is interested in habits that we take responsibility for. Here is Peirce's account of assertion:

If a man desires to assert anything very solemnly, he takes such steps as will enable him to go before a magistrate or notary and take a binding oath to it. Taking an oath is...not mere saying, but is *doing*...[I]t would be followed by very real effects, in case the substance of what is asserted should be proved untrue. This ingredient, the assuming of responsibility, which is so prominent in solemn assertion, must be present in every genuine assertion. For clearly, every assertion involves an effort to make the intended interpreter believe what is asserted, to which end a reason for believing it must be furnished...Nobody takes any positive stock in those conventional utterances, such as "I am perfectly delighted to see you," upon whose falsehood no punishment at all is visited...[E]ven in solitary meditation every judgment is an effort to press home, upon the self of the immediate future and of the general future, some truth...there is contained an element of assuming responsibility, of "taking the consequences."

(*CP* 5. 546; 1908)

Belief, assertion, or judgment is a matter of stating the proposition in question to oneself, and being prepared to act on it. If those actions fail or if the consequences fail to pan out, the believer or assertor is held to account in some way or other. Peirce linked assertion with that of betting, something that will also be of interest when we turn to Ramsey:

What is the difference between making an *assertion* and *laying a wager*? Both are acts whereby the agent deliberately subjects himself to evil consequences if a certain proposition is not true. Only when he offers to bet he hopes the other man will make himself responsible in the same way for the truth of the contrary proposition; while when he makes an *assertion* he always (or almost always) wishes the man to whom he makes it to be led to do what he does.

(*CP* 5. 31; 1903)

Peirce and his fellow co-founders of pragmatism got their inspiration for this dispositional account of belief and assertion from Alexander Bain. The dispositional understanding of belief is so important to the pragmatist project that Peirce says that pragmatism is "scarcely more than a corollary" of it. Roughly, pragmatism holds that beliefs are true if they would lead to successful action and false if they would not. If a belief would be indefeasible, or would always work out by fitting with experience and our other well-grounded beliefs, it is true. I say "roughly" for two reasons. First, this is not designed to be an analytic definition of truth, but an unpacking of one aspect of our concept of truth—the practical aspect. Second, the unpacking of this idea is a site of contention amongst pragmatists. For instance, sometimes James suggests that a belief is true if it leads to making one's life go better here and now, while Peirce insisted that truth be linked to a more robust conception of success, in which a belief could only be taken to be successful if it was put in place by a method "not extraneous to the facts" (W 3. 253; 1877).[4]

Peirce was a crackerjack formal logician. He devised a quantified logic based on diagrams, at the same time as, but independently of, Frege. He was adamant that logic and inquiry are intertwined, and it is in this context that he arrives at his famous division of reasoning into three kinds: deduction, induction, and abduction. The last is an inference to the best explanation. It takes the form:

The surprising fact, C, is observed;
But if A were true, C would be a matter of course.
Hence, there is reason to suspect that A is true.   (CP 5. 189; 1903)

Abduction, or what he early on called "hypothesis," is fundamentally creative[5] or "ampliative." It goes beyond what is in the premises, whereas deduction explicates what is in the premises. Abduction is thus capable of importing new ideas into our body of belief. Peirce put considerable effort into saying what makes an abductive inference good. Factors in its evaluation include simplicity, fitting with our other beliefs, and so on. But like all inference forms, it is good, generally, if it is part of a method that leads to beliefs that are reliable, indefeasible, or do not let us down.[6] That is, simplicity and the like are not what these days get called "pragmatic" criteria, in a sense the classical pragmatists would not have touched. In this new use, pragmatic criteria are distinguished from criteria such as observation and being in line with logical constraints, that are somehow supposed to be linked with truth as correspondence or real truth. The classical pragmatists did not bifurcate grounds for belief in this way. Everything that goes into the method of inquiry that gets us stable, indefeasible beliefs are criteria related to truth.

Peirce might well have been happy with the label "inference to the best explanation," as he says that this mode of inference is "the operation of adopting an explanatory hypothesis" (CP 5. 189; 1903). But he wouldn't have been happy with the modern idea that explanatory power is an indicator that our belief corresponds to reality. For one thing, Peirce thought that the hypothesis that is the best explanation is not one we can

---

    [4] See Misak (2013) for a sustained interpretation of Peirce's position.
    [5] Richardson (2008: 346) uses this term. It is exactly right.
    [6] As we shall see when we turn to Ramsey, we can never put a single belief to the test in isolation from other beliefs and desires, but that is not fatal to the position.

assert. We do not take responsibility for it. The conclusions of abductive inferences are mere conjectures—we must "hold ourselves ready to throw them overboard at a moment's notice from experience" (*CP* 1. 634; 1898). For Peirce, "abduction commits us to nothing. It merely causes a hypothesis to be set down upon our docket of cases to be tried" (*CP* 5. 602; 1903). But if the role of abductive inference is not to enable us to infer that the best explanation is true, or even reasonable to believe, what *is* its role? For Peirce, abduction is part of the logic of discovery, an essential part of the logic of inquiry. The hypotheses of abduction must be confirmed by induction, but they each play a critical role in inquiry aimed at truth—truth as the pragmatist conceives it.[7]

Peirce takes the first step in inquiry to be an abductive inference. A hypothesis or a conjecture is identified that explains some surprising experience—it explains some exception to our belief-habit, or to what was expected. Consequences are then deduced from this hypothesis and are tested by varieties of induction and statistical inference. If the hypothesis passes these tests, then it is accepted—it is stable and believed unless it is upset by new and surprising experience. Inquiry thus proceeds as follows: from abduction, to deduction, to induction. Peirce thinks that because abduction and induction both add to our knowledge, "some logicians have confounded them." But he is clear that he means to describe the two types of inference as separate stages of a tripartite scientific inquiry (*W* 3. 330; 1878). Abduction is part of an inquiry aimed at truth, and we take the upshot of that inquiry to be true. But what we take to be true is fallible, and we can never say with certainty that we know this or that to be true. Not only can the conclusions of abductive inferences not simply be taken to be true, but even the conclusion of our best inquiries here and now can't be inferred to be true.

So on the pragmatist account of truth and inquiry, there is no artificial bifurcation into a logic of discovery (that does not aim at truth) and a logic of confirmation (deduction and straight-rule induction) that does aim at truth. Peirce thinks that we cannot aim at truth unless we utilize all three forms of inference, and then what we get is fallible belief that might well be indefeasible or true, but we cannot know that it is.

To see the point more clearly, let us take a brief excursion into Peirce's idea of a regulative assumption of inquiry. His account of truth rests on the assumption that inquiry will extend into the indefinite future. But that doesn't mean that it will indeed extend indefinitely:

Now, there exist no reasons... for thinking that the human race, or any intellectual race, will exist forever. On the other hand, there can be no reason against it; and, fortunately... there is nothing in the facts to forbid our having a *hope*, or calm and cheerful wish, that the community may last beyond any assignable date.   (*W* 3. 285; 1878)

Abductive inference also rests on a regulative assumption. We assume that whenever we observe C, there will be some hypothesis A that entails C or makes C probable—we

---

[7] See Misak (2004), McKaughan (2008), and Plutynski (2011) for more extended accounts of how Peirce takes abduction to be part of inquiry.

assume that there is an explanation for our surprising observations. Inquiry, too, rests on an assumption: that, for any matter into which we are inquiring, we would find an answer to the question that is pressing in on us: "the only assumption upon which [we] can act rationally is the hope of success" (*W* 2. 272; 1869). Thus the principle of bivalence—for any *p*, *p* is either true or false—rather than being a law of logic, is a regulative assumption of inquiry.

Indispensability arguments—arguments that move from "we need to assume *p* and *x*." to "*p* is true" or "*x* is indispensable to our best theory" to "*x* exists"—were frequently employed in the late nineteenth century. James, as is often the case, is the most extreme on the matter. His thoughts on the will to believe can be seen as an argument that if one finds something indispensable for his or her life, one can believe it.[8] Peirce instead took his start from Kant, and tried to naturalize the position he found there, carving out new and interesting territory on the subject of indispensability arguments. Peirce was clear that not only should the fact that an assumption is indispensable to our practice of inquiry not convince us of its necessary truth, it should not even convince us of its truth. Like the conclusions of abductive inferences, indispensability is not itself a "ground of belief." Peirce says: "It may be indispensable that I should have $500 in the bank—because I have given checks to that amount. But I have never found that the indispensability directly affected my balance, in the least" (*CP* 2. 113; 1902). Peirce's view is that "we are obliged to suppose, but we need not assert" that, for instance, there are determinate answers to our questions. Refusing to make such essential assumptions is to block the path of inquiry and, in Peirce's book, that is the cardinal philosophical sin. Our reason for making the assumptions is driven, Peirce says, by "desperation." If we do not make them, we will "be quite unable to know anything of positive fact" (*CP* 5. 603; 1903). As Hookway puts it, for Peirce, to show that a belief is unavoidable for us does not show that it is true, but it provides a strong reason for hoping that it is true and for regarding it as legitimate in our search for knowledge (1999: 181).

In "The Fixation of Belief," Peirce states that a "fundamental hypothesis" is taken for granted in inquiry. That hypothesis is as follows: "There are real things, whose characters are entirely independent of our beliefs about them" and yet can be discovered through empirical investigation (*W* 3. 254; 1877). The assumption is a fundamental *hypothesis*, or a result of an abductive inference. The best explanation of what we observe is that the world exists independently of us and constrains us. As inquirers, we believe in concrete, independently existing objects. This assumption is, to use a phrase of Arthur Fine's, part of our natural ontological stance. The best explanation of what we observe invokes other minds, real existing objects, causes, and so on. This is Peirce's approach to metaphysics: it is a natural science. Beliefs about independently existing objects and states are reasonable, given how nicely their existence fits with experience and prediction. What exists is what would be found to exist, by the method of abduction, deduction, and induction.

---

[8] See Misak (2011) and (2013) for a sustained account of this issue and the differences between various pragmatists.

Importantly, however, if causes, for instance, are part of the best explanation of our observations, that is not a proof of their existence. Moreover, the fact that causes are part of our best explanations can stand even if philosophy cannot say exactly what causes are. Causal hypotheses, and the more general hypothesis that there are causes, are on the docket to be tried. If their trials prove and continue to prove successful, we are justified in believing them. That is what justification is for Peirce. And truth, for Peirce, is being forever justified.

Inference to the best explanation is thus linked to truth: it gives us hypotheses that, after testing, we are justified in taking to be true. It might be asked how we ever get this justification for a fundamental hypothesis, such as causation, since every test that we conduct will assume the truth of fundamental hypotheses. It seems that there is no way to test the assumptions that lie at the foundation of our beliefs. Ramsey, we shall see, employs a signature pragmatist move. Once we take the step of assuming a hypothesis like causation, it is going to be embedded in our theories and framework. Only a revolution would cause us to abandon such a central belief. But abandon it we could. Peirce, Ramsey, C.I. Lewis, and Quine are all pragmatists who are clear about making this move.

One of the advantages of thinking about all three inference forms as essential to inquiry and truth is that a backstop is supplied for induction. Peirce presents us with a deep and interesting solution to the problem of induction, although he was relatively silent about Hume's way of setting it out (and unfair to Hume when he did mention it). One of Hume's most powerful arguments was to show that the inductive inference from "all observed As are Bs" to "all As are Bs" seems not to stand on solid ground. As I have argued elsewhere,[9] Peirce anticipates Nelson Goodman's *Fact, Fiction and Forecast* (1995) in allowing us to see our way through Hume's problem of induction by reframing it as a problem not for induction, but for hypothesis formation. The seemingly unsolvable problem of induction starts to disintegrate once we acknowledge that regularities abound, but only some of them want explanations. Only unexpected or surprising regularities make a demand on us to make an inference to the best explanation. Once abductive inference has done its work in identifying explanations of the surprises we encounter, then the job of induction is to test those abductive hypotheses. The problem of induction is turned into the problem of which abductively arrived-at hypotheses are good, or should be the ones selected for inductive testing.

3. Ramsey

Ramsey does not deal with abductive inference or inference to the best explanation as explicitly as Peirce. But in this section, I hope to show that Ramsey followed

---

[9] Misak (2004) and (2013).

Peirce to his conclusions about how inference to the best explanation is bound up with truth.

One of the most important things Ramsey picked up from Peirce was the pragmatist account of belief and its connection to action or behavior. Like Peirce, he does not reduce belief to behavior, as he takes belief to have both a subjective and an objective component. It is both an external or observable phenomenon and an internal or mental phenomenon. If a chicken "believes" that a certain caterpillar is poisonous, it abstains from eating that kind of caterpillar on account of the unpleasant experiences associated with eating them:

> The mental factors in such a belief would be parts of the chicken's behaviour, which are somehow related to the objective factors, viz. the kind of caterpillar and poisonousness. An exact analysis of this relation would be very difficult, but it might well be held that in regard to this kind of belief the pragmatist view was correct, i.e. that the relation between the chicken's behaviour and the objective factors was that the actions were such as to be useful if, and only if, the caterpillars were actually poisonous.   (FP: 40)

Like Peirce, Ramsey is interested primarily in the habits involved in assertion or judgment, rather than those involved in instinct. "It's a fly" is a judgment; brushing it off is not (OT: 50). Judgment is expressible in words and is marked by an "affirmative attitude," hence it is "capable of truth and falsity" (OT: 52). This dispositional account of belief leads Ramsey to his most famous result—that we can measure partial belief in terms of how people act, especially in betting contexts. Ramsey sees that this will be a complex matter:

> there is no uniform action which believing "p" will always produce. It may lead to no action at all, except in particular circumstances, so that its causal properties will only express what effects result from it when certain other conditions are fulfilled. And, again, only certain sorts of causes and effects must be admitted; for instance, we are not concerned with the factors determining, and the results determined by, the rhythm of the words.   (FP: 44)

Ramsey's technical contributions to showing how this complex matter might be resolved go beyond the scope of this chapter. What shall concern me are the implications for inference to the best explanation. I shall suggest that Ramsey, with Peirce, thought that inference to the best explanation was key to understanding truth and knowledge, without glorifying the conclusions of those inferences. Since Ramsey grew up in the logical analyst culture of Russell and the early Wittgenstein, his way of framing the issues fits more snugly into our contemporary discussions of inference to the best explanation, than does Peirce's.

In a famous paper, titled "Theories,"[10] Ramsey distinguishes primary and secondary systems. The primary system is that envisioned by Russell, the Tractarian

---

[10] As so often with Ramsey's final pieces of work, this was not a finished paper. After his death, his friend Richard Braithwaite selected some drafts of Ramsey's and put them together with his few published pieces, as *The Foundations of Mathematics*.

Wittgenstein, and the Vienna Circle. It is a notation describing immediate experience—it is the language of sense data or, as Wittgenstein later put it, the language with no owner.[11] The secondary system is the language of physical objects, the language of theories, and ordinary language. Ramsey argues that the primary language is of limited use, and that almost all of what we want to explain requires a "theoretical construction" (*T*: 114). He also thinks that a single-minded focus on the primary language leads to solipsism: "Solipsism in the ordinary sense in which as e.g. in Carnap the primary world consists of my experiences past present and future will not do. For this primary world is the world about which I am now thinking" (*NPPM*: 66). If we are to make sense of what James called the "blooming buzzing confusion" of experience, it is the secondary system, with its theories, laws, and claims about what exists, to which we must turn. Ramsey thinks that the secondary system might not be uniquely determined by the primary system. It does not follow, with certainty, from the primary system. Ramsey's project is thus not that of Russell, Carnap, or Wittgenstein in the 1920s, which was to either reduce the secondary system to the primary system; or to construct the secondary system from the primary system; or to demarcate what was sayable in the primary language but not in the secondary language. Stathis Psillos summarizes Ramsey's alternative view:

We treat our theory of the world as a *growing existential statement*. We do that because we want our theory to express a judgement: to be truth-valuable. In writing the theory, we commit ourselves to the existence of things that make our theory true and, in particular, to the existence of unobservable things that cause or explain the observable phenomena. We don't *have to* do this. But we think we are better off doing it, for theoretical, methodological and practical reasons. So we are *bold*. Our boldness extends a bit more. We take the world to have a certain structure (to have *natural joints*)…We don't want our theory to be true just in case it is empirically adequate. We want the structure of the world to act as an *external constraint* on the truth or falsity of our theory. So we posit the existence of a natural structure of the world (with its natural properties and relations). We come to realise that this move is *not* optional once we have made the first bold step of positing a domain of unobservable entities. These entities are powerless without properties and relations, and the substantive truth of our theories requires that these are real (or natural) properties and relations.   (2006: 85–6)

Psillos does not identify Ramsey's argument as particularly pragmatist. But it is a perfect summary of Peirce's modest, fallibilist, low-profile justification of the assumption of those entities and principles that we seem to require. We must be bold in committing ourselves to the existence of other minds, causes, and things in an external world because that is the best explanation of the observable phenomena, although Peirce, we saw, takes some of the load off of abductive inference and has it carried by deductive and inductive inference as well.

In making this move, Ramsey invented and utilized a technical innovation that defines a concept or translates a sentence in terms of the causal roles it plays. Many

---

[11] See 2009: §398 and Boncompagni (2017).

philosophers have utilized what Hempel later named these "Ramsey sentences" for their own ends. In his 1958 "Observation Language and Theoretical Language," Carnap argued that theoretical entities could be defined away by them (and noted that he had first found the idea in Ramsey). David Lewis (1972) later applied this reductionist idea to the philosophy of mind, arguing that Ramsey sentences could allow us to analyze or reduce mental terms in behavioral terms. Ramsey, however, was not interested in defining away problematic terms, such as the theoretical terms of science. He was interested in a very different stance, one which Psillos calls Ramseyan humility,[12] to mark the thought that we must not foreclose on the possibility that our theory will not be uniquely determined, or that it might be determined in ways we do not anticipate.

Ramsey makes a similar point in another late draft paper, "General Propositions and Causality." There is "nothing ... beyond the regularity [of succession the world exhibits] to be called causality"—there is only "this conduct of ours" of making "sentences called causal laws" on the basis of this regularity (*GC*: 160). Ramsey takes himself to exhibit a "realistic spirit" about causality, while rejecting the kinds of realism that say that there are causal facts. By "realistic" Ramsey means to remind us that philosophy must *matter* to us. A philosophical theory must not neglect the facts of our own experience in favor of an elegant theoretical construction; it must not be such that we can only believe it to be true if we divorce ourselves from the very features of our experience that we wished to examine in the first place. This realistic spirit is what animated Peirce (and, I would argue, Hume if he is read correctly).

From the little that has been said thus far, it will be unsurprising that Ramsey argued against the correspondence view of truth, as he found it in Russell and Wittgenstein. He was suspicious of the position that a proposition could correspond to simple objects in the world. Even as an 18 year old, he held that there simply are no such "mysterious entities" as propositions, "so unlike anything else in the world" (*NP*: 111–12), and introducing more problems than they solve. In the book manuscript he was writing when he died, he asks us to consider the belief that Jones is a liar or a fool (*OT*: 11). It seems that a correspondence theorist (at least the kind of logical atomist of Ramsey's day), who reduces everything to the primary language that then hooks onto the world, must explain the truth of this belief by appeal to its correspondence to a specious disjunctive "fact" that Jones is either a liar or a fool. Ramsey runs similar arguments with respect to universal, causal, and negative beliefs.

In the face of the failure of the correspondence theory, Ramsey turns to Peirce's account of belief and truth. We have habits of belief, and those beliefs are evaluated in terms of whether they serve us well, in an appropriate sense of "serve." We assume that one system of belief will fit the facts and will serve us best, and we aim at getting to that best system of belief. Perhaps that aim is in vain—perhaps there is no one

---

[12]   He is building on David Lewis' paper "Ramseyan Humility," posthumously published in 2009.

uniquely determined true system, or even if there is, perhaps we will not arrive at it. But if we are to have any hope of attaining that aim, we had better make inferences to the best explanations so that our beliefs or assertions are as strong as they can be, in the sense that they answer the questions that matter to us, and serve us well.[13] To bring Peirce and Ramsey together, the idea is that experience gives rise to puzzling questions and we can provide answers to those questions by inferring the best explanations. This is something we do because we want to have a complete account of all we care about. So, while inference to the best explanation is fallible, it is something we cannot do without. It is part of our inquiry aimed at truth (as the pragmatist sees it).

Ramsey is also in step with Peirce about how to deal with buried secrets, or questions whose answer seems unavailable to inquiry. He considers the view in philosophy of science that we would now call holism about the meaning of theoretical terms: what provides for the verification of a theoretical statement is our whole interconnected system of belief, not some isolated bundle of empirical content. This position is often meant to show how statements about unobservable entities, such as the very small and the very large, are meaningful and truth-apt, despite the fact that they seem not to be verifiable by observation. Ramsey sees that the argument might be deployed in additional ways:

[T]his view can be extended to include not merely what appear to be statements about facts which could not be observed, but also all statements apparently about facts which have not or will not actually be observed. So that questions about cosmogony, or the back of the moon, or anything no one has ever seen may not have any independent meaning, but only be about what it would be best for us to say in order to get a satisfactory scientific system. If this were so, "The back of the moon is made of green cheese" might be both "true" and "false," equally satisfactory "theories" having been found, one of them containing that sentence or rather allowing it to be deduced, and the other containing its contradictory.   (*OT*: 34)

Ramsey rejects the view contained in the last sentence, which results in a relativism on which baldly inconsistent beliefs could both be true. He also rejects the whole idea upon which this kind of philosophy of science rests: that some of the statements of science can be defined "by means of a 'dictionary' in such a way that they can be proved true and false by observation" and that others have more "aesthetic" merits (such a simplicity, elegance, or other "pragmatic" criteria) (*OT*: 33). On the contrary, Ramsey thinks "our ordinary statements about the external world express definite judgments, which are true or false" (*OT*: 34). But notice that he has alerted us to the Peircean solution to the problem of buried secrets. If our best theory would have it that the backside of some far-flung planet is not made of green cheese, then we are justified in saying that it is not made of green cheese. Beliefs about far-flung planets, theoretical entities, and other

---

[13] This is a highly condensed version of the argument I make in Misak (2016).

hard or impossible-to-verify matters play a role in our system of belief, and so that system imparts meaning and truth-values to them.

We saw that Peirce called assumptions, such as bivalence, regulative assumptions of inquiry. At the end of "Theories," Ramsey says something similar. The assumptions, for instance, that nature changes gradually and not by leaps, or that nature is simple

can only be laid down if we are sure that they will not come into conflict with future experience combined with the causal axioms...To assign to nature the simplest course except when experience proves the contrary is a good maxim of theory making, but it cannot be put *into* the theory in the form "Natura non facit saltum" except when we see her do so.   (*T*: 135)

On my reading of Ramsey, a principle such as the simplicity of nature is a principle that we are reasonable to invoke when making our theories, except when nature proves to the contrary. It may be that our best theory has it that nature (or a part of it) is simple, and if so, nature is simple. Ramsey took ill as he was working through this position, and it turned out that he was to run out of time.[14]

I hope to have shown in this chapter how there is an interesting and novel pragmatist position on inference to the best explanation, begun by Peirce, and completed by Ramsey. First, they offer a pragmatist, teleological justification for inference to the best explanation. This mode of inference is part of an inquiry aimed at understanding, and since such an inquiry is something we value, we have reason to employ such a mode of inference. What's more, the telos or ultimate goal of inference to the best explanation (as a part of inquiry) is indefeasible understanding or truth. Inference to the best explanation is justified because it is part of an inquiry aimed at truth. That justification is very different from standard justifications that look to present or past facts to justify inference to the best explanation.

Second, Peirce and Ramsey speak to the status of the conclusions of inferences to the best explanation. When we infer the existence of other minds, causes, and external entities to explain what we observe, we are not proving the existence of those things. But as inquiry proceeds and as we test the hypotheses we get from inference to the best explanation, we bet we are likely to rest with the existence of those entities. Indeed, our system is already on that path, and hence only highly disruptive evidence is likely to shake it. If our beliefs about the existence of these things turn out to be indefeasible, then they are true. None of this provides us with certainty, or with proof that what ends up in our best system of belief corresponds to the world as it is independently of us. This is the subtle pragmatist move about inference to the best explanation, and it is made by two of the most subtle of all the pragmatists, Peirce and Ramsey.

---

[14] See Misak (2015) for my latest attempt to make sense of situations in which our best theories are silent about some claim.

References

Boncompagni, Anna (2017). "The 'Middle' Wittgenstein (and the 'Later' Ramsey) on the Pragmatist Conception of Truth." In *The Practical Turn: Pragmatism in the British Long 20th Century*. Eds. Cheryl Misak and Huw Price. Oxford: Oxford University Press.

Carnap, Rudolp (1958). "Observation Language and Theoretical Language." Translated and reprinted in J. Hintikka (Ed.), *Rudolf Carnap, Logical Empiricist* (1975). Dordrecht: Reidel.

Fowler, Karen Joy (2013). *We Are All Completely Beside Ourselves*. New York: Penguin.

Goodman, Nelson (1955). *Fact, Fiction and Forecast*. Cambridge, MA: Harvard University Press.

Hookway, Christopher (1999). "Modest Transcendental Arguments and Sceptical Doubts." In *Transcendental Arguments: Problems and Prospects*. Ed. R. Stern. Oxford: Clarendon Press, 173–88.

Lewis, David (1972). "Psychophysical and Theoretical Identifications," in N. Block (1980), *Readings in the Philosophy of Psychology*, Volumes 1 and 2, Cambridge, MA: Harvard University Press, 207–15.

Lewis, David (2009). "Ramseyan Humility." In *Conceptual Analysis and Philosophical Naturalism*. Eds. David Braddon-Mitchell and Robert Nola. Cambridge, MA: MIT Press, 203–22.

McKaughan, Daniel J. (2008). "From Ugly Duckling to Swan: C.S. Peirce, Abduction, and the Pursuit of Scientific Theories." *Transactions of the Charles S. Peirce Society* 44/3: 446–68.

Methven, Steven (2015). *Frank Ramsey and the Realistic Spirit*. London: Palgrave Macmillan.

Misak, Cheryl (2004 [1991]). *Truth and the End of Inquiry*. 2nd edition. Oxford: Oxford University Press.

Misak, Cheryl (2011). "American Pragmatism and Indispensability Arguments." *Transactions of the Charles S. Peirce Society* 47/3: 261–73.

Misak, Cheryl (2013). *The American Pragmatists*. Oxford: Oxford University Press.

Misak, Cheryl (2015). "Pragmatism and the Function of Truth." In *Meaning without Representation*. Ed. Stephen Gross. Oxford: Oxford University Press.

Misak, Cheryl (2016). *Cambridge Pragmatism: From Peirce and James to Ramsey and Wittgenstein*. Oxford: Oxford University Press.

Peirce, Charles Sanders (*CP*) (1931–58). *The Collected Papers of Charles Sanders Peirce*. Eds. C. Hartshorne and P. Weiss (Volumes I–IV), A. Burks (Volumes VII and VIII). Cambridge, MA: Belknap Press.

Peirce, Charles Sanders (*W*) (1982–). *The Writings of Charles S. Peirce: A Chronological Edition*. Gen. Ed. E. Moore. Bloomington: Indiana University Press.

Plutynski, Anya (2011). "Four Problems of Abduction: A Brief History." *HOPOS: Journal of the International Society for the History and Philosophy of Science*: 1–22.

Psillos, Stathis (2006). "Ramsey's Ramsey-Sentences." In *Cambridge and Vienna: Frank P. Ramsey and the Vienna Circle*. Ed. Maria Carla Galavotti. Dordrecht: Springer, 67–90.

Ramsey, F.P. (1990). *F.P. Ramsey: Philosophical Papers*. Ed. D.H. Mellor. Cambridge: Cambridge University Press.

Ramsey, F.P. (*FP*) (1990 [1927]). "Facts and Propositions." Reprinted in Ramsey (1990), 34–51.

Ramsey, F.P. (*GC*) (1990 [1929]). "General Propositions and Causality." Reprinted in Ramsey (1990), 145–63.

Ramsey, F.P. (*T*) (1990 [1929]). "Theories." Reprinted in Ramsey (1990), 112–36.

Ramsey, F.P. (*NPPM*) (1991). *Notes on Philosophy, Probability and Mathematics*. Ed. Maria Carla Galavotti. Naples: Bibliopolis.

Ramsey, F.P. (*NP*) (1991 [1921]). "The Nature of Propositions." Reprinted in Ramsey (1991 [1930]), 107–19.

Ramsey, F.P. (*OT*) (1991 [1930]). *On Truth*. Eds. Nicholas Rescher and Ulrich Majer. Dordrecht: Kluwer.

Richardson, Alan (2008). "Philosophy of Science in America." In *The Oxford Handbook of American Philosophy*. Ed. Cheryl Misak. London: Oxford University Press, 339–74.

Wittgenstein, Ludwig (2009 [1953]). *Philosophical Investigations*, 4th ed. Trans. G.E.M. Anscombe, P.M.S. Hacker, and Joachim Schulte. Oxford: Blackwell.

4

Inference to the Best Explanation, Cleaned Up and Made Respectable

Jonah N. Schupbach

Inference to the Best Explanation (IBE) is a form of uncertain inference in which one reasons to a hypothesis based upon the premise that it provides a better potential explanation of some given evidence than any other available, competing hypothesis. When inferring the best explanation, one regards the explanatoriness of a hypothesis as good reason to favor that hypothesis. In this way, IBE links the explanatory value of a hypothesis to its epistemic value.

Philosophers and psychologists alike emphasize the widespread use and intuitive appeal of IBE in human reasoning (Harman, 1965; Lipton, 2004; Keil, 2006; Lombrozo, 2006; Douven, 2011; Douven and Schupbach, 2015b). In everyday affairs, people often reason to hypotheses based on their explanatory value; I might, for example, infer that my train has not yet come through the station because this hypothesis better explains the large number of people standing on the platform than any other plausible, competing hypothesis. And the applicability of IBE stretches far beyond the mundane. Scientists often infer to the best explanation; geologists may infer the occurrence of an earthquake millions of years ago because this event would, more than any other plausible hypothesis, explain various deformations in layers of bedrock. Court cases and forensic studies are decided to various degrees using IBE. This is true also of diagnostic procedures, whether performed by clinicians or auto mechanics. Philosophers themselves often rely on IBE when debating some of the most venerable topics in the history of philosophy.[1] In all of these cases across domains, people favor hypotheses on account of their ability to explain evidence.

---

[1] To take a small but informative sample: In the philosophy of religion, several well-known arguments for and against the existence of God are instances of IBE (e.g., Swinburne, 2004, p. 20). Some epistemologists claim that IBE provides us with our best response to various forms of skepticism (e.g., Vogel, 1990). In the philosophy of science, arguments to the existence of unobservables as well as arguments for scientific progress generally have the form of IBE (e.g., Putnam, 1975 and Psillos, 1999). And the same can be said of debunking arguments in ethics, and arguments for realist positions in metaethics and metaphysics—witness Lewis's (1986) central argument to possible worlds realism.

Despite its ubiquity and apparent cogency, IBE has a stormy history. It is difficult to think of another form of inference that has been, at once, so heartily defended by its champions and disparaged by its critics. Harman (1965, p. 88) boldly claims that IBE is the "basic form of nondeductive inference," having normative and conceptual priority over other forms of uncertain inference. Fumerton (1980) argues for the opposite claim that IBE is no more than an incomplete description of simpler forms of induction, having no independent epistemic merit. Van Fraassen (1989, pp. 142–3) famously offers the "bad lot" objection against IBE: IBE *assumes without argument* that the true hypothesis is likely to be one of the hypotheses under consideration. The upshot is that it can hardly be said to give us a reliable vehicle for inferring to conclusions that are more probably true.[2]

Of all the objections put to IBE, however, there is one that is most fundamental. The worry, expressed by the proponents and opponents of IBE alike, is that despite decades of serious philosophical investigation, the specific nature of IBE is still up for grabs. In the words of one of IBE's foremost supporters (Lipton, 2004, p. 2), "[IBE] is more a slogan than an articulated philosophy." This worry is of primary importance because it needs to be addressed before IBE's more specific vices and virtues may be explored; who is to say whether Harman, Fumerton, van Fraassen, and others are correct in their evaluations of IBE so long as this inference form has no clear articulation?

This chapter, first of all, attempts to rectify this situation by specifying more precisely the nature of IBE. The most significant roadblock currently standing in the way of a clear account of IBE is our lack of understanding regarding the concept(s) of explanatoriness. The key premise of any instance of IBE claims a difference in explanatoriness between available potential explanations. Yet, the notion of explanatoriness is ambiguous. Section 1 accordingly distinguishes one particular version of IBE by first explicating precisely one prevalent sense of explanatoriness.

This chapter is not merely interested in the clear *articulation* of IBE, however, but also in its *evaluation*. To this end, Section 2.1 argues that the specific version of IBE introduced in Section 1 is cogent, meaning that its premise always lends epistemic support to its conclusion. Section 2.2 goes further and defends, through a series of computer simulations, IBE as a respectably reliable mode of inductive inference (at least when compared to the somewhat less contentious case of Bayesian inference).

1. IBE, Cleaned Up

The key premise of any particular inference to the best explanation refers to a difference in explanatoriness—or explanatory goodness—between considered hypotheses. But explanatoriness is famously evaluated along different dimensions, corresponding to

---

[2] See (Schupbach, 2014) for a recent response to the bad lot objection. Douven and Schupbach (2015a) additionally offer a brief response to van Fraassen's claim that IBE is a poor form of inference insofar as it commits the probabilistic, epistemic agent to diachronically incoherent updates.

the various acclaimed explanatory virtues. Potential explanations may be prized for their great simplicity, unification, generality, power, or some combination of these (or other) virtues. One immediate consequence of this, often overlooked by IBE's commentators, is that the nature of an inference to the best explanation will depend upon the notion of explanatoriness at work therein. As a general category, IBE is polymorphous. There are at least as many distinct forms of IBE as there are distinct senses in which a hypothesis may be judged more explanatory than others; Inference to the Most Unifying Potential Explanation, for example, differs (prima facie, quite substantially) from Inference to the Simplest Potential Explanation.[3]

This basic point gives rise to a concern for generalist accounts and evaluations of IBE (i.e., much of the extant work on IBE). Such accounts gloss over potentially crucial differences between versions of IBE, confounding any attempt to evaluate seriously any specific version—the normative upshot of Inference to the Most Unifying Potential Explanation plausibly differs from that of Inference to the Simplest Potential Explanation. Any careful articulation and evaluation of IBE must rather build upon a precise account of the notion of explanatoriness determining what it takes for a potential explanation to be best. In the remainder of this chapter, I heed this advice by focusing my sights on one particular acclaimed explanatory virtue and the corresponding version of IBE.[4]

1.1 Explanatoriness as power

Our aim is to distinguish a particular version of IBE by first explicating an important notion of explanatory goodness. The result will be more interesting to the extent that the notion of explanatory goodness we focus on is one that reasoners indeed have in mind on some of the occasions in which they infer best explanations. With that in mind, we take a cue from C. S. Peirce's (1935, 5.189) description of explanatory inference (or "abduction"):

Long before I first classed abduction as an inference it was recognized by logicians that the operation of adopting an explanatory hypothesis—which is just what abduction is—was subject to certain conditions. Namely, the hypothesis cannot be admitted, even as a hypothesis, unless it be supposed that it would account for the facts or some of them. The form of inference, therefore, is this:

---

[3] If one is a pluralist about the nature of explanation itself, then varieties of IBE may further multiply, with Inference to the Best Causal-Mechanical Explanation, for example, differing from Inference to the Best Covering Law Explanation, and so on. Whether these are differences that make a difference to the logic of IBE not already captured by the distinct notions of explanatoriness is an important question. Regardless, my focus in this chapter will be on one particular brand of IBE distinguished by a single explanatory virtue at work in its central premise.

[4] None of this is meant to suggest that all precisely articulated notions of explanatoriness—and corresponding species of IBE—will only refer to one explanatory virtue. Plausibly, many instances of IBE involve a notion of explanatoriness that effectively strikes a balance between several distinct virtues. Any informative evaluation of this brand of IBE must build upon a precise account of what these virtues are and how they are balanced.

The surprising fact, C, is observed;
But if A were true, C would be a matter of course;
Hence, there is reason to suspect that A is true.

According to Peirce, an inference in which one adopts an explanatory hypothesis begins when a "surprising fact" calls out for explanation. A hypothesis is put forth then, which must render the surprising fact a "matter of course." The key idea here is that a hypothesis explains some surprising fact well if it is able to render that fact unsurprising (i.e., expected). Let us call Peirce's notion of explanatoriness, having to do with a hypothesis's ability to make evidence unsurprising, "power."

Reasoners often assess how explanatory a hypothesis is with respect to some evidence by gauging its power over that evidence (Schupbach and Sprenger, 2011, p. 108). Indeed, this particular notion of explanatoriness is so prevalent in instances of IBE that Peirce just seems to *identify* power with the general notion of explanatoriness in the above passage. While Peirce is surely wrong to suggest that we *always* adopt explanatory hypotheses on the basis of their power over explananda,[5] it does seem that this virtue adequately describes the notion of explanatory goodness at work in many applications of IBE. Accordingly, we focus in the rest of this chapter on applications of IBE in which explanatoriness is evaluated purely as power.

To develop a precise explication of power, we start with Peirce's idea that an explanatory hypothesis has power over some surprising explanandum if it is able to render that explanandum unsurprising. This thought naturally lends itself to a subtler condition for an explication of power: a hypothesis has power over a proposition *to the extent that* it makes that proposition less surprising—or more expected—than it otherwise was. So, a geologist will favor a prehistoric earthquake as a powerful explanation of certain observed deformations in layers of bedrock to the extent that deformations of that particular character, in that particular layer of bedrock, and so on would be less surprising given the occurrence of such an earthquake. This condition is not a mere restatement of Peirce's idea. For one thing, given this condition, a hypothesis may provide a powerful explanation of a surprising proposition and still not render it a matter of course in any sense; i.e., a hypothesis may make a proposition much less surprising while still not making it unsurprising. Additionally, this subtler condition does not suggest that a proposition must be surprising in order to be explained; a hypothesis may make a proposition much less surprising (or more expected) even if the latter is not so surprising to begin with.

This condition may be used to motivate further conditions for an account of power. First, just as (positive) power comes with a decrease in surprise, one might say that a hypothesis has "negative power" over some proposition to the extent that it makes that proposition *more* surprising. I would judge the hypothesis that my train has already come through the station to be a terrible explanation of the large number of people

---

[5] After all, sometimes we infer best explanations based on their having virtues best describable as monadic properties (as opposed to relational properties between these hypotheses and evidence), simplicity being the most obvious example.

standing on the platform; this is because I know that the majority of people in the station at this time of day are there to catch this particular train; my train typically leaves behind an empty platform. This hypothesis thus has negative power; if adopted, the crowd that I observe before me is even *more* surprising than it already was.

Given the above, a hypothesis lacks all (positive or negative) power whatever relative to some given explanandum if the latter is neither more nor less surprising in light of that hypothesis. The perceived motions of the planet Uranus are less surprising in light of the hypothesized existence of Neptune, but they are neither more nor less surprising given that my train has not yet passed through the station. The latter hypothesis is simply impotent with respect to that explanandum.

Insofar as a hypothesis has power over a proposition to the extent that it renders the latter unsurprising, one might additionally conclude that a hypothesis provides a *maximally* powerful explanation of some proposition just when it would lead one to expect that proposition to be true with certainty; this occurs when the hypothesis implies the truth of that proposition. On the other hand, a *minimally* powerful explanation of some known proposition is one that renders the latter maximally surprising, and this occurs when the hypothesis implies that the proposition in question is false.

Finally, the less surprising a proposition's truth is in light of a hypothesis, the more surprising is its falsity. Given the above, this means that the more power a hypothesis has over a proposition, the less it has over the negation of that proposition. To summarize then, the intuitive starting point provided by Peirce can naturally be extended so that it provides the following compelling conditions for an explication of power:

Condition 1:   A hypothesis has *(positive) power* over a proposition to the extent that it decreases the degree to which that proposition is surprising (i.e., increases the degree to which we expect that proposition to be true).

Condition 2:   A hypothesis has *negative power* over a proposition to the extent that it increases the degree to which that proposition is surprising.

Condition 3:   A hypothesis has *no power* over (i.e., is *impotent* with respect to) a proposition if and only if the latter is neither more nor less surprising in light of that hypothesis.

Condition 4:   A hypothesis has *maximal power* over a proposition if and only if it leads us to expect with certainty that the proposition is true.

Condition 5:   A hypothesis has *minimal power* over a proposition if and only if it leads us to expect with certainty that the proposition is false.

Condition 6:   The more power a hypothesis has relative to a proposition, the less it has relative to the negation of that proposition.

1.2 The measure of power $\mathcal{E}$

The task of this section of the chapter will be to apply the above considerations in order to arrive at a precise explication of power. If one makes use of the probability calculus to clarify and interpret these conditions, then only one measure of power with a certain desirable mathematical structure satisfies a subset of *Conditions 1–6*. Hence, the intuitions pertaining to power presented in the previous section already suffice to pin down a formal account of this concept. This account then clarifies, in the precise language of the probability theory, what it takes for a hypothesis to provide the best available explanation, when explanatoriness is evaluated purely in terms of power.

The key interpretive move of this section is to formalize a decrease in surprise (increase in expectedness) as an increase in probability. This move may seem dubious depending upon one's interpretation of probability. Given a physical interpretation (e.g., a relative frequency or propensity interpretation), it would indeed be difficult to saddle such a psychological concept as surprise with a probabilistic account. However, when probabilities are themselves given a more psychological interpretation (whether in terms of simple degrees of belief or the more normative degrees of *rational* belief), this move makes sense. In this case, probabilities map neatly onto degrees of (rational) expectedness. Accordingly, given the inverse relation between surprise and expectedness (the more surprising a proposition, the less one expects it to be true), surprise is straightforwardly related to probabilities: the observation that h decreases the degree to which e is surprising corresponds with the judgment that h increases the degree to which e is expected, expressed probabilistically by the inequality $Pr(e) < Pr(e|h)$.[6]

As part of its "desirable mathematical structure" (which we specify exactly with two purely formal conditions of adequacy in the appendix), we require that the degree of power that hypothesis h has over evidence e, $\mathcal{E}(e,h)$, be real-valued on the closed interval $[-1,1]$. In explanatory contexts, $\mathcal{E}(e,h) = 1$ ($\mathcal{E}$'s maximal value) is the value at which h is interpreted as a maximally powerful potential explanation of e. $\mathcal{E}(e,h) = -1$ indicates the minimal degree of power for h relative to e, where h is interpreted as providing a maximally powerful potential explanation for e being *false*. $\mathcal{E}(e,h) = 0$ is the "neutral point" at which h lacks any power relative to e and its negation.

What are the corresponding formal conditions under which $\mathcal{E}$ takes these values? Here is where *Conditions 1–6* become relevant. As noted, $\mathcal{E}(e,h)$ should take the value 0 precisely when h lacks any power relative to e (and $\neg e$). *Condition 3* specifies that this occurs if and only if e (and $\neg e$) is neither more nor less surprising in light of h. Given the inverse relation between surprise and probability, this condition is explicated as h and e being statistically irrelevant to one another: $Pr(e|h) = Pr(e)$, or equivalently (remembering that Pr is a regular probability measure and that e and h are contingent propositions), $Pr(h \wedge e) = Pr(h) \times Pr(e)$.

CA1 (Neutrality): $\mathcal{E}(e,h) = 0$ if and only if $Pr(h \wedge e) = Pr(h) \times Pr(e)$.

$\mathcal{E}(e,h)$ takes a maximum value of 1 if and only if h is maximally powerful with respect to e. *Condition 4* clarifies that such will be the case precisely when h leads us to

---

[6] The background knowledge term k always belongs to the right of the solidus "|" in Bayesian formalizations (e.g., $Pr(e|k) < Pr(e|h \wedge k)$). Nonetheless, here and in the remainder of this chapter, I leave k implicit in all formalizations for ease of exposition.

expect with certainty that e is true. Such a notion is straightforwardly formalized with the equality $Pr(e \mid h) = 1$.

CA2 (Maximality): $\mathcal{E}(e,h) = 1$ if and only if $Pr(e \mid h) = 1$.

Condition 6 above requires that as the power of h relative to e increases, that of h relative to $\neg e$ decreases. When explanatoriness is assessed as power, this amounts to the idea that the more h explains the truth of e, the less it explains its falsity. Maximality and Neutrality provide us with further rationale for this condition. Maximality tells us that $\mathcal{E}(e,h)$ should be maximal only if $Pr(e \mid h) = 1$. Importantly, in such a case, $Pr(\neg e \mid h) = 0$, and this value intuitively corresponds to the point at which we should expect $\mathcal{E}(\neg e, h)$ to be minimal (see *Condition 5* above). In other words, given *Maximality*, we see that $\mathcal{E}(e,h)$ takes its maximal value 1 precisely when $\mathcal{E}(\neg e, h)$ takes its minimal value -1 and vice versa. Also, we know from *Neutrality* that $\mathcal{E}(e,h)$ and $\mathcal{E}(\neg e, h)$ should always equal zero at the same point given that $Pr(h \wedge e) = Pr(h) \times Pr(e)$ if and only if $Pr(h \wedge \neg e) = Pr(h) \times Pr(\neg e)$. These considerations lead to the following requirement:

CA3 (Symmetry): $\mathcal{E}(e,h) = -\mathcal{E}(\neg e, h)$.

The final condition of adequacy appeals to a scenario in which degree of power is unaffected. If a hypothesis h_2 is impotent with respect to another hypothesis h_1, to some proposition e, and to any logical combination of h_1 and e, then *Condition 3* tells us that it does nothing to increase or decrease the degree to which these are surprising. In such a case, conjoining h_2 to h_1 will do nothing to increase or decrease the degree to which e is surprising in light of h_1. Given *Neutrality*, we can state this in other words: if h_2 has no *power* whatever relative to e, h_1, or any logical combination of e and h_1, then its presence will not affect the overall power of h_1 relative to e. This gives us the following condition:

CA4 (Irrelevant Conjunction): If $Pr(e \wedge h_2) = Pr(e) \times Pr(h_2)$ and $Pr(h_1 \wedge h_2) = Pr(h_1) \times Pr(h_2)$ and $Pr(e \wedge h_1 \wedge h_2) = Pr(e \wedge h_1) \times Pr(h_2)$, then $\mathcal{E}(e, h_1 \wedge h_2) = \mathcal{E}(e, h_1)$.

These four adequacy conditions conjointly determine a unique measure of power as stated in the following theorem (proof in the appendix).[7]

---

[7] Measure $\mathcal{E}$ is structurally equivalent to Kemeny and Oppenheim's (1952) measure of "factual support,"

$$F(h,e) = \frac{Pr(e \mid h) - Pr(e \mid \neg h)}{Pr(e \mid h) + Pr(e \mid \neg h)},$$

which itself is ordinally equivalent to the log-likelihood measure of incremental confirmation $L(h,e) = \log[Pr(e \mid h) / Pr(e \mid \neg h)]$ (Good, 1983; Fitelson, 1999). The key difference between $\mathcal{E}$ and these measures is in their interpretation and application; $\mathcal{E}(e,h)$ is $F(h,e)$ with h and e interchanged. This difference is significant, as the conditions of adequacy used to motivate the measures differ. It is easy to verify that F and L at least fail to satisfy *CA2* and *CA3*, making them unsuitable for measuring power—though both are among the most plausible measures of incremental confirmation. This is appropriate, since these conditions properly constrain measures of power, but they make little sense as constraints on measures of incremental confirmation.

Theorem 1. The only measure with a desirable mathematical structure that satisfies CA1–CA4 is

$$\mathcal{E}(e,h) = \frac{Pr(h|e) - Pr(h|\neg e)}{Pr(h|e) + Pr(h|\neg e)}.$$

Note that this measure also satisfies the conditions from Section 1.1 that were not needed in order to prove Theorem 1. *Conditions 1* and *2* require that power increases (decreases) as the degree to which e is surprising decreases (increases) in light of h. Put more formally, these conditions require that $\mathcal{E}(e,h) > 0$ to the extent that $Pr(e) < Pr(e|h)$. These conditions are satisfied by $\mathcal{E}$ given that $\mathcal{E}(e,h) > 0$ to the extent that $Pr(h|e) > Pr(h|\neg e)$, which in turn is true just to the extent that $Pr(e|h) > Pr(e)$. [8] *Condition 5* requires that power is minimal if and only if e is certainly false in light of h. This fact also follows necessarily from $\mathcal{E}$ given that $\mathcal{E}(e,h) = -1$ if and only if $Pr(e|h) = 0$. [9] Thus, these conditions determine for us an intuitively well-grounded, unique measure of power. [10]

With $\mathcal{E}$ in hand, we may formally articulate an important version of IBE. In cases where the premise that h provides the best available potential explanation of the evidence e can be restated as the claim that this hypothesis has more power over e than any competing hypothesis, we have that $\mathcal{E}(e,h) > \mathcal{E}(e,h_i)$ for any and all of h's explanatory competitors h_i. The corresponding full version of IBE, which we can denote IBE_p ("p" designating the notion of explanatoriness as power), has the following form:

$$e$$

$$(IBE_p) \quad \mathcal{E}(e,h) > \mathcal{E}(e,h_i), \text{ for any } h_i \text{ competing with } h$$

$$\therefore h$$

The question of whether or not this species of IBE is a cogent inference form is now more tractable. We investigate this question in the next section.

2. IBE, Made Respectable

The nature of IBE changes depending on the precise sense of explanatoriness at work in its central premise. And the evaluation of IBE naturally follows suit. Any

---

[8] This is easy to see in light of the fact that

$$\frac{Pr(h|e)}{Pr(h|\neg e)} = \frac{Pr(e|h)}{Pr(e)} \times \frac{1 - Pr(e)}{1 - Pr(e|h)}.$$

[9] $\mathcal{E}(e,h) = -1$ just in case $\mathcal{E}(e,h) = -Pr(h|\neg e) / Pr(h|\neg e)$. But this equality holds only if $Pr(h) \neq 0$ and $Pr(h|e) = 0$ which implies that $Pr(e|h) = 0$.

[10] Alternative uniqueness theorems providing different axiomatic foundations for $\mathcal{E}$ may be found in (Schupbach and Sprenger, 2011) and (Cohen, 2015). That $\mathcal{E}$ can be defended via several distinct uniqueness theorems helps alleviate the worry that our result is driven by a faulty condition of adequacy. Schupbach and Sprenger (2011) also provide further support for $\mathcal{E}$ via several theorems, which show that $\mathcal{E}$ matches clear intuitions about power. $\mathcal{E}$ gains yet another line of support as an accurate measure of (explanatory) power in (Schupbach 2011), where I show experimentally that $\mathcal{E}$ is a good predictor of actual human judgments of explanatoriness.

informative evaluation of IBE will attend to a precisely explicated species of IBE. Correspondingly, any attempt to evaluate (defend or criticize) IBE in general without first precisely articulating the version of IBE will be at least as confused as the general category of explanatoriness itself. Once different versions of IBE are disentangled, it may well turn out that some of these are epistemically defensible and others not. This will depend most obviously on whether the notion of explanatoriness at work in a particular version of IBE carries any genuine epistemic force.

This section evaluates IBE$_p$, the version of IBE instantiated when explanatoriness is evaluated as power. The strategy is as follows: Section 2.1 first defends IBE$_p$ as cogent, arguing that there is a clear sense in which its premises always support its conclusion. Section 2.1 also suggests that IBE$_p$ is useful as an informal heuristic allowing us to approximate sound probabilistic reasoning. Section 2.2 thus asks just how reliable this inference form is when compared to Bayesian inference. It turns out that IBE$_p$ stacks up quite well. Indeed, under certain (arguably common) conditions, IBE$_p$ provides a *more* reliable mode of inference than that based on sound probabilistic reasoning.

2.1 Some implications of power

As a first step toward evaluating IBE$_p$, it is enlightening to spell out the probabilistic implications of a single hypothesis h having positive power over evidence e, $\mathcal{E}(e,h) > 0$. Filling in the details of $\mathcal{E}$, this judgment can be shown to have the following probabilistic consequences (where '$\Leftrightarrow$' symbolizes interderivability):[11]

$$\frac{Pr(h\,|\,e) - Pr(h\,|\,\neg e)}{Pr(h\,|\,e) + Pr(h\,|\,\neg e)} > 0$$

$$\Leftrightarrow Pr(h\,|\,e) > Pr(h\,|\,\neg e)$$

$$\Leftrightarrow \frac{Pr(e\,|\,h)}{Pr(e)} > \frac{Pr(\neg e\,|\,h)}{Pr(\neg e)}$$

$$\Leftrightarrow Pr(e\,|\,h) - Pr(e\,|\,h)Pr(e) > Pr(e) - Pr(e\,|\,h)Pr(e)$$

$$\Leftrightarrow Pr(e\,|\,h) > Pr(e)$$

$$\Leftrightarrow Pr(e\,|\,h) > Pr(e\,|\,\neg h) \qquad\qquad \text{(L)}$$

$$\Leftrightarrow Pr(h\,|\,e) > Pr(h) \qquad\qquad \text{(C)}$$

(L) and (C) are especially interesting; these results tell us that positive power can be probabilistically represented using either a likelihood comparison or the notion of incremental confirmation, respectively. We will have more to say, in the rest of this section, about the likelihood comparisons indicated by certain explanatory judgments. (C) reveals that, to the extent that a hypothesis is able to provide a powerful explanation of the evidence in question, that evidence confirms (raises the probability

---

[11] Recall that *Pr* is a regular probability measure and that *e* and *h* are contingent propositions.

of) that hypothesis. This suggests a particular sense in which the judgment that a hypothesis is positively explanatory of the evidence does constitute a reason to favor that hypothesis.

IBE$_p$'s central premise does not claim, however, that h has positive power over e. Instead, it makes the comparative claim that h offers a more powerful potential explanation of e than does any competing hypothesis h_i, $\mathcal{E}(e,h) > \mathcal{E}(e,h_i)$. Filling in the probabilistic details of $\mathcal{E}$, this explanatory judgment is explicated as follows:

$$\frac{Pr(h\,|\,e) - Pr(h\,|\,\neg e)}{Pr(h\,|\,e) + Pr(h\,|\,\neg e)} > \frac{Pr(h_i\,|\,e) - Pr(h_i\,|\,\neg e)}{Pr(h_i\,|\,e) + Pr(h_i\,|\,\neg e)}$$

$$\Leftrightarrow \frac{Pr(h\,|\,e)}{Pr(h\,|\,\neg e)} > \frac{Pr(h_i\,|\,e)}{Pr(h_i\,|\,\neg e)}$$

$$\Leftrightarrow \frac{Pr(e\,|\,h)Pr(\neg e)}{Pr(\neg e\,|\,h)Pr(e)} > \frac{Pr(e\,|\,h_i)Pr(\neg e)}{Pr(\neg e\,|\,h_i)Pr(e)}$$

$$\Leftrightarrow Pr(e\,|\,h) - Pr(e\,|\,h)Pr(e\,|\,h_i) > Pr(e\,|\,h_i) - Pr(e\,|\,h)Pr(e\,|\,h_i)$$

$$\Leftrightarrow Pr(e\,|\,h) > Pr(e\,|\,h_i)$$

$\mathcal{E}$ thus reveals that, in multiple-hypothesis settings, the hypothesis that offers the most powerful potential explanation of some proposition will be the one that makes that proposition the most likely. In Bayesian terms, the hypothesis judged to provide the best explanation will have the greatest corresponding *likelihood* of any explanatory hypothesis considered. This result clarifies the nature of the reason that favors the most explanatory hypothesis over those that are explanatorily inferior. A hypothesis's likelihood ($Pr(e|h)$) is positively related to its overall probability in light of the evidence ($Pr(h|e)$), as can be seen via Bayes's Theorem:

$$Pr(h|e) = \frac{Pr(h) \times Pr(e|h)}{Pr(e)}$$

Holding all else constant, the greater a hypothesis's corresponding likelihood, the greater its probability given e.

Furthermore, when comparing various hypotheses with respect to the same evidence e (as in instances of IBE), $Pr(e)$ is the same regardless of which hypothesis one has in mind. Accordingly, we can say that if h offers the most powerful of the available potential explanations of e, then it is also the most probable hypothesis given e so long as it is at least as plausible as its competitors apart from considerations of e—i.e., so long as $Pr(h) \geq Pr(h_i)$, for all rival hypotheses h_i. Of course, the most explanatory hypothesis may be less plausible apart from considerations of e as compared to other hypotheses; in this case, it is possible for h to provide the best explanation and *not* be the most probable available hypothesis overall. Nonetheless, it is also true that the power of h over e may be greater than that of rival hypotheses to such an extent that it overcomes the fact that $Pr(h)$ is comparatively low and makes it the case that *h is* the most probable competing hypothesis.

In general then, the judgment that a hypothesis provides the most powerful explanation of the evidence provides us with a good reason to favor that hypothesis. This is because comparative judgments of power bear witness to relative degrees of statistical relevance between *e* and considered hypotheses. The hypothesis with the greatest power over *e* corresponds to that which is the most statistically relevant to *e*, implying that this hypothesis has the greatest corresponding likelihood. A hypothesis's likelihood is positively related to its overall probability in light of the evidence. The judgment that a hypothesis provides the best available explanation of the evidence *does* therefore constitute reason to favor that hypothesis over its explanatory competitors, because this judgment reflects probabilistic information that has a positive bearing on *h*'s overall probability in light of *e*. In this sense, IBE_p is manifestly a cogent form of nondeductive inference.

At this point, it is important to bear in mind what a general defense of a nondeductive inference *form* can and cannot provide. Precisely in virtue of its nondeductive nature, such a form cannot fairly be criticized for not always guiding us from true premises to a true conclusion. Instead, the most that we can *generally* require of such an inference form is that, whenever we instantiate it, we do end up with premises that—in some way, to some extent—positively support the conclusion. The above claim that IBE_p is cogent thus amounts to the claim that any inference to the most powerful explanation's premises will provide positive support for the corresponding conclusion.[12]

2.2 What computers can teach us about IBE

The picture that arises out of the above defense of IBE_p's cogency is that considerations of power have epistemic value on account of the role they play in reflecting important probabilistic information. When a person recognizes that a hypothesis has the most power over the evidence, that person has taken account of a fact with probabilistic ramifications in favor of that hypothesis. In this way, IBE_p enables us to account for relevant probabilistic information when reasoning without necessarily having explicit awareness of the individual probabilities involved or even any working knowledge of probability theory. The foregoing investigation into the epistemic implications of power thus sheds new light on Peter Lipton's oft-repeated dictum that "explanatory loveliness is a guide to judgments of likeliness" (2004, p. 121).

---

[12] Note that this is a far cry from claiming that the conclusion of any particular inference of this form is *justified*. Whether an inference *form* is cogent is determined at a general level—based upon whether there is a logical sense in which the sort of premises required by that form provide positive evidence for the sort of conclusion described. Whether a particular conclusion of an inference is *justified*, on the other hand, is not generally decidable. There must be at least some reason in favor of the conclusion of any particular instance of an inference form, if that form is cogent. However, other epistemic considerations may bear upon this conclusion in such a way that it is overall unjustified. Whether or not the conclusion of a particular such inference is justified is determined by the full epistemic details of one's context; whether or not IBE_p is a cogent form of inference is not determined by such contextually specific factors.

IBE_p describes a cogent inference form because the power of a hypothesis is a genuine epistemic virtue; all else being equal between competing hypotheses, the most powerful hypothesis will also be the most probable. But all else is seldom equal in real life. Consequently, in contexts where people typically make inferences to the most powerful explanations, it might be that, despite its cogency, this inference form is not very *useful*; though considerations of power reflect important probabilistic information in such contexts, IBE_p may commonly misguide us because of the probabilistic information that these considerations ignore (viz., prior probabilities). Just how useful IBE_p is depends *inter alia* on its potential for guiding us to true hypotheses despite selectively attending only to some of the relevant probabilistic information.

In this section, I use computer simulations—based closely upon those devised and reported by Glass (2012)—to model and compare the performance of IBE_p versus probabilistic reasoning for the sorts of everyday contexts in which people are inclined to infer most powerful explanations. The general methodology that these simulations employ is summarized in the following steps:

1. For each of a specified number n of competing (mutually exclusive) explanatory hypotheses, assign values of the prior probabilities ($Pr(h_i)$) and likelihoods ($Pr(e|h_i)$). Priors and likelihoods are drawn randomly from a normal and uniform distribution, respectively (see discussion below for more details).

2. Using weights corresponding to the respective values of $Pr(h_i)$, randomly select the "true" hypothesis h_j from $h_1, h_2, \ldots, h_n$. Each h_i has a $Pr(h_i)$ chance of being selected.

3. Using the value of $Pr(e|h_j)$ (the likelihood associated with the true hypothesis), check whether e "occurs." If e occurs, continue with steps 4–6; otherwise, end this iteration.

4. Check which of the n hypotheses has the greatest power; i.e., find h_k where $\mathcal{E}(e, h_k) > \mathcal{E}(e, h_i)$ for all $i \neq k$.

5. Check which of the n hypotheses is the most probable in light of e; i.e., find h_l where $Pr(h_l|e) > Pr(h_i|e)$ for all $i \neq l$.

6. If $h_k = h_j$, count this as a case where the most explanatory hypothesis matches the true hypothesis; if $h_l = h_j$, count this as a case where the most probable hypothesis matches the true hypothesis.

Steps 1–6 constitute one iteration of the simulation. After a large number of repeated iterations, the simulation provides estimates of how often the hypothesis with the greatest power (relative to e) corresponds to the true hypothesis and how often the hypothesis with the greatest probability (conditional on e) corresponds to the true hypothesis. In either case, this is calculated as the number of times that one gets such a match divided by the number of instances in which e occurs.

The goal is for this procedure to model *real-world* contexts in which people are inclined to infer most powerful explanations, and thereby to give us an estimate of IBE_p's average, actual accuracy in such contexts. Whether one is able to accomplish this end (and precisely which real-world contexts are modeled) is contingent upon several assumptions built into the simulation. Two important decisions in particular constrain the model's proper application: (1) whether one includes a "catch-all" hypothesis, and (2) how exactly one assigns prior probabilities (values of $Pr(h_i)$) to the hypotheses.

Regarding (1), in general, if explanatory hypotheses h_1 through h_n are not only assumed to be mutually exclusive but also jointly exhaustive, then one's model will represent a situation in which it is known that one of these competing hypotheses must be true. In such a case, there is no need to include a "catch-all" hypothesis to represent all unimagined hypotheses. To take a simple example, one might be interested in inferring whether a particular coin is fair or biased by examining how well these respective hypotheses explain a series of observed coin flips. Given that the coin must either be fair or biased, there is no room to include a third, catch-all hypothesis.

However, there are many contexts in which it is not known with certainty that one of the considered hypotheses is true. In order to represent this scenario, a model must include a catch-all hypothesis. Within the above simulation procedure, a catch-all hypothesis can be chosen as the true hypothesis h_j in step 2, but it cannot be chosen as the most explanatory (h_k in step 4) or probable (h_l in step 5) of the available competing hypotheses for the simple reason that it is not considered by—and therefore not available to—the reasoner.

Decisions pertaining to (2) are more difficult. How should one go about assigning prior probabilities to the explanatory hypotheses in these simulations if the goal is to model contexts in which people are inclined to infer most powerful explanations? Such probabilities must always sum to one,[13] but is there more to say than this? At least the following seems clear: the set of hypotheses reasoners are willing to entertain in such contexts will be determined in part by how plausible those hypotheses are to begin with. When faced with evidence in need of explanation, a person may be able to conjure up any number of alternative, explanatory hypotheses having various degrees of power over that evidence. But the fact that a given hypothesis is conjurable and powerful is not enough to place that hypothesis within the ranks of those that a reasoner is willing to infer. No matter how well I think that an ancient extraterrestrial visitation, for example, would explain the patterned deformations that I observe in layers of bedrock, I will not consider this hypothesis when inferring the best explanation. This is because, to my mind, that hypothesis is so implausible to begin with that it's not worth consideration. By contrast, insofar as someone believes that the extraterrestrial hypothesis *is* plausible, that person will find it appropriate to consider for potential inference.

This is particularly true when reasoners are inclined to rest all inferential weight on considerations of power. In such cases, considerations of prior plausibility are neglected.

---

[13] This is true in either case regarding decisions about (1). If no catch-all hypothesis is required, then h_1 through h_n are mutually exclusive and jointly exhaustive, their prior probabilities thus necessarily summing to one. If a catch-all is required, then h_1 through h_n *plus the catch-all hypothesis* are mutually exclusive and jointly exhaustive, with prior probabilities thus summing to one.

But people are not inclined to neglect such considerations when they weigh heavily for or against considered hypotheses. That is, it is plausible to think that people only allow power alone to do the inferential heavy lifting in cases where there is no substantial difference in prior plausibility that also weighs in favor of one of the hypotheses.

The upshot is that the hypotheses considered when people infer most powerful explanations will all typically be comparably plausible (though they might all have low probability—e.g., if there are a sufficiently large number of mutually exclusive hypotheses to consider). For the sake of modeling the usual IBE$_p$ context, then, the prior probabilities of the considered hypotheses are chosen in such a way that they tend to be closer in value to one another. This is only enforced for the *considered* hypotheses though; when a catch-all hypothesis is included in a simulated context, the prior probability of this catch-all hypothesis is allowed to stray from the values of the prior probabilities corresponding to the considered hypotheses.[14]

This basic simulation design was run for two distinct scenarios corresponding to the choice of whether or not to include a catch-all hypothesis. Within each of these two scenarios, a specific simulation was run for a particular number n of competing explanatory hypotheses (n ranging from 2 to 10). Any individual simulation included 1 million repetitions to secure accuracy.

Results are shown in Figures 4.1 and 4.2. For a given number of hypotheses, these figures display the percentage of cases in which the most powerful hypothesis is true as

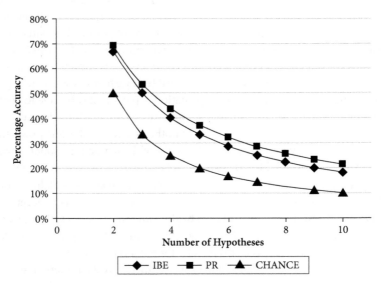

Figure 4.1 Percentage accuracies in contexts that do not include a catch-all.

[14] This is achieved by sampling prior probabilities randomly from a normal distribution ($\mu = 0.5$, $\sigma = 0.15$), choosing the prior probability of the catch-all randomly from a uniform distribution between 0 and 1, and then renormalizing so that the probabilities sum to 1.

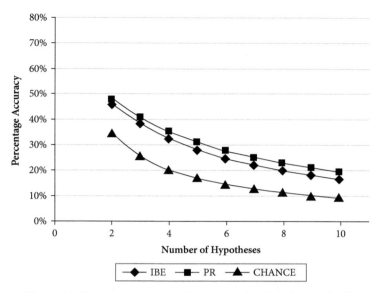

Figure 4.2 Percentage accuracies in contexts that include a catch-all.

compared to the percentage of cases in which the most probable hypothesis is true. For reference, the percentage accuracies of a random guess ("chance") from the lot of available hypotheses is also displayed. Figure 4.1 shows these results for contexts that do not include a catch-all hypothesis, while Figure 4.2 shows the results corresponding to contexts that do.

Both figures reveal that percentage accuracies decrease as the number of hypotheses increases. This validates the intuitive idea that as the number of competing hypotheses increases, so does the number of ways in which one's inferred conclusion could go wrong. Hence, accuracy decreases when there are more hypotheses to which one can infer. Note, however, that IBE$_p$ and probabilistic reasoning are both unsurprisingly much more accurate in contexts with no catch-all hypothesis. This fact allows us to clarify one sense in which increasing the number of considered hypotheses could actually *increase* the respective accuracies of these inference rules. Each new hypothesis added to the lot of those considered decreases the probability of (i.e., the need for) a catch-all hypothesis; each such addition brings us a step closer to the special case where our considered hypotheses *partition* the space of possibilities, leaving the catch-all with zero probability. And as one moves closer to a context in which there is no space left for a catch-all in this way, the result may be an overall increase in accuracy. Thus, comparing Figures 4.1 and 4.2, the addition of an explanatory hypothesis that exhausts the remaining possibility space (so that there is no longer any need for a catch-all hypothesis) slightly *improves* the average accuracy of IBE$_p$ and probabilistic reasoning in all cases.

Table 4.1 Relative percentage accuracies of
IBE_p (percentage accuracy of IBE_p/percentage
accuracy of probabilistic reasoning).

n	No catch-all	Catch-all
2	0.9639	0.9642
3	0.9398	0.9409
4	0.9174	0.9205
5	0.9024	0.9000
6	0.8882	0.8881
7	0.8772	0.8800
8	0.8711	0.8646
9	0.8584	0.8571
10	0.8505	0.8462

As can be seen from Figures 4.1 and 4.2, IBE_p approximates probabilistic reasoning very well indeed, the average accuracy of the former being consistently only slightly less than that of the latter. More specifically, both in contexts that do and those that do not include a catch-all hypothesis, IBE_p's accuracy is consistently, on average, only about 3 per cent below that of probabilistic reasoning. To compare IBE_p's reliability to that of probabilistic reasoning more directly, we can calculate its *relative* percentage accuracy (i.e., the percentage accuracy of IBE_p divided by that of probabilistic reasoning). These results are displayed in Table 4.1. Again, the results suggest that IBE_p's reliability is not much worse than that of probabilistic reasoning. Whether or not a context includes a catch-all, IBE_p identifies the true hypothesis about 90 percent as often as probabilistic reasoning—averaging over the simulated contexts.

Thus far, our results suggest that IBE_p's epistemic import is parasitic upon Bayesianism's. IBE_p is cogent insofar as it gives us an informal handle on some, but not all, of the probabilistic information needed for Bayesian inference, and it is useful because it is nearly as reliable as the latter (and much more reliable than chance). Practically speaking, we might point out that IBE_p seems eminently more useful to human reasoners than Bayesian inference insofar as it serves reasoners who are, for whatever reason, not able to apply probabilistic reasoning directly; still, if this is right, IBE_p may be thought merely heuristically useful as a poor man's Bayesianism.

However, the above simulations incorporate an unrealistic, simplifying assumption that gives Bayesianism a substantial advantage. Specifically, these assume that an agent's prior probabilities perfectly match the objective chances of the various hypotheses being true. Thus, h_i's chance of being selected as the true hypothesis in any iteration of the simulation is determined straightforwardly by the value of $Pr(h_i)$. Relaxing this assumption by allowing agents to have inaccurate priors accordingly results in Bayesian reasoning having a worse reliability. By contrast, IBE_p neglects priors and ultimately puts all inferential weight on likelihood comparisons. And so, relaxing this assumption has no effect on IBE_p's reliability. The predicted upshot is that, as an agent's

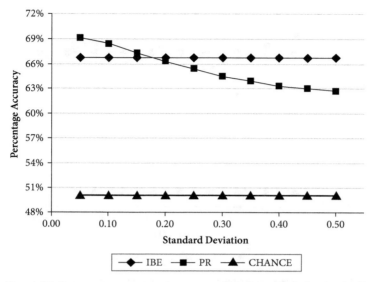

Figure 4.3 Percentage accuracies in contexts that do not include a catch-all.

priors are allowed, on the average, to diverge from objective chances, IBE_p may become more reliable than probabilistic reasoning.

This is easily verified by complicating the above simulations as follows. Steps 1–3 remain the same, although the "prior probabilities" referred to in those steps are now interpreted as the objective chances that the various hypotheses are true. After these initial steps, each prior is calculated by adding to the corresponding chance the value of a normally distributed random variable with mean 0 and specified standard deviation (and then renormalizing to ensure that they sum to 1). This standard deviation explicates the average error of the agent's prior probabilities. While the true hypothesis (and whether e occurs) is determined on the basis of the objective chances, the remaining steps calculate greatest power and posterior probability using the (erroneous) prior probabilities.

The above predictions are verified in the results of all variations—average accuracies for the specific case where $n = 2$, for standard deviations varying from 0.05 to 0.50, are shown in Figures 4.3 and 4.4. Both in contexts that do and those that do not include a catch-all hypothesis, the average reliability of probabilistic reasoning dips below that of IBE_p already with rather modest allowances for error in the priors—though it never dips below that of chance.

3. Conclusions

Past work on the nature and value of IBE largely treats this inference form as one unified category. However, once one remembers that explanatory goodness is evaluated

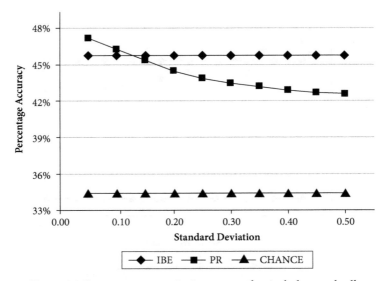

Figure 4.4 Percentage accuracies in contexts that include a catch-all.

on distinct dimensions that can (and often do) vary from case to case, this generalist perspective looks dubious and misleading. Different versions of IBE can be distinguished by the notions of explanatoriness at work in their respective central premises. And these are differences that plausibly matter a great deal to IBE's normative evaluation. Depending on how explanatory goodness is evaluated, IBE may or may not describe a respectable form of uncertain inference.

In Section 1 of this chapter, I put forward a Bayesian explication of one specific sense of explanatory goodness, and I articulated precisely the corresponding version of IBE. Then, in Section 2, I defended this version of IBE as inductively cogent and respectably reliable (at least when compared to Bayesian reasoning). At the start of his most well-known attack on IBE, van Fraassen (1989, p. 131) writes, "As long as [IBE] is left vague, it seems to fit much rational activity. But when we scrutinize its credentials, we find it seriously wanting." This chapter demonstrates, to the contrary, that once we clearly articulate the nature of IBE via an explication of explanatoriness, this inference form can gain a sound new defense.

Appendix A: Uniqueness of $\mathcal{E}$

The mathematical structure that we require of our explicatum is specified in the following two formal conditions of adequacy:

Normality.   For any probability space $(\Omega, \mathcal{A}, Pr(\cdot))$—where Pr is a regular probability measure—$\mathcal{E}$ is a measurable function from two contingent propositions $e, h \in \mathcal{A}$ to a real number $\mathcal{E}(e, h) \in [-1, 1]$.

Structure. $\mathcal{E}$ is the ratio of two functions of $Pr(e \wedge h)$, $Pr(\neg e \wedge h)$, $Pr(e \wedge \neg h)$ and $Pr(\neg e \wedge \neg h)$, each of which are homogenous in their arguments to degree $k \geq 1$, where k is the smallest integer permitted by *Normality* and CA1–CA4.[15]

These conditions require that $\mathcal{E}(e,h)$ be probabilistic in nature and simple in a well-defined sense.

Theorem 1.[16]   *The only measure that satisfies* Normality, Structure, *and* CA1–CA4 *is*

$$\mathcal{E}(e,h) = \frac{Pr(h \mid e) - Pr(h \mid \neg e)}{Pr(h \mid e) + Pr(h \mid \neg e)}.$$

Notation.   Let $x = Pr(e \wedge h)$, $y = Pr(e \wedge \neg h)$, $z = Pr(\neg e \wedge h)$ and $t = Pr(\neg e \wedge \neg h)$ with $x + y + z + t = 1$. Then, by *Structure*, $\mathcal{E}(e,h)$ has the form

$$f(x,y,z,t) = \frac{f_n(x,y,z,t)}{f_d(x,y,z,t)},$$

where $f_n(x,y,z,t)$ and $f_d(x,y,z,t)$ are homogeneous in their arguments to the same least degree $k \geq 1$.

Lemma 1.   *There is no f with f_n, f_d of degree 1 that satisfies* Normality, Structure, *and* CA1–CA4; *i.e., $k \neq 1$.*

Proof.   Let $k = 1$. Then $f_n(x,y,z,t)$ has the form $ax + by + cz + dt$ ($a, b, c,$ and d are coefficients). By CA1, $f(x,y,z,t) = 0$ (and so $ax + by + cz + dt = 0$) if and only if $x = Pr(h \wedge e) = Pr(h) \times Pr(e) = (x + z)(x + y)$. Now we can show that this biconditional cannot be generally satisfied by locating four different parameter settings of (x,y,z,t) that each satisfy $x = (x + z)(x + y)$ but across which there are no (non-zero) coefficients that satisfy $ax + by + cz + dt = 0$. The following four parameter settings suffice: (1/2, 1/4, 1/6, 1/12), (1/2, 1/3, 1/10, 1/15), (1/2, 3/8, 1/14, 3/56), and (1/4, 1/4, 1/4, 1/4). Since these vectors are linearly independent (i.e., their span has dimension 4), the only way to satisfy $ax + by + cz + dt = 0$ across these cases is if $a = b = c = d = 0$. QED.

Lemma 2.   *CA4 entails that for any value of $\beta \in (0,1)$,*

---

[15] A function is homogenous to degree k iff multiplying its arguments all by the same factor c multiplies its value by c^k. The homogeneity requirement ensures that the functional form of $\mathcal{E}$ itself does not determine which of the terms ($Pr(e \wedge h)$, $Pr(\neg e \wedge h)$, $Pr(e \wedge \neg h)$, $Pr(\neg e \wedge \neg h)$) should have more weight. Representing $\mathcal{E}$ as the ratio of two functions serves the purpose of normalization. $Pr(e \wedge h)$, $Pr(\neg e \wedge h)$, $Pr(e \wedge \neg h)$, and $Pr(\neg e \wedge \neg h)$ fully determine the probability distribution over the truth-functional compounds of e and h, so it is appropriate to represent $\mathcal{E}$ as a function of them. Finally, the requirement that $\mathcal{E}$ be the ratio of two functions, each having "the least possible degree $k \geq 1$," reflects a minimal and well-defined simplicity assumption akin to those advocated by Carnap (1950, chapter 1) and Kemeny and Oppenheim (1952, p. 315). Any reader skeptical of simplicity's place in these conditions of adequacy is referred to (Schupbach and Sprenger, 2011), which contains an alternative uniqueness proof from different conditions of adequacy (not including *Structure*).

[16] This theorem and its proof are closely related to, and were inspired by, Kemeny and Oppenheim's (1952) discussion and proof of their Theorem 17.

$$f(x, y, z, t) = f(\beta x, y + (1 - \beta)x, \beta z, t + (1 - \beta)z). \tag{1}$$

Proof. This lemma is a consequence of **CA4**, which describes conditions under which degrees of power must be the same. For any $x, y, z, t \in [0,1]$ such that $x + y + z + t = 1$, allow that there could be an e and h_1 such that $x = Pr(e \wedge h_1)$, $y = Pr(e \wedge \neg h_1)$, $z = Pr(\neg e \wedge h_1)$, and $t = Pr(\neg e \wedge \neg h_1)$. For any β, allow that there may be an h_2 that satisfies the antecedent conditions of **CA4** and such that $Pr(h_2) = \beta$.[17]

With regards to such an e, h_1, and h_2, **CA4** requires that $\mathcal{E}(e, h_1 \wedge h_2) = E(e, h_1)$. We can show that this is equivalent to (1) by establishing the following:

$$\beta x = Pr(e \wedge (h_1 \wedge h_2))$$
$$\beta z = Pr(\neg e \wedge (h_1 \wedge h_2))$$
$$y + (1 - \beta)x = Pr(e \wedge \neg(h_1 \wedge h_2))$$
$$t + (1 - \beta)z = Pr(\neg e \wedge \neg(h_1 \wedge h_2)).$$

These equations are demonstrated straightforwardly, making use of the antecedent conditions of **CA4**. For example, these require that $Pr(e \wedge (h_1 \wedge h_2)) = Pr(h_2)Pr(e \wedge h_1) = \beta x$ (establishing the first equation above). This condition entails that $Pr(e \wedge h_1 \wedge \neg h_2) = Pr(\neg h_2) Pr(e \wedge h_1)$, allowing us to demonstrate the second equation:

$$Pr(e \wedge \neg(h_1 \wedge h_2)) = Pr[(e \wedge \neg h_1) \vee (e \wedge \neg h_2)]$$
$$= Pr(e \wedge \neg h_1) + Pr(e \wedge \neg h_2) - Pr(e \wedge \neg h_1 \wedge \neg h_2)$$
$$= Pr(e \wedge \neg h_1) + Pr(e \wedge h_1 \wedge \neg h_2)$$
$$= Pr(e \wedge \neg h_1) + Pr(\neg h_2)Pr(e \wedge h_1) = y + (1 - \beta)x$$

The other two equations are demonstrated *mutatis mutandis*. QED.

Proof of Theorem 1 (Uniqueness of $\mathcal{E}$).   Lemma 1 shows that there are no f_n, f_d of degree 1 that satisfy our desiderata. Here, I show that there is exactly one ratio of such functions of degree $k = 2$, which completes the proof (given the formal requirements set out in *Structure*). If $k = 2$, $f(x, y, z, t)$ takes the form

$$\frac{f_n(x, y, z, t)}{f_d(x, y, z, t)} = \frac{ax^2 + bxy + cy^2 + dxz + eyz + gz^2 + ixt + jyt + rzt + st^2}{\bar{a}x^2 + \bar{b}xy + \bar{c}y^2 + \bar{d}xz + \bar{e}yz + \bar{g}z^2 + \bar{i}xt + \bar{j}yt + \bar{r}zt + \bar{s}t^2}. \tag{2}$$

As previously noted, **CA1** tells us that f's numerator has to be zero if and only if $x = (x + y)(x + z)$. Making use of $x + y + z + t = 1$, we conclude that this is the case if and only if:

---

[17] In Bayesian terms, this amounts to allowing that an agent could have credences x, y, z, and t in the corresponding conjunctions and β in a proposition that is statistically independent of e, h_1, and $e \wedge h_1$. More generally, it amounts to not restricting the sorts of probability spaces to which $\mathcal{E}$ might apply.

$$x - (x + y)(x + z) = x - x^2 - xy - xz - yz$$
$$= x(1 - x - y - z) - yz$$
$$= xt - yz = 0.$$

The obvious way to satisfy **CA1** (i.e., to ensure that $f_n(x, y, z, t) = 0$ iff $xt - yz = 0$) is to set $e = -i$ and all other coefficients (but i) in the numerator to zero. That this is the *only* way to satisfy **CA1** is a straightforward consequence of Hilbert's Nullstellensatz—a fundamental theorem in and to algebraic geometry. In this context, the Nullstellensatz says that, given that the two polynomials $ax^2 + bxy + cy^2 + dxz + eyz + gz^2 + ixt + jyt + rzt + st^2$ and $xt - yz$ have exactly the same zeros, they are constant multiples of each other. Accordingly, f can be reduced to

$$f(x, y, z, t) = \frac{i(xt - yz)}{\bar{a}x^2 + \bar{b}xy + \bar{c}y^2 + \bar{d}xz + \bar{e}yz + \bar{g}z^2 + \bar{i}xt + \bar{j}yt + \bar{r}zt + \bar{s}t^2}.$$

Turning now to the denominator, **CA2** requires that $f(x, y, z, t) = 1$ iff $Pr(e \mid h) = Pr(e \wedge h) / Pr(h) = x / (x + z) = 1$. Thus, if $z = 0$, $f(x, y, z, t) = 1$. Accordingly, for any case in which $y = z = 0$, **CA2** yields $f(x, 0, 0, t) = 1 = ixt / (\bar{a}x^2 + \bar{i}xt + \bar{s}t^2)$, and by a comparison of coefficients, we get $\bar{a} = \bar{s} = 0$ and $\bar{i} = i$. **CA3** ($\mathcal{E}(e, h) = -\mathcal{E}(\neg e, h)$) is equivalent to

$$f(x, y, z, t) = -f(z, t, x, y). \tag{3}$$

Combining (3) with **CA2**, we have $f(x, 0, 0, t) = 1 = -f(0, t, x, 0) = ixt / (\bar{c}t^2 + \bar{e}xt + \bar{g}x^2)$. Comparing coefficients again, we obtain $\bar{c} = \bar{g} = 0$ and $\bar{e} = i$, reducing f to

$$f(x, y, z, t) = \frac{i(xt - yz)}{\bar{b}xy + \bar{d}xz + i(xt + yz) + \bar{j}yt + \bar{r}zt}.$$

Assume now that $\bar{j} \neq 0$. Let $x, z \to 0$. We know by **CA2** that in this case, $f \to 1$. Since the numerator vanishes, the denominator must vanish too, but by $\bar{j} \neq 0$ it stays bounded away from zero, leading to a contradiction ($f \to 0$). Hence $\bar{j} = 0$. In a similar vein, we can argue for $\bar{b} = 0$ by letting $z, t \to 0$ and for $\bar{r} = 0$ by letting $x, y \to 0$ — making use of (3) again: $-1 = f(0, 0, z, t)$.

Thus, letting $\alpha = \bar{d} / i$, f can be written as

$$f(x, y, z, t) = \frac{i(xt - yz)}{\bar{d}xz + i(xt + yz)}$$
$$= \frac{(xt - yz)}{(xt + yz) + \alpha xz}. \tag{4}$$

To fix the value of α, we make use of **CA4**, which requires $f(x, y, z, t) = f(\beta x, y + (1 - \beta)x, \beta z, t + (1 - \beta)z)$—see **Lemma 2**, equation (1). Applying this constraint to (4), we obtain

$$\frac{xt - yz}{xt + yz + \alpha xz} = \frac{\beta x(t + z - \beta z) - (y + x - \beta x)\beta z}{\beta x(t + z - \beta z) + (y + x - \beta x)\beta z + \beta^2 xz}$$

$$= \frac{xt - yz}{xt + yz + (2 - 2\beta + \alpha\beta)xz}.$$

For this to be true in general, we have to demand that $\alpha = 2 - 2\beta + \alpha\beta$, which implies that $\alpha = 2$. Hence,

$$f(x,y,z,t) = \frac{xt - yz}{xt + yz + 2xz}.$$

After replacing x, y, z, and t by their corresponding joint probabilities, some algebraic manipulations show that this ratio is equivalent to the following:

$$\mathcal{E}(e,h) = \frac{Pr(h \mid e) - Pr(h \mid \neg e)}{Pr(h \mid e) + Pr(h \mid \neg e)}$$

which is therefore the unique function satisfying all of the conditions. QED.

Acknowledgments

I owe special thanks to David Danks, John Earman, David Glass, Edouard Machery, Kevin McCain, Lydia McGrew, Ryan Muldoon, John Norton, Ted Poston, Jan Sprenger, and Rev. Michael van Opstall for helpful comments pertaining to this project. I am doubly grateful to Jan Sprenger, who co-authored an earlier draft of the appendix.

References

Carnap, R. (1950). *Logical Foundations of Probability*. University of Chicago Press, Chicago.

Cohen, M. P. (2015). On Schupbach and Sprenger's measures of explanatory power. *Philosophy of Science*, 82(1): 97–109.

Douven, I. (2011). Abduction. In Zalta, E. N., editor, *The Stanford Encyclopedia of Philosophy*. Spring 2011 edition.

Douven, I. and Schupbach, J. N. (2015a). Probabilistic alternatives to Bayesianism: The case of explanationism. *Frontiers in Psychology*, 6(459): 1–9.

Douven, I. and Schupbach, J. N. (2015b). The role of explanatory considerations in updating. *Cognition*, 142: 299–311.

Fitelson, B. (1999). The plurality of Bayesian measures of confirmation and the problem of measure sensitivity. *Philosophy of Science*, 66: S362–S378.

Fumerton, R. A. (1980). Induction and reasoning to the best explanation. *Philosophy of Science*, 47: 589–600.

Glass, D. H. (2012). Inference to the best explanation: Does it track truth? *Synthese*, 185: 411–27.

Good, I. J. (1983). *Good Thinking: The Foundations of Probability and Its Applications*. University of Minnesota Press, Minneapolis.

Harman, G. H. (1965). The inference to the best explanation. *Philosophical Review*, 74: 88–95.

Keil, F. C. (2006). Explanation and understanding. *Annual Review of Psychology*, 57: 227–54.

Kemeny, J. G. and Oppenheim, P. (1952). Degree of factual support. *Philosophy of Science*, 19: 307–24.

Lewis, D. (1986). *On the Plurality of Worlds*. Blackwell, Oxford.

Lipton, P. (2004). *Inference to the Best Explanation*. Routledge, New York, 2nd edition.

Lombrozo, T. (2006). The structure and function of explanations. *Trends in Cognitive Sciences*, 10(10): 464–70.

Peirce, C. S. (1931–5). *The Collected Papers of Charles Sanders Peirce*, volumes I–VI. Harvard University Press, Cambridge, MA.

Psillos, S. (1999). *Scientific Realism: How Science Tracks Truth*. Routledge, London.

Putnam, H. (1975). *Mathematics, Matter, and Method*, Volume I of *Philosophical Papers*. Cambridge University Press, Cambridge.

Schupbach, J. N. (2011). Comparing probabilistic measures of explanatory power. *Philosophy of Science*, 78(5): 813–29.

Schupbach, J. N. (2014). Is the bad lot objection just misguided? *Erkenntnis*, 79(1): 55–64.

Schupbach, J. N. and Sprenger, J. (2011). The logic of explanatory power. *Philosophy of Science*, 78(1): 105–27.

Swinburne, R. (2004). *The Existence of God*. Oxford University Press, Oxford, 2nd edition.

van Fraassen, B. C. (1989). *Laws and Symmetry*. Oxford University Press, New York.

Vogel, J. (1990). Cartesian skepticism and Inference to the Best Explanation. *Journal of Philosophy*, 87(11): 658–66.

PART II

The Fundamentality of Inference to the Best Explanation

5

Reasoning to the Best Explanation

Richard Fumerton

1. Introduction

The primary purpose of this chapter is to explore the viability of reasoning to the best explanation as a *fundamental* source of epistemically rational beliefs and a potentially useful weapon for use in responding to the skeptic.[1] After making some important distinctions among various forms of explanationism,[2] I'll reach a somewhat pessimistic conclusion about the prospects for explanationist epistemologies solving fundamental epistemological problems.

2. The Concept of Explanation

One might suppose that in order to properly explore the power of reasoning to the best explanation, one would first need to figure out what an explanation is. But we can't possibly do that in this context. Entire books (and lives) have been devoted to the difficult question of how to understand what it is to explain successfully some phenomena. The term "explanation" is itself extraordinarily ambiguous. We talk about philosophical explanations of various phenomena, where we often have in mind something like an analysis, a theory, or an account of some kind of how to understand the phenomena in question. In other quite removed contexts we even talk of excuses as a kind of explanation (e.g. I have a good explanation of why I didn't meet you as promised). But here we will be primarily concerned with causal or scientific (broadly understood) explanations of various phenomena.

I have, of course, already begged questions in the preceding sentence by presupposing that the search for paradigmatic scientific explanations is primarily the search for causes. How much I begged the question probably depends on how expansive our view is of causal explanation. It is an understatement to suggest that there is no consensus

---

[1] I would like to thank Ted Poston and Kevin McCain for comments on an earlier draft of this chapter.

[2] I'll use this term more loosely than does McCain (2014, chapter 6). McCain has in mind a very explicit form of evidentialism within which reasoning to the best explanation plays a dominant role. I use the term to describe any view that takes reasoning to the best explanation to be a fundamental sort of reasoning that can play an important role in refuting the skeptic.

on how to understand causation or causal explanation. The gamut runs from those who take the relation to be unanalyzable[3] and generality theorists who take the concept of causal *law* to be the key to understanding causation.[4] Regularity theorists are a species of generality theorists who think that one can understand lawful regularity in terms of contingent universals of material implication.[5] But there are additional internal debates among generality theorists as to whether the laws required for causation need to be universal or might, instead, describe relevant non-accidental statistical regularities (see, for example, Anscombe 1993).

Again, I'm not about to try to settle any of these philosophical controversies here. I think that for the most part my arguments won't depend on any particular account of what makes a given proposed explanans the correct explanation of some particular explanandum. I will, however, presuppose that at the very least knowledge of what causally explains what is a posteriori. One cannot discover a priori the truth of a proposed causal explanation.

3. Reasoning to the Best Explanation and the Internalism/Externalism Debate in Epistemology

Internalist/externalist controversies cut across just about every interesting debate in contemporary epistemology. Without attempting to define in a general way internalism or externalism, it seems plausible to suggest at the outset that the mere fact that E is the correct explanation for O wouldn't *by itself* make it epistemically rational for someone S to believe E, even when that person knows that O has occurred. And that is so even if we distinguish (as we must) doxastic from propositional justification. This latter distinction is itself not easy to define, but as a rough gesture let us say that there is propositional justification for S to believe E when there is reason for S to believe E (whether or not S actually bases his belief that E on the reason he possesses to believe E). S has doxastic justification for believing E when S has propositional justification for believing E and properly bases his belief that E on that justification. It isn't hard to imagine a situation in which E is the correct explanation for O, but where neither S nor anyone else has the slightest reason to suppose that E even obtains let alone is an explanation for O. In such a scenario, S clearly has no propositional justification for believing E and, *eo ipso*, has no doxastic justification for believing E.

It is important to note that the above argument, by itself, has no significant implications for whether one should be an internalist or an externalist about justification in general. One might concede that S would need to have good reason to believe that E is the explanation (or, a candidate explanation) for some known phenomenon O before S

---

[3] See, among many others, Davidson (1967) and Fales (1990).

[4] I say "causal" law to acknowledge the obvious fact that lawful connections include non-causal relationships between such things as barometers and storms and present states of the world and past states of the world.

[5] Sophisticated regularity theorists will always allow that the analysis must at least take into account hierarchies of such generalizations.

would get a reason to believe E, but go on to give a thoroughly externalist account of how one gets good reason (justification) for believing the relevant causal claim. The issue is, however, a bit complicated. I have, for example, defended in a number of places (1988, 1995, 2006) a thesis I called *inferential internalism*. The inferential internalist argues that for S to be justified in inferring P from E, S would need justification for believing that E makes probable P (where E's entailing P would be the upper limit of E's making probable P). But despite the label "inferential *internalism*," that view, also, is compatible with a radical externalism about epistemic justification. Most externalists are not inferential internalists, but should they change their mind about that, they might still hold onto an externalist account of how one gets justification for believing that E makes probable P. So my claim here is highly limited, and it is one to which I shall return shortly. While it is entirely plausible to suppose that we often use beliefs about causation (and explanation) in reaching conclusions, such beliefs need to be *justified* beliefs in order to generate any further justified beliefs.

If what I have just argued is correct, then if reasoning to the best explanation is to help solve an epistemological problem for some subject S, the reasoning will need to be of the following form: (1) O and (2) E is the most plausible explanation of O; therefore, (3) E. And as is true of any argument, the premises of the argument would need to be justifiably believed if one is to justifiably believe any conclusion based (solely) on the premises.

4. Forms of Explanationism

There are extreme and moderate forms of explanationism.[6] I suppose the most extreme would be the thesis that *every* justified belief is justified by its place in an explanatory schema. Given this idea, our justification for believing E might be that it at least partially explains O. And our justification for believing O would be that it is explained at least partially by something else we believe E, and, perhaps, that it also explains some other phenomenon. Our justification for believing E would be that it is either a believed explanans of something else we believe or a believed explanandum of something else we believe. If we don't insist that we have *prior* justification for believing our explanans, our explanandum, or the relation between them, we probably have what amounts to a coherence theory of justification, one that insists that the relevant "glue" of coherence is the place of a hypothesis in a believed explanatory schema. Assuming that we don't embrace a coherence theory of truth (who does?), we are simply endorsing a coherence theory of justification with an addendum specifying the relevant sort of coherence that we think generates justification.

About a coherence theory of explanatory justification I don't have much to say that doesn't apply equally to a coherence theory of justification more generally. There

---

[6] There are many more forms of explanationism than I consider here. For a more comprehensive discussion of different theses associated with explanationism, see Lycan (2002).

are all sorts of technical problems with coherence theories,[7] but their most glaring defect to me is that they place implausibly high standards on what it takes to have a justified belief. As I get older I find that I have very odd aches and pains for which I have no explanation whatsoever (other than the trivial—that's what happens to you when you get older). The point is that my justification for believing that I am in pain, when I am, seems utterly unaffected by the paucity of explanations available to me. I have no explanation for some of the pains that I am quite certain I experience, and I like to think that I am stoic enough that they play very little role in explaining other phenomena of which I am aware. This last claim, to be sure, might be a little cavalier. I probably realize that the pain of which I am aware is more likely to cause me to complain to someone who I think might be a tiny bit sympathetic. It is probably more likely to cause me to toss and turn in bed at night, and there are probably indefinitely many small ways in which the pain further affects my behavior and dispositions to behave. But even when I think about the causal effects of the pain, it seems patently obvious to me that my *justification* for believing that I have the pain is quite unrelated to these further *different* beliefs about the pain's effects. Lastly, the explanationist might argue that even if nothing is involved in my justification that I am in pain but my awareness of the pain, surely I would still think of the pain as the best explanation of the awareness. But here again, it seems to me that this would mistakenly over-intellectualize the nature of my justification. It doesn't seem to me that awareness is always, or even usually, accompanied by meta-awareness. It might even require considerable conceptual sophistication to acquire the concept of direct awareness. But even children have perfectly good justification for believing that they are in pain when they are.

The above is not much of an argument against someone determined to locate the source of epistemic justification for a belief with its place in a web of beliefs about explanatory connection, but I must confess it thoroughly convinces me that the very idea is misguided.

I find a similar argument utterly convincing against Klein's infinitism. Klein (1999) thinks that the epistemic status of a belief is a function of our ability to conjure up arguments in favor of what we believe. And, consistently, he argues that the premises of such arguments must gain their positive epistemic status from the fact that we can generate still further arguments in their favor—and so on, ad infinitum. There are again technical problems with the view,[8] but to be honest, the most straightforward problem I find with it is similar to the one described above. If I gave myself some time, I could probably generate some sort of argument whose premises make probable for me that I am in severe pain (when I am), but the fact of the matter is that I know with complete certainty that I am in pain right now (damn back!) and I know that this epistemic certainty has nothing to do with the capacity I have to exercise my imagination in generating indefinitely many arguments culminating ultimately in the conclusion that I am in pain.

---

[7] See Fumerton (1994).   [8] See Fumerton (2014).

There is a closely related problem in thinking of reasoning to the best explanation as the source of all justified belief.[9] We get interested in explanations when we run across something we want to explain. We are not in a position to even look for explanations until we have found our explanandum. But our justified belief that there is some state of affairs that needs explaining surely doesn't derive from our having found a place for the phenomenon in our explanatory scheme of things. That pain is before my consciousness. I know it exists as surely as I know anything else. I would probably *like* an explanation of its existence, and I might care to know what other effects it will have on my life. But my certainty of the pain's existence has nothing to do with my success at getting answers to these other epistemic questions. My noticing the pain is prior to my seeking explanations.

So global explanationism doesn't seem to be a contender as a general account of the nature of justification. But that shouldn't be surprising and it doesn't tell against the less grandiose claim that we should recognize reasoning to the best explanation as *one* promising way in which we might "bridge" the epistemic gaps between our available evidence and our various common-sense beliefs. When we come home and see our door smashed and our valuables missing we quickly conclude that we have been robbed. While we haven't laboriously intellectualized it this much, it is tempting to suggest that we have quickly come to the conclusion that a robbery explains the available data, and for that reason we regard ourselves as having a justified belief that a robbery took place. Hume sometimes seemed to suggest that when we move past what is given to us in experience it is virtually *always* through what he calls causal reasoning.[10] We hear the sounds characteristic of a carriage and infer that the carriage pulled up. We hear voices in the hall and infer that there are people outside our room. These and indefinitely many other examples suggest that whether or not reasoning to the best explanation is the *only* form of reasoning that takes us beyond objects of which we are presently aware, it is a powerful and familiar way to increase through inference our stock of common-sense beliefs.

All that *seems* right. We undeniably do rely on our implicit knowledge of causal connections (or statistical correlations) when reaching conclusions about the world around us. But there are two importantly different ways in which that could be true, and if we don't distinguish them we are courting confusion.

5. Fundamental and Derivative Epistemic Principles

The terminology in what follows is somewhat arbitrary. We may either distinguish fundamental from derivative epistemic principles or, alternatively (as I would actually

---

[9] I'm not suggesting that all self-identified explanationists think that it is. In comments on a draft of this chapter, Kevin McCain suggests that the most one should claim is that all *explicit* reasoning involves reasoning to the best explanation. We may be able to form all sorts of foundationally justified beliefs that don't involve any sort of reasoning at all.

[10] Hume (1888, Book I, part IV, sec. II, p. 212).

prefer), we could distinguish *genuine* epistemic principles from *spurious* epistemic principles. What would be an example of the latter? Well consider an example that I often use. Many of us have learned that, in some sense or other, it is appropriate to "infer" that a solution is acidic from the fact that the litmus paper in the solution turned red. It is usually appropriate to "infer" that someone's name is "Fred" from the fact that he says it is. What sounds like a clap of thunder strongly "implies" that there is lightning not that far away. A baby's crying often indicates that the baby is either hungry or tired, and the list could go on and on. All of these commonplace predictions might plausibly be described as relying on hypotheses about what best explains what. So who would want to deny the obvious—that there is such a thing as reasoning to the best explanation and that such reasoning pervades much of our thinking about the world around us?

The question, however, is not whether we use information about causal connections and causal explanations in reaching conclusions about the world—we do. The question, at least the one I want to ask, is whether these forms of reasoning are *sui generis* kinds of reasoning that do not owe their epistemic power to other more fundamental kinds of reasoning.

Perhaps the best way to raise the relevant question is to ask whether the person employing reasoning to the best explanation is relying on the critical explanatory claim as a *principle* in drawing a conclusion from premises, or whether the person is using the critical information in formulating a *premise* from which to draw the relevant conclusion. If I know what usually causes that distinctive sound I associate with thunder, then I can certainly use the fact that I heard that sound as part of my enthymematic reasoning to reach the conclusion that lightning just struck. But if the causal hypothesis appears as a *premise*, there is no reason yet to suppose that the *reasoning* is anything other than a more mundane sort of statistical reasoning. I antecedently know that this sort of sound is *usually* produced by lightning; I can inductively infer from my awareness of the sound, and my knowledge of its usual source, that in all probability lightning occurred. This is just classic enumerative induction where the correlation described in the premise involves causes and effects (see Fumerton 1980).[11]

---

[11] Interestingly enough, some philosophers would reverse the direction in which one should attempt the reduction. Harman (1965) and Fales (1990) both argue that inductive reasoning is best construed as reasoning to the best explanation. On this view observed correlations are best explained by regularities and that's how we get justification for believing that the relevant regularities exist. This dispute eventually turns on what one takes the fundamental epistemic principles to be. If any sort of non-deductive reasoning is legitimate, it seems to me, there must be relations of making probable that hold between the premises of such reasoning and its conclusion. The solution to the problem of induction is just to recognize that there is a principle of induction that is true (indeed necessarily true), that can be known a priori, and that sanctions the relevant inference. Any other *fundamental* principles of non-deductive reasoning would have the same modal and epistemic status. This chapter doesn't purport to defend a fully developed solution to the problem of induction. Any such effort would need to respond to Goodman's so-called new riddle of induction. To solve the riddle, perhaps one needs to restrict the kind of properties that can be "projected" (in Goodman's terminology, 1979) to something like natural properties. Those sympathetic to the idea that we should explain inductive reasoning in terms of reasoning to the best explanation might go on to argue that the restriction to natural properties is best explained by the fact that only natural properties stand in causal connection. It still seems to me that we would need induction to reach that conclusion, but the issue is, to be sure, complex.

These days there is much more controversy over whether critical "connecting" principles belong as premises of an argument or belong outside the argument as inference rules licensing a given inference. Once, almost everyone in the analytic tradition assumed that Hume was right and that if a man told you that such and such is the case, your reason for believing what he said was only as good as your reason for thinking that he was telling the truth and was in a good position to know what he was talking about. Partly because of paranoia (or, perhaps, justified belief) about the difficulty of avoiding skeptical conclusions when so much is demanded of rational acceptance of testimony, a significant movement began to recognize testimony as its own *fundamental* and legitimate source of justification. Just as there are rules licensing deductively valid inference and, perhaps, rules licensing inferences of enumerative induction, so also, it was argued, there are independent, self-standing rules licensing trust in what others say (see, for examples, Coady 1992; Zagzebski 2012; Goldberg 2010). I still stand with Hume. We may not bring to the fore of consciousness all that we rely upon in forming beliefs based on what others tell us, but the relevant beliefs are there lurking in the background, and it doesn't take much to remind ourselves that we are critically relying upon them. Furthermore, even as they hovered in the background as non-occurrent beliefs, or even merely beliefs we *would* form were we prompted in the right way, it is not clear that the ground of those dispositional beliefs isn't active in producing a prodigious array of output beliefs.[12] And in so far as they are active, it seems to me that they would themselves need to be justified.

I have tried to give examples that illustrate the distinction, but we haven't yet *defined* clearly the distinction between fundamental and derivative epistemic principles. The distinction is easier to define given certain metaepistemological views rather than others. On the view I defend, it is critical to distinguish epistemic principles that can be known *a priori* and those than can be known only *a posteriori*. It might be true that the color of litmus paper is a reliable guide to the acidity of a solution, that the reading on a thermometer outside one's window is a reliable guide to the temperature outside, that the gas gauge on the car is a reliable guide to the amount of fuel in the car, but all of these truths are obviously discoverable only *a posteriori*. To reach a rational conclusion about the gas in the car we need to have good reason to trust the gauge as a reliable indicator of the gas. To rationally rely on the color of litmus paper to reach a conclusion about the acidity of the liquid, we need independent empirical reason to think that the color correlates reliably with the acidity. All this seems to me pretty obvious, but as soon as we realize that it is true, we will have lost all temptation to think that there are independent rules of *reasoning* governing the color of litmus paper and what one should infer from it, the readings of thermometers, gas gauges, and the like and what one should infer from those readings. Gas gauge "inferences" (if that's what you want

---

[12] There are difficult questions here. I am presupposing that dispositions have grounds and that grounds can be causes. Whether we say that the ground of a dispositional belief is something upon which we can base another belief is terminological.

to call them) are just inferences that utilize information about gas gauges in *premises*. The relevant facts about indicator relations are just additional *premises* that we rely upon in reaching our conclusion about the level of gas in the tank.

Now let me repeat that the terminology we use here isn't particularly important. In the preceding remarks I sometimes talked about the "inferences" we would make from observations about indicators and gauges of various sorts. If we so choose, we can classify kinds of inferences based on the kinds of *premises* we use in our reasoning. Alternatively, we can classify kinds of legitimate inferences based on the structure of arguments and the ways in which we think their premises entail or make probable their conclusions. If we make the former decision, we can talk about litmus paper inferences, gas gauge inferences, inferences from testimony, and, more generally, causal reasoning—reasoning that involves premises describing causal connections.

Let us return to reasoning to the best explanation. When we do (and we obviously do) use information about causation and causal correlations in reaching conclusions, are we reasoning in a special way, or are we simply employing premises about causation and causal correlations in reaching conclusions (where the pattern of reasoning is most often best characterized as enumerative induction)? One could, of course, insist it is misleading to suggest that an inductive argument whose premises describe causal connections is really an inductive argument. As I admitted above, there is nothing to stop one from distinguishing inductive arguments based on the kinds of correlations that are described in premises. But I don't see any reason for thinking that the content of the premises affects the right way to think about the *structure* of the argument and the *relation* of its premises to its conclusions. So as far as I can see we haven't much of a reason to suppose that we are employing anything other than inductive reasoning when employing the kind of argument discussed above.

Consider Peirce's (1903, Vol. 2, p. 231) description of abductive reasoning (reasoning to the best explanation).[13] We observe something striking—something that surprises us or puzzles us. We discover, for example, what certainly look like the fossilized remains of fish skeletons in rocks that are far from any body of water. It strikes us that this needs an explanation. Note that we have already employed an implicit premise that would require further justification—that most natural phenomena *have* explanations. But let that go for now. Perhaps we have a half-way decent inductive argument that when we look hard enough for causal explanations we eventually find them.[14] In the case of our desert rock with remains of fish, it probably wouldn't have been that

---

[13] There is a controversy over whether Peirce intended abduction to be a kind of inference, or whether he thought of it as something that belongs more to the context of discovery (as people used to think). On the latter interpretation, abduction is a process that simply suggests hypotheses, hypotheses that would still need to be independently confirmed if they were to be justifiably believed. See Plutynski (2011) for an excellent discussion of debates related to the interpretation of Peirce.

[14] This is problematic. For certain sorts of phenomena we have been looking for a very long time for the relevant explanations and have come up empty. We typically don't give up on the search (any more than we assume that the missing sock in the wash just disappeared *in nihilo*).

hard for ancient people to have reached the conclusion that the water once covered the land where the remains were found. They might have made a further epistemic leap by also inferring that the most common way that water ends up covering land that is usually dry is through flooding—and when the dry land is a long way from any water the flood must have been impressive indeed. Eureka!—we have an explanation of the prevalence of flood myths in ancient cultures.

Assuming that it occurred, the above "flood" reasoning wasn't irrational—it actually came pretty close to the truth. After all the water did once cover land where one finds fossilized remains of fish. Their explanans only got in trouble with its further suppositions of how that water came to cover that land. But our interest here is with the *pattern* of reasoning. Our reasoners had, we may presume, knowledge that the skeletal remains of fish would often be found in areas where water covered land. In the wake of another finding of fish skeletons, why shouldn't one *inductively* infer that what has happened so often before has happened again? It isn't even difficult to defend the claim that the reasoning in question is simply enthymematic inductive reasoning. And when the enthymeme is filled in, our reasoning to the best explanation, as a separate and independent sort of reasoning, has simply disappeared.

5.1 Objection

Straw men burn easily. I took *one* example of something that we might casually describe as reasoning to the best explanation and showed how with its critical suppressed premises made explicit, the reasoning could be made to look like a familiar sort of inductive reasoning. But, one might worry, I deliberately chose an example where I could do just that. The philosophical interest in reasoning to the best explanation is to see how it might answer an epistemological question where other more familiar modes of reasoning are hopeless. As Hume (1888, p. 212) pointed out, memory and induction aren't taking us anywhere close to knowledge of a mind-independent, enduring physical world. At least we are not making the journey on that path if our starting point is limited to what we know through direct knowledge about subjective experiences and the way they come and go. Setting aside the question of how we get knowledge of past correlations between experiences, if we are limited to direct knowledge of subjective experience, we won't be able to establish the needed correlations between appearance and reality in order to inductively establish the former as reliable indicators of the latter.

It is in these dire circumstances that we might hope that reasoning to the best explanation will ride to the rescue. And it does sometimes give that appearance. I've heard that Europeans discovered the gorilla only in the late eighteenth century. I daresay that before they ever had their first opportunity to dissect a dead gorilla, the Europeans were willing to wager it had a heart, lungs, kidneys, and a host of other organs. They clearly weren't in a position, one might argue, to inductively establish such conclusions. Inductive reasoning would require correlating the outside appearance of the gorillas with their internal organs and, by hypothesis, they hadn't done that. But the prima facie appearance of a problem using inductive reasoning is just that—appearance. It is true that the Europeans hadn't dissected gorillas, but gorillas are not just gorillas. Every animal is a member of many different kinds—the gorilla is also a mammal, and a vertebrate, and a creature with eyes, and a creature with eyes and hair, and a creature with eyes and hair and ears, and ... Biologically we might not bother to make all of the distinctions that are there to be made, but epistemologically we can pay attention to whatever we find useful. The Europeans had observed countless examples of creatures with eyes, ears, a mouth, and fur that were also found to have hearts, lungs, and kidneys. Assuming that is so, they would have ample ground to infer inductively that this animal also had the relevant internal organs.

The point of the example is simply to remind ourselves that enumerative induction is a potentially powerful tool—at least as a way of concluding that a given conclusion has prima facie plausibility. In discussing the alleged distinction between theoretical and observable entities, Grover Maxwell (1962) described the quasi-fictional example of a scientist trying to figure out how disease was transmitted from patients in one part of a hospital to patients in a quite different part of the same hospital. The scientist in question eventually suggested "crobe" theory. People were already generally aware that animals like rats could carry disease across distances, and our crobe theorist suggested that there were creatures too small to see (theoretical posits) that were "carriers" of disease. In Maxwell's story other scientists (representing radical empiricists) ridiculed crobe theory for positing something that was "unobservable" and hence unverifiable. With the invention of the compound microscope the crobe theorist was eventually vindicated.

Maxwell told the story to cast doubt on the question of whether there really is an intelligible and useful distinction between the theoretical (unobservable) and the observable in science. But I repeated the story for a different reason. Once again, one could use the reasoning employed by the "crobe" theorist as an example of the potential power of inductive reasoning. In the story the scientist had correlated disease transmission with creatures carrying disease. He found another case of disease transmission and (tentatively, to be sure) concluded that we have another case of creatures carrying disease—this time creatures too small to be seen.

5.2 Best explanation

The assimilation of reasoning to the best explanation to inductive reasoning can also plausibly illuminate the criteria philosophers often discuss in trying to say something interesting about what makes a given explanation the epistemically "best" explanation available. On many accounts of explanation, it is useful to distinguish the formal characteristics a good explanation must have, from the substantive requirement of truth that will always be required for an explanans (even one that is epistemically unlikely) to be correct. So on Hempel's (1965) classic D-N model of explanation, a successful explanans of some event would need to contain a description of antecedent conditions and laws that together entail a description of the explanandum.[15] These are the *formal* conditions a successful explanation must meet. For the explanation to be correct, the statements describing antecedent conditions and lawful regularities also need to be true. On anything like the D-N model there are literally infinitely many explanations that satisfy the formal conditions. So the epistemic problems will at least include the problem of choosing from among those possible explanations.

It is at this point that some philosophers have talked about such epistemic virtues as simplicity, consiliency (range of kinds of phenomena explained), and analogy (or familiarity).[16] Other things being equal, choose the simplest of the possible explanations of some phenomenon, the explanation that would explain the most, the explanation that is most familiar (most like other explanations we know to be correct), and so on. But here we must ask about the modal status of the relevant criteria. Is it a necessary truth that simpler, more comprehensive, or more familiar explanations are likely to be true, or is it a contingent fact learned inductively from experience? And once again, it seems plausible to me that it is the latter. So take a familiar, though relatively mundane attempt to choose from among possible explanations of something we want explained. You are on a jury and the prosecutor is trying to present a convincing argument that Jones killed his ex-wife. It looks as if it is a pretty good case. Jones was seen in the neighborhood shortly before his ex-wife was killed, his blood was found at the scene, his skin under his ex-wife's fingernails, his fingerprints on the knife that was the murder weapon. Defense lawyers don't give up, though. It is their job to paint alternative scenarios. The cops had it in for Jones and planted the evidence. Jones had a doppelgänger who elaborately tried to stage all of the relevant evidence and commit the murder during a period of time where Jones had no alibi, and so on. We find Jones guilty because we think that the prosecutor's explanation of what happened is far simpler than the defense alternatives. But what does simplicity consist in here? In the end, aren't we just relying on such obvious background knowledge that in the vast majority of cases like this people haven't engaged in the behavior postulated by the defense? In general conspiracy theories are implausible because we have antecedent knowledge that the kind of behavior postulated by the conspiracy theorist doesn't usually happen and, in the few cases where it does, the conspiracy usually gets out—as Franklin famously said, two people can keep a secret if one of them is dead.

I'm not suggesting that this is the only way that appeal to simplicity can be justified. There probably is a sense in which if two proposed incompatible explanations E1 and E2 for an explanadum O satisfy all of the formal requirements of a good explanation, but E2 involves more claims at the same level of generality as does E1, then, other

---

[15] Antecedent conditions aren't necessary in explaining lawful regularities. One can in principle explain lawful regularities by deducing them from more general laws. The explanans for some laws (like Kepler's and Galileo's) would also need a description of specific bodies like the earth and the sun. It is, of course, an understatement to suggest that Hempel's D-N model of explanation faces enormous difficulties, not the least of which is the need to distinguish somehow causal laws from other sorts of laws.

[16] See Thagard (1978) for a discussion of these.

things being equal E1 is more likely to be correct than E2. But that is only because the probability of conjunctions with independent conjuncts is lower than the probability of each conjunct. And the more conjuncts there are in a conjunctive explanation, the lower the probability. All of this would need much more work—there are formal problems with trying to figure out when two claims have a similar level of specificity, and the conjuncts of conjunctions employed in explanations will rarely be independent, but the basic idea is still sound.

It is plausible to suppose that we can use induction to justify other criteria for choosing among possible explanations (again assuming that we have reason to believe that there is an explanation in the first place). Consider, for example, familiarity. If a proposed explanation is like another already known to be a successful explanation, then there will be a more general level of description of the two *kinds* of explanation such that the known explanation will be an instance of this kind of explanation. The more instances of this kind of explanation known to be successful, the stronger will be our evidence for thinking that a new explanation of this same kind will also be successful.

I have been emphasizing the power of induction to find reasons for choosing from among alternative potential explanations. But that probably betrays my views about what I take to be genuinely fundamental epistemic reasoning. Another philosopher with other sympathies might justify the relevant choice differently. It strikes me, for example, that in his recent book, McCain (2014) is ultimately sympathetic to a kind of phenomenal conservatism of a sort defended by Huemer (2001). McCain seems to think that a proposed explanation becomes plausible because it "seems" to you to be true. Again, if seeming evidentialism were a plausible view, that might be a way of getting justification for the needed premises. But it is the seeming evidentialism that is doing all of the epistemic hard lifting.

Before leaving the topic of criteria for choosing among explanations it is worth underscoring a point I made years ago (Fumerton 1992). For you to rationally believe a proposed explanation E of some known phenomenon O, it is not enough that you have reason to believe that there is an explanation of O and that E is more likely for you than each of E's competitors. E would also need to be more likely for you than the *disjunction* of all of E's competitors. And again, if anything like a covering law model of explanation were correct, there are literally infinitely many competitors that will satisfy the formal requirements of a good explanation. The disjuncts of this disjunction might not be independent, but even if they are not, the probability of the disjunction is in danger of becoming very significant.

5.3 The problem of perception and causal reasoning

I have allowed that we do engage in causal reasoning *if* we characterize kinds of reasoning by the kinds of *premises* that reasoning employs. I have argued, however, that in a host of examples one can plausibly construe the actual inference involved in causal reasoning as just a species of inductive reasoning. But still, even the "crobe" theory we discussed earlier posited a *kind* of thing whose *properties* can be, and have been, observed (properties like being a living thing and moving through space). And as we noted earlier, we might hope that reasoning to the best explanation will step up when we are trying to reach conclusions about a kind of thing whose defining properties have never been, and perhaps *could never be* observed. It really would be nice if reasoning to the best explanation could provide an alternative to inductive reasoning. The difficulty is that we may not be able to give any convincing examples of reasoning of this sort that is paradigmatically legitimate. Let's return again to the problem of perception.

Again, we are obviously begging questions in characterizing the problem of perception as I did earlier. We are supposing that our available evidence really is restricted to what we know directly about the character of subjective and fleeting experience. We are giving ourselves knowledge of correlations between appearances, but we can't figure out how to correlate appearance with something other than appearance. We can't, therefore, figure out how a straightforward *inductive* argument would go for the conclusion that there are physical objects with characteristically *physical* properties. Nor is it easy to see how more subtle sorts of inductive reasoning would go. With our examples of gorillas and "crobes" we presupposed that we had unproblematic access to at least some physical objects and their properties. But if our epistemic base really is restricted to subjective experience, it is hard to see how we can usefully subsume the unobserved and unobservable physical world under more epistemically useful *general* categories with which we are already familiar. With what information can we work?

Well, I suppose one might argue that we have discovered non-accidental (non-coincidental) correlations among sensations. Certain sequences of visual and kinesthetic sensations have been almost always (not *always*) followed by certain kinds of tactile sensations. We could generate the concept of causal connection (or at least non-accidental connection) through our awareness of these correlations. Given a limited enough sample, we could start working on the inductively established conclusion that there are usually explanations for many of the sensations that we have, explanations that cite the occurrence of prior sensations. But that doesn't take us very far. When we wake up from a sleep not remembering where we are and have the experiences as of opening our eyes, we can inductively predict that we will have some sort of visual experience. But we wouldn't have been able to predict that *specific* character of the visual experience we have. We could give up on the hypothesis that experiences have causal explanations, of course, or we could hypothesize something else as the cause of the experience—again, I mean the specific *character* of the experience. And that is where the theoretical posit of physical objects might save the generalization we were working on—the generalization that everything has a cause, or at least is non-accidentally correlated with prior conditions.

But now we encounter Berkeley's objection to Hylas's attempt to posit matter as the cause of sensations—unknown in terms of its intrinsic properties, but known as whatever is the cause of the phenomenal world of appearance. Berkeley (1713, pp. 202–4) wanted to know why we would posit material objects rather than a mind. After all, we already know that minds can cause (subjective) ideas of the imagination (or so Berkeley argued). So employing a criterion of familiarity should lead us to posit a mental cause for subjective sensations. Moreover, if we do have some reason for preferring the simpler of two hypotheses, it looks as if Berkeley's idealism wins. Berkeley's idealism posited only two kinds of things—minds and ideas (of which sensations are a species). Materialists recognize three kinds of things—minds, ideas, and matter. Moreover, in terms of numbers of things posited, Berkeley could make the case again that he has in his ontology fewer things. He just has a God, finite minds, and the sensations that God causes. To be sure, with respect to both kinds and raw numbers it is not easy to count. Berkeley may have commitment only to minds and ideas, but one of his minds is a great deal different from the others. Moreover, one might argue that the internal complexity of that mind (the various causally efficacious acts of will) needed to explain the huge variety of sensations people have, might stand in a kind of one-to-one correspondence with the huge variety of material objects that the materialist posits.[17]

The plausibility of positing physical objects as the cause of sensations is directly related to how ambitious one is in characterizing physical objects. Even Hume seemed to admit that if we are content to understand the physical simply as *whatever* it is that plays various causal roles, the hypothesis that there are physical objects looks promising (1888, p. 68), though in the *Enquiry* (1777, p. 155), he goes on to suggest that this idea of matter as "an unknown, inexplicable *something*" is "a notion so imperfect that no skeptic will think it worthy to contend against it."

By contrast, David Chalmers (2003) suggests that our commitment to physical objects really *is* nothing other than a commitment to something with the capacity to cause experiences to come and go in the way with which we are familiar. If he is right, then the problem of justifying belief in the physical world is just the problem of justifying belief that there are causes of the structural connections between experiences. I'm still not sure that we can justify that hypothesis through a special kind of reasoning that we might call reasoning to the best explanation. But I'm also not sure that we can *inductively* establish the relevant hypothesis either.

References

Anscombe, Elizabeth. 1993. "Causality and Determinism." In *Causation*, eds. Ernest Sosa and Michael Tooley. Oxford: Oxford University Press, pp. 88–104.

Berkeley, George. 1713. *Three Dialogues between Hylas and Philonous*. In *Philosophical Works: Including the Works on Vision*, ed. M. R. Ayers (ed.). Totowa, NJ: Rowman and Littlefield, 1975, pp. 129–207. (Page references follow the standard of referring to the original pagination, which most versions note on the side margin.)

[17] McCain (2014, chapter 6) discusses this issue in more detail.

Chalmers, David. 2003. "The Matrix as Metaphysics," <consc.net/papers/matrix.html>.

Coady, C. A. J. 1992. *Testimony*. Oxford: Oxford University Press.

Davidson, Donald. 1967. "Causal Relations." *Journal of Philosophy*, 64, 691–703.

Fales, Evan. 1990. *Causation and Universals*. New York: Routledge.

Fumerton, Richard. 1980. "Induction and Reasoning to the Best Explanation." *Philosophy of Science*, December, 589–600.

Fumerton, Richard. 1988. "The Internalism/Externalism Controversy." *Philosophical Perspectives*, 2, 443–59.

Fumerton, Richard. 1992. "Skepticism and Reasoning to the Best Explanation." In *Philosophical Topics*, ed. Enrique Villanueva. Chichester: Wiley.

Fumerton, Richard. 1994. "The Incoherence of Coherence Theories." *Journal of Philosophical Research*, XIX, 89–102.

Fumerton, Richard. 1995. *Metaepistemology and Skepticism*. Boston, MA: Rowman and Littlefield.

Fumerton, Richard. 2006. "Epistemic Internalism, Philosophical Assurance, and the Skeptical Predicament." In *Knowledge and Reality: Essays in Honor of Alvin Plantinga*, eds. Thomas M. Crisp, Matthew Davidson, and David Vander Laan. Dordrecht: Springer, 2, 179–92.

Fumerton, Richard. 2014. "Infinitism." In *Ad Infinitum: New Essays on Epistemological Infinitism*, eds. John Turri and Peter Klein. Oxford: Oxford University Press, 75–86.

Goldberg, Sandy. 2010. *Relying on Others*. Oxford: Oxford University Press.

Goodman, Nelson. 1979. *Fact, Fiction and Forecast*, 3rd ed. Indianapolis, IN: Hackett.

Harman, Gilbert. 1965. "The Inference to the Best Explanation." *Philosophical Review*, 74, 88–5.

Hempel, Carl. 1965. *Aspects of Scientific Explanation*. New York: Free Press.

Huemer, Michael. 2001. *Skepticism and the Veil of Perception*. Lanham, MD: Rowman and Littlefield.

Hume, David. 1777. *Enquiries*, ed. L. A. Selby-Bigge. Oxford: Clarendon Press.

Hume, David. 1988. *A Treatise of Human Nature*, ed. L. A. Selby-Bigge. Oxford: Clarendon Press.

Klein, Peter. 1999. "Human Knowledge and the Infinite Regress of Reasons." *Philosophical Perspectives*, 13, 300–25.

Lycan, William. 2002. "Explanation and Epistemology." In *The Oxford Handbook of Epistemology*, ed. Paul Moser. Oxford: Oxford University Press.

McCain, Kevin. 2014. *Evidentialism and Epistemic Justification*. New York: Routledge.

Maxwell, Grover. 1962. "The Ontological Status of Theoretical Entities." In *Scientific Explanation, Space, and Time*, Vol. 3, *Minnesota Studies in the Philosophy of Science*, eds. Herbert Feigl and Grover Maxwell. Minneapolis: University of Minnesota Press, 3–15.

Peirce, Charles Sanders. 1903. *The Essential Peirce: Selected Philosophical Writings*. Vol. 2, *1893–1913*, ed. the Peirce Edition Project. Bloomington: Indiana University Press.

Plutynski, Anya. 2011. "Four Problems of Baduction: A Brief History." *HOPOS: Journal of the International Society for the History and Philosophy of Science*, 1–22.

Poston, Ted. 2014. *Reason and Explanation: A Defense of Explanatory Coherentism*. New York: Palgrave Macmillan.

Thagard, Paul. 1978. "The Best Explanation: Criteria for Theory Choice." *Journal of Philosophy* 75: 76–92.

Zagzebski, Linda. 2012. *Epistemic Authority*. Oxford: Oxford University Press, 2012.

6

Inference to the Best Explanation
Fundamentalism's Failures

Kareem Khalifa, Jared Millson, and Mark Risjord

Many epistemologists take Inference to the Best Explanation (IBE) to be "fundamental." For instance, Lycan (1988, 128) writes that "all justified reasoning is fundamentally explanatory reasoning." Conee and Feldman (2008, 97) concur: "fundamental epistemic principles are principles of best explanation." Call them *fundamentalists*. They assert that nothing deeper could justify IBE, as is typically assumed of rules of deductive inference, such as *modus ponens*.

However, logicians account for *modus ponens* with the valuation rule for the material conditional. By contrast, fundamentalists account for IBE with an ill-defined set of relations that happen to furnish their favorite set of inductive inferences. To our eye, this seems a little too convenient—there is too much room for *ad hoc*, just-so stories about the "striking" correspondence between our explanatory and inductive practices.

We will argue that the (*explanatory*) *pluralism* adopted by the leading theorists of the best explanation—philosophers of science—undermines fundamentalism. Section 1 clarifies fundamentalism's key tenets. Section 2 presents pluralism's challenge to fundamentalism. Section 3 considers a potential fundamentalist reply to this challenge. Sections 4 through 6 canvass the leading candidates for developing this fundamentalist reply, showing each to be unsatisfactory.

1. Fundamentalism

While most self-described explanationists are fundamentalists, a precise statement of the latter is hard to find. Weintraub (2013, 203) succinctly states fundamentalism's core thesis, "IBE is an autonomous (indispensable) form of inference." Similarly, Lycan (2002, 417) describes "Sturdy Explanationism" as the thesis that "explanatory inference can do its justifying intrinsically, that is, without being derived from some other form of ampliative inference, such as probability theory, taken as more basic."

Both Weintraub and Lycan take fundamentalism to be a claim about IBE's autonomy relative to other inductive inferences, but they disagree about what count as "other inductive inferences." Weintraub (2013, 204) contrasts IBE only with *enumerative induction*, i.e. "inference from a sample to the entire population or to the next case." However, fundamentalists should argue that IBE is autonomous relative to every alternative rule of induction, which also includes Mill's Methods, hypothetico-deductivism, error-statistical inference, and Bayesian Conditionalization.[1]

IBE's alleged autonomy amounts to its *irreducibility* to these other rules of induction. Reductionists claim that any warranting (i.e. inductively good) instances of IBE can be recast as warranting instances of one or more non-explanatory inferences, while denying the converse. More precisely:

IBE is reducible to another inductive argument schema(s) X if:
(R1)   Some warranting instances of X do not require the *best explains* operator, and
(R2)   For all warranting instances of IBE, I_E, wherein H *best explains* E, there is some warranting instance of X, I_X, such that:
    (a) H is a logical consequence of I_X's conclusion; and
    (b) E is a logical consequence of I_X's premises.

This requires three clarifications. First, the *argument schema* for IBE is:

e

$\underline{h \text{ best explains } e}_{\text{[probably]}}$

h

The lowercase letters are variables that take statements as their values. The remaining formal vocabulary is the *best explains* operator. Other argument schemas have variables and distinctive operators. For instance, Bayesians' paradigmatic formal vocabulary invokes conditional probability.

Second, *instances* of IBE[2] obtain when the schema's variables assume specific values, e.g.

IBE Mono
The patient has a high number of atypical lymphocytes.
That the patient has mononucleosis best explains why he has a high number of $\underline{\text{atypical lymphocytes.}}_{\text{[probably]}}$
The patient has mononucleosis.

Hence, in (R2), H and E are h and e's respective values.

Third, beginning in Section 3, we will discuss how an argument schema could "require the *best explains* operator" in greater detail. In the interim, we rely on the reader's intuitions.

---

[1] Norton (2003) surveys the induction literature.
[2] We use "IBE" as a proper name when referring to its argument schema; as a count noun when referring to its instances.

Conditions (R1) and (R2) characterize a form of reduction that is fair to the fundamentalist and to the reductionist. As an illustration, consider the following *Bayesian Reduction*:

B1.   Whenever h_i explains e better than h_j, $P(e|h_i) > P(e|h_j)$.
B2.   Whenever h_i is more explanatory virtuous (e.g. simpler, more powerful, more fruitful, etc.) than h_j, $P(h_i) > P(h_j)$.
B3.   Degrees of belief can be mapped onto IBE, e.g. via detachment or acceptance rules for probabilities.
B4.   No other principles link explanation to induction.[3]

Ceteris paribus, a straight Conditionalization Rule—the Bayesian's paradigmatic argument schema—would then favor h_i over h_j:

$$P(h \mid e) = \frac{P(e \mid h) \times P(h)}{P(e)}$$

If B1 through B4 are true, then IBE is reducible to Bayesian inference. Per (R1), proponents of this reduction must argue that some warranting Bayesian inferences do not appeal to the best explanation. For instance, they might argue that P(the coin exists| the coin landed heads) = 1, yet the coin landing heads does not explain why the coin exists. Likewise, any IBE will be an instance of Conditionalization. Presumably, explanationists take all IBEs to preserve warrant. Hence (R2) would also be satisfied.

Suppose this reduction is correct. Then although IBE would be warranting, it would be dispensable, for Conditionalization could be used without any explanatory detours. Consequently, fundamentalists must deny reductionism as defined by (R1) and (R2).

2. Reductive Pluralism

In the balance of this chapter, we argue for reductionism. We are not the first to do so (Fumerton 1980; Salmon 2001). However, our position differs substantially from our predecessors', as we harness a long-budding consensus in the philosophy of science. According to this consensus, scientific explanations come in many shapes and sizes.[4] Caution about overgeneralizing is the norm. For instance, even Woodward's (2003) ambitious and influential theory of explanation is restricted to causal explanation, but countenances the legitimacy of some non-causal explanations. Indeed, whole books have been written on specific kinds of explanations in specific fields, e.g. asymptotic explanations in physics (Batterman 2002) and mechanistic explanations in neuroscience (Craver 2007). Importantly, few contemporary philosophers of

---

[3] Since this is only offered as illustration, we will not engage the various authors who have challenged these assumptions, e.g. Lipton (2004).
[4] Achinstein (1981), McCauley (1996), and Díez et al. (2013) offer general arguments for explanatory pluralism. Many other authors presuppose it, e.g. Douglas (2009), or argue for pluralism within particular scientific disciplines, e.g. Risjord (2000).

science are offering general analyses of explanation that *compete* with each other—they're not offering general analyses of explanation *at all*.

2.1 *Thick relations*

Among these philosophers of science, the working assumption is that *explanation* can be unpacked in many different ways. This suggests the following:

> *Pluralist Thesis*:   If "*h* best explains *e*" is true, then *h* describes *e*'s causes or *h* nomologically entails *e* or *h* describes the mechanisms underlying *e* or . . .

A few clarifications are needed. Call the causal, nomological, mechanistic, etc. relations *thick relations*. Taking the prevailing consensus among philosophers of science as our cue, the ellipsis in the Pluralist Thesis only includes these thick relations. Provisionally, let *thick inferences* be inferences with at least one premise invoking a thick relation, and having no references to the *best explains* operator.

We will argue that IBE is reducible to thick inference, i.e.

(T1)   Some warranting thick inferences do not require the *best explains* operator, and
(T2)   For all warranting instances of IBE, I_E, wherein H *best explains* E, there is some warranting thick inference, I_T, such that:
　　(a) H is a logical consequence of I_T's conclusion; and
　　(b) E is a logical consequence of I_T's premises.

Consider the following:

Oxygen
The house caught fire.
That the house was in an oxygenated atmosphere is a cause of the house catching fire. [probably]
The house was in an oxygenated atmosphere.

Clearly, causation is one of our thick relations, and this causal inference preserves warrant. So, this is a warranting thick inference. However, oxygen is almost never the best explanation of a fire. Hence *Oxygen* does not require the *best explains* operator.[5] So (T1) is true.

2.2 *Thick virtues*

While (T1) shows that some thick inferences are not IBEs, (T2) requires us to show that all IBEs are thick inferences. Hence, while *Oxygen* validates (T1) precisely because it trades in a warranting but less-than-best explanation, (T2) requires us to show that any reference to the best explanation can be replaced by thick relations without loss.

---

[5] Sections 3 through 6 provide a sustained argument that our thick reduction base does not require the *best explains* operator. For now, we rest on the assumption that a causal explanation can be correct but not the best.

Fundamentalists typically assume that the best explanation is one that strikes the right balance between several theoretical virtues (e.g. scope, simplicity, conservatism, testability, and fruitfulness). So, in order to reduce IBE to thick inference, we will argue for the following:

> *The Replacement Thesis*:   For all h, e, and v, if h_1 is a better explanation of e than h_2 with respect to virtue v, then v can be explicated using thick-relation-concepts and without using the *explains*-concept.

Once we establish this, then some amalgam of thick-relations-concepts can replace any appeal to "best explains." Consequently, any warranting inference invoking "best explains" will also be a warranting thick inference—just as (T2) requires.

To illustrate how explanatory pluralists can capture the theoretical virtues via the Replacement Thesis, consider Thagard's (1978, 79) account of consilience:

> Let FT_i be the set of classes of facts [i.e. phenomena] *explained by* theory T_i. Then...(1) T_1 is more consilient than T_2 if and only if the cardinality of FT_1 is greater than the cardinality of FT_2; or (2)...FT_2 is a proper subset of FT_1.

Since the Pluralist Thesis already replaces any instance of "explains" with a thick relation, we can simply repeat that move here, e.g.

> (*Thick Consilience*)   Let FT_i be the set of classes of facts *causally modeled, nomologically entailed, mechanistically modeled, or...* by theory T_i. Then...(1) T_1 is more consilient than T_2 if and only if the cardinality of FT_1 is greater than the cardinality of FT_2; or (2) FT_2 is a proper subset of FT_1.

Any analysis of a virtue that appeals to explanatory relationships can be treated in analogous fashion.[6]

A fortiori, virtues invoking explanatory and *non-explanatory* relationships will also be fodder for reduction. This requires two steps. First, we identify a (barbarously) disjunctive relationship:

> R is a relationship of entailment, causal representation, mechanistic representation, prediction, analogy, embedding, or ...

Some of these relations are thick relations that replace instances of "explains," while others (e.g. prediction) were never proffered as explanatory in the first place.

Second, we recast all of the theoretical virtues in terms of this disjunctive relationship:

- *Thick Simplicity*: Ceteris paribus, if h_1 is a simpler explanation than h_2, then h_1 stands in at least as many relationships R to the evidence as h_2, but posits fewer entities, processes, etc. than h_2.

---

[6] Examples include Thagard's (1978, 1989, 1992) differing treatments of consilience, simplicity, and analogy; and Psillos' (2009) treatments of completeness, importance, parsimony, unification, and precision.

- *Thick Conservatism*: Ceteris paribus, if h_1 is a more conservative explanation than h_2, then h_1 stands in more relationships R to what is already accepted/believed.
- *Thick Fruitfulness*: Ceteris paribus, if h_1 is a more fruitful explanation than h_2, then h_1 suggests or discloses new relationships R, or new phenomena that stand in R to other accepted statements.
- *Thick Testability*: Ceteris paribus, if h_1 is a more testable explanation than h_2, then h_1 stands in a relation R to a set of empirical conditions that are more readily accessible and/or that more readily determine whether or not h_1 is true/acceptable.

Thus, by replacing explanations with our disjunctive relationship, we can jettison explanations from our inductive inferences without loss.

This concludes our argument for the Replacement Thesis. While this already suffices to establish (T2), pluralism puts further reductive pressures on the theoretical virtues. Like explanation, pluralism about theoretical virtues is the norm among philosophers of science. Moreover, this latter pluralism has many dimensions. First, many individual virtues have multiple interpretations. For instance, "simplicity" may denote quantitative parsimony (fewer entities), qualitative parsimony (fewer kinds of entities), or commitment to a parameter-minimizing information criterion (e.g., Akaike). Any given interpretation applies to only a subset of our best explanations. Second, explanations of different phenomena weight virtues differently. For instance, fundamental physics emphasizes simplicity, biology, mechanism. Frequently, simpler explanations eschew mechanistic detail, so tradeoffs are inevitable.[7] Third, some of our best explanations jettison time-honored theoretical virtues. For instance, a natural corrective to oversimplification is greater complexity; novelty trumps conservatism when background assumptions are skewed; etc. (Longino 1995).

These points allow us to refine the Pluralist Thesis. Let a *thick bundle* be a thick relation plus a list of virtues, each of which is given specific interpretation and weight relative to the other members on this list. ("Thick inference" now refers to inferences involving thick bundles.) Here is a simple thick bundle:

Thick relation: deductive-nomological
List of virtues: "power" and "simplicity"
Interpretation of virtues:
(1)   "power" = Thick Consilience.
(2)   "simplicity" = Thick Simplicity.
Weighting of virtues: Among the statements that nomologically entail e, use the following, lexically ordered rules to determine which to accept:
(1)   Maximize Thick Consilience.
(2)   Maximize Thick Simplicity.

---

[7] Kuhn (1977) makes similar points, though he emphasizes that different *inquirers* can reasonably differ about how they interpret and weight the virtues when explaining the same phenomenon. Less controversially, we only claim that *explanations* of different phenomena can differ in these ways.

Since writing out these bundles is cumbersome, let $W_i he$ indicate that h is the "winner" of the weighting stage. Then the Pluralist Thesis becomes:

Thick Bundle Thesis: If "*h* best explains *e*" is true, then $W_1 he$ or $W_2 he$ or $W_3 he$ or...

Importantly, *Oxygen* is still a perfectly good example of (T1), e.g., if the following thick bundle is among our disjuncts:

Thick relation: causal
List of virtues: ∅.

Nor is this bundle *ad hoc*. Like *Oxygen*, many causal inferences allow us to drop the *best*-component of the *best explains* operator. For instance, consider:

Thick Mono
The patient has a high number of atypical lymphocytes.
The patient's mononucleosis caused him to have a high number of atypical lymphocytes. [probably]
The patient has mononucleosis.

Since our paradigmatic explanations are causal, this means that many IBEs are reducible to causal inferences without reference to the virtues.[8] Hence, the pluralist challenge to fundamentalism is not limited to *recherché* examples.

To summarize, our reductive pluralism readily assimilates the theoretical virtues to thick bundles. Moreover, replacing theoretically virtuous explanations with thick bundles does nothing to alter the inferences in question; it is simply more cumbersome. Hence, any IBE can be turned into a thick inference, so (T2) is on firm ground. Furthermore, (T1) remains unaffected by our more catholic notion of a thick inference. Hence, fundamentalism should be rejected.

2.3 Induction

We've largely taken it for granted that thick inferences are warranting in a way that makes for a suitable reduction base. However, reducing IBE to thick inferences might seem confused, as these are not argument schemata. There are two replies to this.

First, in principle, thick inferences are reducible to traditional alternatives to IBE.[9] We have already shown that replacing "*h* best explains *e*" with its thicker correlate will not make an inferential difference. As before, represent the latter as $W_i he$. Now plug *e* and $W_i he$ into your preferred form of (non-explanatory) induction, along with any additional non-explanatory assumptions *a* needed to get *h* as a conclusion. Presto! You've reduced IBE to your favorite form of non-explanatory inference. For instance,

---

[8] This is slightly more complicated. Causal inferences typically introduce causal vocabulary in their conclusions. However, since *h* is a logical consequence of *h causes e*, the reduction still goes through.

[9] Thanks to Leah Henderson and Brad Weslake for raising this point.

Bayesians could use the preceding arguments to effect reductionism so long as the following holds:

$$P(h \mid e, W_i he, a) \gg P(h).$$

Second, explanatory pluralism might occasion a kind of inductive pluralism in which there is a proliferation of different argument schemas—perhaps to the point where the intuitive distinction between schemas and instances breaks down. Proponents of "material" theories of induction have defended inductive pluralism of this sort (Sellars 1980; Brandom 1994; Norton 2003). On this view, inductive inferences are not warranting in virtue of their formal vocabulary, but through the remaining material vocabulary of specific instances. For instance, consider Norton's (2003) example:

Bismuth	_Wax_
Some samples of bismuth melt at 271°C. [probably]	Some samples of wax melt at 91°C. [probably]
All samples of bismuth melt at 271°C.	All samples of wax melt at 91°C.

Both inferences are instances of the following argument schema for enumerative induction:

Schema EI
Some As are Bs. [probably]
Therefore, all As are Bs.

The inductive "formalist" who includes _EI_ in her stable of accepted argument schemata must claim that _all_ instances of _EI_ are warranting. And yet, while _Bismuth_ preserves warrant, _Wax_ does not. Materialists about induction claim that the non-formal vocabulary, e.g. _bismuth_, _wax_, and _melting point_, ultimately accounts for the differences between these two inferences. By contrast, formalists must seek argument schemata with more stringent formal vocabulary to account for this difference.

Similarly, on the current proposal, instances of IBE that fail to preserve warrant are not the result of two propositions failing to stand in the formal _best explains_ relation. Rather, thick bundles explain these breakdowns. An inference such as _IBE Mono_ is good only because of the thick _causal_ relationship between _atypical lymphocytes_ and _mononucleosis_. On this view, thick-bundle vocabulary is material relative to the formal vocabulary in IBE, i.e. _best explains_.

Thus, thick inferences can be warranting in two ways: (1) as instances of traditional non-explanatory argument schemata, or (2) within the context of a material theory of induction. For the purposes of this chapter, we are content with either approach, for both are compatible with (T1) and (T2).[10]

---

[10] Having said that, we prefer the latter. See note 12.

3. Multiple Realizability?

So far, we have argued against fundamentalism by showing that some thick inferences do not require the *best explains* operator (T1), and that any warrant-preserving instance of IBE can be replaced by a thick inference (T2). From this it follows that IBE is reducible to thick inference. In short, because thick bundles do all of the inferential work in IBE, the *best explains* operator is dispensable.

In response to the foregoing argument, it might be pointed out that many fundamentalists seem congenial to some version of explanatory pluralism (Harman 1965; Lycan 1988; Thagard 1992; Poston 2014; Psillos 2009). Presumably, they reverse our preferred order of analysis. In other words, they hold that thick inferences are warranting precisely because they are IBEs, and that causal, nomological, mechanistic, etc. relations are inextricably bound to the *best explains* operator. According to this fundamentalist position, (T1) is false on the grounds that thick bundles multiply realize the *best explains* role. Furthermore, the Replacement Thesis—which undergirded our argument for (T2)—rested on our argument for (T1). Consequently, this fundamentalist rebuttal poses a deep challenge to our reductionist arguments. Call it the *Multiple Realizability Defense*:

(MR1)   For all h and e, if there is a thick inference from e to h, then h stands in R to e.

(MR2)   For all h and e, if h stands in R to e, then "h best explains e" is true.

(MR3)   For all h and e, if h stands in relation R to e, then e inductively warrants h.

(~T1)   All warranting thick inferences require the *best explains* operator.

Here R is the common role realized by different thick bundles. If any relation satisfies these three conditions, then all thick inferences are R-mongering inferences, and all R-mongering inferences require the *best explains* operator. Consequently, if the fundamentalist-*cum*-multiple-realizer can find such a relation, then (T1) is false, and our proposed reduction is thwarted.

Let's discuss why each of these three conditions is necessary for the fundamentalist's defense. The Multiple Realizability Defense claims that all of the thick bundles realize a common role. If we treat the relation R as this role, then (MR1) captures this anti-reductionist commitment. Indeed, even if (MR2) and (MR3) are true when (MR1) is false, then those thick bundles that do not realize this common role can eschew the *best explains* operator. Consequently, without (MR1), (T1) remains unthreatened.

Turn now to (MR2). Quite clearly, h's standing in relation R to e must be sufficient for making "h best explains e" true. Otherwise, the claim that all thick inferences realize a common role, i.e. (MR1), will not guarantee that all thick inferences require the *best explains* operator (~T1).

Finally, suppose that (MR1) and (MR2) are true, but (MR3) is false. Then although all thick inferences realize a common role, and that role tracks with our best explanations, our best explanations ultimately do not track with good inference. Hence, the thick bundles would be doing all of the relevant inferential work, and explanatory considerations would be playing some other, non-inferential role. Presumably, fundamentalists claim that our inductive inferences are good *because* of the relevant explanatory considerations. Consequently, (MR3) is also necessary to dislodge (T1).

Thus, fundamentalists need all of the thick bundles to realize some best-explains role. We will examine three leading candidates: a primitive best-explains relation, counterfactual dependence, and the relationship between a why-question and its answer. We will argue that none can underwrite the Multiple Realizability Defense. Nevertheless, the failures of these three proposals are instructive, for they suggest a more promising "nominalist" alternative to fundamentalism. On this view, "best explains" is merely a convenient label for picking out thick bundles that serve certain human interests.

4. Primitivism

The Multiple Realizability Defense is schematic, as the relation R can be instantiated in different ways. The "primitivist" takes the *best explains* operator to be unanalyzable. Therefore, the primitivist fills in (MR1) through (MR3) as follows:

(P1)   For all h and e, if there is a thick inference from e to h, then h best explains e.

(P2)   For all h and e, if h best explains e, then "h best explains e" is true.

(P3)   For all h and e, if h best explains e, then e inductively warrants h.

The chief problem with primitivism is (P1). After all, *Oxygen* is a thick inference that does not invoke the best explanation. So, primitivism leaves the class of thick inferences that motivate our reductionism untouched.

Furthermore, other theories of inference say far more about their formal vocabulary than primitivism says about the *best explains* operator. We have already mentioned how logicians account for *modus ponens*. However, this is not confined to deduction, as Bayesians account for Conditionalization using Dutch books, representation theorems, and/or scoring rules. Consequently, a primitivist version of the Multiple Realizability Defense threatens to lapse into obscurantism.

Are there any arguments for primitivism in general, and (P1) in particular? Poston (2014, 69–80) offers the most promising arguments of this sort,[11] but they are arguably unsound. Poston's first datum is that philosophers of science have failed to provide an "analysis of explanation having this form: a group of statements φ explains another statement ψ if and only if X, where X is some set of non-trivial conditions," despite sixty years of trying to do so. Here, he draws parallels with Williamson's (2000) primitivism about knowledge, which arose in light of post-Gettier epistemology's similar frustrations in furnishing an analysis of knowledge.

---

[11] Poston also offers the "Argument from Better Explanations in Math," and the "Epistemological Argument." The first is essentially an argument for pluralism. The second assumes fundamentalism, and hence is question-begging in the current discussion. It should also be noted that Poston is only a primitivist about *explains*; like us, he is a pluralist about the virtues (Poston 2011).

Let us grant that the failure to analyze *explains* is because this concept is unanalyzable, and not because of a lack of philosophical ingenuity. While the unanalyzability of *explains* is consistent with there being a single, primitive best explains role realized by all thick bundles—as Poston is suggesting—it does not entail this primitivist stance, for it is also compatible with there being *multiple* best explains roles realized by different thick bundles—as we are suggesting.

Indeed, over twenty-five years ago, Salmon—one of the last great purveyors of the analyses that Poston criticizes—had already suggested this multiplicity of explanatory roles. At the conclusion of his survey of the explanation literature, Salmon suggests that there are several conceptions of explanation, each with a different role. For instance, according to Salmon (1989, 182), unificationist explanations, such as Kitcher's (1989), "serve to organize and systematize our knowledge in the most efficient and coherent possible fashion," while causal-mechanical explanations provide "knowledge of the hidden mechanisms by which nature works." While Salmon notes that these roles are often complementary, he does not assert that they have some common core—much less a common core that can punch an inference ticket, as the Multiple Realizability Defense requires. Moreover, others have argued that these roles are distinct: some causal explanations provide dis-unified (unsystematic, loosely organized) knowledge (Cartwright 1999), while some unifying explanations are non-causal (Kitcher 1989). Consequently, the fact that *explanation* is unanalyzable does not entail that thick-bundle vocabulary—such as causal or nomological vocabulary—realizes a common role.

Poston's next argument is the "Argument from Cognition." The claim hinges on a piece of linguistic data that "because" is the ninety-fourth most common word in English (and similar patterns emerge in other languages) and the words that appear ahead of it are not likely to provide an analysis of *explanation*. Similarly, simple explanatory concepts appear in early childhood, and the common practice of removing a mystery seems fraught with explanatory reasoning. According to Poston (2014, 76), this evidence suggests that "The idea that there are facts of the form '*this because that*' [is] the kind of notion that is not learned by a mastery of more fundamental concepts."

However, *theorizing* or *analyzing* the concept of explanation differs from *possessing* that concept. Indeed, formal semantics rests on precisely this distinction. For instance, the words "will," "would," "about," "time," "when," and "can" are all among the top 100 words used in English. All of them are analyzed using possible-worlds semantics. Yet, we can rest assured that "possible world" is not a part of the everyman's lexicon. Hence, the Argument from Cognition might show that possession of the concept *because* does not require much else. However, this does not preclude analyzing that concept using other, thicker concepts.

Indeed, reductive pluralism explains away the intuitions that drive others towards primitivism. If the only universal conditions for the best explanation are trivial, then the concept, *best explains*, may *seem* to lack a deeper substrate and hence *seem* primitive. However, on the reductive pluralist line, this is only because nothing unifies this substrate, which consists of thick inferences. Essentially, the primitivist's guiding intuition that explanatory inferences run *deep* is replaced with the pluralist's idea that thick inferences run *wild*.

5. Counterfactual Dependence

Primitivism's failures suggest that fundamentalists need thick bundles to multiply realize an explanatory role that is at least partially analyzable. On this front, a promising candidate is counterfactual dependence. Regardless of whether they are causal, nomological, or otherwise, all explanations must be "difference-makers." So, the Multiple Realizability Defense could be instantiated in the following way:

(CD1)   For all *h* and *e*, if there is a thick inference from *e* to *h*, then *e* would have been *e'*, had *h* been *h'*.

(CD2)   For all *h* and *e*, if *e* would have been *e'*, had *h* been *h'*, then "*h* best explains *e*" is true.

(CD3)   For all *h* and *e*, if *e* would have been *e'*, had *h* been *h'*, then *e* inductively warrants *h*.

(CD1) through (CD3) correspond to (MR1) to (MR3), respectively. Here the counterfactual claim indicates that *h* made a difference to *e*, or that *e* counterfactually depends on *h*.

This proposal faces two problems. First, consider our earlier example:

Counterfactual Oxygen
The house caught fire.
Had the house been in a deoxygenated (rather than oxygenated) atmosphere, it would not have caught fire (rather than catching fire). [probably]
The house was in an oxygenated atmosphere.

As before, this is a warranting (counterfactual) inference, but not an IBE. Hence, there is no reason to grant (CD2).

Indeed, while *Counterfactual Oxygen* invokes a less-than-best explanation, other warranting counterfactual inferences invoke no explanations.

For instance, consider the following:

Moore
(Circuits per chip) = $2^{(\text{year}-1975)/1.5}$
There are 8,192 circuits per chip.
Had it been 2011 (rather than 2014), then the number of circuits per chip would have been 4,096 (rather than 8,192). [probably]
It is the year 2014.

The first premise, Moore's law, is a well-confirmed regularity, but it is not an explanatory relationship. For example, the fact that last year was 2014 does not explain why there were 8,192 circuits per chip. Presumably, this number is explained instead by technological innovations in the computing industry. Nevertheless, there is a difference-making relationship here, and it can be used in an inductive inference—even if we don't have information about the relevant innovations in the computing industry.

Generally, some predictive or descriptive models support counterfactuals, yet lack explanatory purport. Since their corresponding counterfactual inferences warrant but have no corresponding IBE, difference making is not a sufficient condition for our best explanations. Hence, in addition to *Oxygen*, Moore provides further evidence that (CD2) is false.

Additionally, as Fumerton (1980, 591–2) argues, (CD3) faces counterexamples, e.g.

Cow
There are footprints on the beach.
If a cow wearing shoes [rather than not wearing shoes] had walked the beach recently, there would be such footprints [rather than hoof-prints]. [probably]
A cow wearing shoes walked the beach recently.

Despite the counterfactual relation holding between the minor premise and the conclusion, this putative inference is not warranting. So, *contra* (CD3), counterfactual dependence is not sufficient for inductive warrant.

Most fundamentalists would retort that the cow explanation lacks theoretical virtues such as conservatism and simplicity. This would mean that these and other virtues do some inferential heavy lifting independently of their difference-making capacities. Hence, *Cow* is not an IBE. Yet, Section 2.2 shows that the virtues' inferential roles are reducible to their thick bundling capacities. So, ultimately, *Cow* is a bad inference not because it fails to be an IBE, but because it fails to be a thick inference.

Still, for all that we've shown, (CD1) may be true. This would mean that difference making is a *necessary* feature of our best explanations. This suggests that thick bundles should be equiprimordial with counterfactual dependence requirements, i.e.

> *Thick Counterfactual Thesis*:   If "*h* best explains *e*" is true, then (i) $W_1 he$ or $W_2 he$ or $W_3 he$ or..., and (ii) *e* counterfactually depends on *h*.

In this way, thick bundles capture any distinctive, virtue-driven inductive roles not exhausted by difference making.

This simply elaborates Section 2's Thick Bundle Thesis. Examples such as *Counterfactual Oxygen* still make it incompatible with fundamentalism. Furthermore, as the falsehood of (CD3) underscores, even if all thick bundles are difference makers, counterfactual dependence is still not substantive enough to account for IBE's warrant. Hence, it cannot live up to its billing as the fundamentalist's hidden role.

6. Why-Questions

Consider our last candidate for a multiply realized role:

(WQ1)   For all *h* and *e*, if there is a thick inference from *e* to *h*, then *h* is a correct answer to "Why *e*?"

(WQ2)   For all h and e, if h is a correct answer to "Why e?", then "h best explains e" is true.

(WQ3)   For all h and e, if h is a correct answer to "Why e?", then e inductively warrants h.

(WQ1) through (WQ3) correspond to (MR1) to (MR3), respectively.

Fundamentalists of this stripe face a dilemma. To see this, consider the following:

Interrogative Oxygen
The house caught fire.
That the house was in an oxygenated atmosphere is a correct answer to "Why did the house catch fire?" [probably]
The house was in an oxygenated atmosphere.

The second premise is not nearly as intuitive as in our *Oxygen* or *Counterfactual Oxygen*. Yet, if *Interrogative Oxygen*'s second premise is true, then the presence of oxygen answers the relevant why-question, but we still have no reason to think that it is the best explanation of why the house caught fire. Hence, (WQ2) is unsubstantiated. If, on the other hand, *Interrogative Oxygen*'s second premise is false, then (WQ1) is false, for *Oxygen* is a good thick inference, yet it does not answer the relevant why-question.

Regardless of how fundamentalists resolve this dilemma, (WQ3) faces further difficulties. First, if a "correct" answer to a why-question is a *true* answer, then (WQ3) reduces IBE to a trivial deductive inference. So, fundamentalists need something weaker. But if "correct" means "best," fundamentalists backslide into primitivism. But if "correct" is neither "true" nor "best," then it is liable to appeal to thick-bundle vocabulary—just as we predict.

Second, even if correct answers must be true, (WQ3) is false on any intuitive reading of "inductive warrant." Consider any event that has a very unlikely and atypical explanation. For instance, suppose that a grass-stained baseball is bleached to make it white again. This tells us why the baseball is white, yet we could not infer that the baseball was bleached from the fact that it was white.

The preceding considerations suggest that explanation is one thing, and inductive warrant is another. This is given further credence when we consider that the leading accounts of why-questions (Sintonen 1989; van Fraassen 1980) rely heavily on contextual mechanisms that are largely orthogonal to inductive considerations. Indeed, the pragmatics of why-questions accounts for why the second premise in *Interrogative Oxygen* is far less obvious than its analogue in *Oxygen*. For van Fraassen, the same why-interrogative sentence can express different why-questions, depending on the context in which the former is uttered. In *typical* contexts, a person asking, "Why did the house burn down?" presupposes that oxygen is in the atmosphere, and is interested in the fire's other causes. But, in *atypical* contexts, where interests and presuppositions are different, we ask a different question, which may well be answered by the assertion that oxygen is in the atmosphere. Hence, because of an underspecified context, *Interrogative Oxygen*'s second premise invokes interrogative vocabulary ("answer," "why?" etc.) with an ambiguous referent. By contrast, the vocabulary characteristic of thick bundles is context invariant: whether oxygen is a cause of the house's burning down is not sensitive to speakers' presuppositions or interests. Hence, unlike *Interrogative Oxygen*, *Oxygen*'s second premise is unambiguous.

However, this comparison between *Interrogative Oxygen* and *Oxygen* should be cold comfort to fundamentalists. By integrating the pragmatics of why-questions into our account, we can make this vivid:

> *Thick Interrogative Counterfactual Thesis*:   "*h* best explains *e*" is true in context *C iff* (i) W_1he or W_2he or W_3he or... (ii) *e* counterfactually depends on *h*, and (iii) given the relevant presuppositions and interests in *C*, *h* is a correct answer to "Why *e*?" in *C*.[12]

We are willing to claim that all and only our best explanations have these three features. By contrast, we only claim the following of inductive warrant:

> If (i) and (ii), then *e* inductively warrants *h*.

In other words, thick *inferences* only involve counterfactual dependence and thick bundles, but only *explanations* are held hostage to speakers' interests and presuppositions by way of the pragmatics of why-questions (iii). On this view, "best explains" merely picks out the thick bundles that suit our current purposes. This accounts for the differences between *Oxygen* and *Interrogative Oxygen*, and remains consistent with the reductionist arguments in Section 2.

7. Conclusion

In summary, we have rejected the popular epistemological view that IBE is a fundamental rule of inductive inference. We were motivated by the idea, popular among philosophers of science, that our best explanations are a diverse lot of thick bundles sharing little structure. From this, we have found that thick bundles underwrite far more good inferences than those involving our best explanations. Consequently, IBE is not fundamental, but instead is reducible to thick inference.

Moreover, thick inference is such an unruly reduction base that there is little to unify IBE. Hence, thick bundles are not multiple realizers of the *best explains* role, since there could be almost as many *best explains* roles as there are thick bundles. The leading candidates for a single *best explains* role—its roles in counterfactual reasoning and in why-questions—merely reinforced the idea that IBE is not a fundamental rule of inference.

---

[12] Material theorists of inference, discussed in Section 2.3, reduce (ii) counterfactual (Sellars 1957) and (iii) interrogative (Millson 2014) inferences to (i) thick, material inferences. We will develop these ideas in future work.

From this view, an alternative picture of IBE has emerged. Using only thick bundles, we can specify all of the target relationships that obtain between various statements, including the inferential relationships. When it piques our interest, we ask questions about one relatum of a thick bundle, and answer it with another. Deploying some pragmatics, we earmark some of these questions as "why-questions," and some of their answers as "best explanations." However, there is nothing more to them. All of their semantic and epistemological properties—including their inferential properties—are in place before this baptism, using only thick bundles. In short, explanatory considerations run shallow.

References

Achinstein, Peter (1981), "Can there be a model of explanation?" *Theory and Decision* 13 (3): 201–27.

Batterman, Robert W. (2002), *The devil in the details: asymptotic reasoning in explanation, reduction and emergence.* Oxford: Oxford University Press.

Brandom, Robert (1994), *Making it explicit: reasoning, representing, and discursive commitment.* Cambridge, MA: Harvard University Press.

Cartwright, Nancy (1999), *The dappled world: A study of the boundaries of science.* Cambridge: Cambridge University Press.

Conee, Earl and Richard Feldman (2008), "Evidence," in Quentin Smith (ed.), *Epistemology: New essays.* Oxford: Oxford University Press, 83–104.

Craver, Carl F. (2007), *Explaining the brain: Mechanisms and the mosaic unity of neuroscience.* Oxford: Clarendon Press.

Díez, José, Kareem Khalifa, and Bert Leuridan (2013), "General theories of explanation: Buyer beware," *Synthese* 190 (3): 379–96.

Douglas, Heather E. (2009), "Reintroducing prediction to explanation," *Philosophy of Science* 76 (4): 444–63.

Fumerton, Richard (1980), "Induction and reasoning to the best explanation," *Philosophy of Science* 47 (4): 589–600.

Harman, Gilbert (1965), "The inference to the best explanation," *Philosophical Review* 74: 88–95.

Kitcher, Philip (1989), "Explanatory unification and the causal structure of the world," in P. Kitcher and W. C. Salmon (eds), *Scientific explanation.* Minneapolis: University of Minnesota Press, 410–506.

Kuhn, Thomas S. (1977), *The essential tension: Selected studies in scientific tradition and change.* Chicago: University of Chicago Press.

Lipton, Peter (2004), *Inference to the best explanation*, 2nd ed. New York: Routledge. Original edition, 1991.

Longino, Helen E. (1995), "Gender, politics, and the theoretical virtues," *Synthese* 104 (3): 383–97.

Lycan, William G. (1988), *Judgement and justification.* Cambridge: Cambridge University Press.

Lycan, William G. (2002), "Explanation and epistemology," in Paul K. Moser (ed.), *The Oxford handbook of epistemology.* Oxford: Oxford University Press, 408–33.

McCauley, Robert N. (1996), "Explanatory pluralism and the coevolution of theories in science," in Robert N. McCauley (ed.), *The churchlands and their critics.* Oxford: Blackwell, 17–47.

Millson, Jared (2014), *How to ask a question in the space of reasons: Assertions, queries, and the normative structure of minimally discursive practices*. PhD dissertation, Emory University Philosophy Department, Ann Arbor: ProQuest/UMI (Publication No. AAT 3634363).

Norton, John D. (2003), "A material theory of induction," *Philosophy of Science* 70: 647–70.

Poston, Ted (2011), "Explanationist plasticity and the problem of the criterion," *Philosophical Papers* 40 (3): 395–419.

Poston, Ted (2014), *Reason and Explanation: A Defense of Explanatory Coherentism*. Basingstoke: Palgrave MacMillan.

Psillos, Stathis (2009), *Knowing the structure of nature: Essays on realism and explanation*. Basingstoke: Palgrave Macmillan.

Risjord, Mark (2000), *Woodcutters and witchcraft: Rationality and interpretive change in the social sciences*. Albany: State University of New York Press.

Salmon, Wesley C. (1989), "Four decades of scientific explanation," in Philip Kitcher and Wesley Salmon (eds), *Scientific Explanation*, Minneapolis: University of Minnesota Press, 3–219.

Salmon, Wesley C. (2001), "Explanation and confirmation: A Bayesian critique of inference to the best explanation," in G. Hon and S. S. Rakover (eds), *Theoretical approaches to explanation*, Dordrecht: Kluwer, 121–36.

Sellars, Wilfrid (1957), "Counterfactuals, dispositions, and the causal modalities," in Herbert Feigl, Grover Maxwell, and Michael Scriven (eds), *Minnesota studies in the philosophy of science*. Minneapolis: University of Minnesota, 225–308.

Sellars, Wilfrid (1980), *Pure pragmatics and possible worlds: The early essays of Wilfrid Sellars*, ed. J. Sicha. Reseda, CA: Ridgeview.

Sintonen, Matti (1989), "Explanation: In search of a rationale," in Philip Kitcher and Wesley C. Salmon (eds), *Scientific explanation*. Minneapolis: University of Minnesota Press, 253–82.

Thagard, Paul (1978), "The best explanation: Criteria for theory choice," *Journal of Philosophy* 75: 76–92.

Thagard, Paul (1989), "Explanatory coherence," *Behavioral and Brain Sciences* 12 (3): 435–502.

Thagard, Paul (1992), *Conceptual revolutions*. Princeton, NJ: Princeton University Press.

van Fraassen, Bas C. (1980), *The scientific image*. New York: Clarendon Press.

Weintraub, Ruth (2013), "Induction and inference to the best explanation," *Philosophical Studies* 166 (1): 203–16.

Williamson, Timothy (2000), *Knowledge and its limits*. Oxford: Oxford University Press.

Woodward, James (2003), *Making things happen: A theory of causal explanation*. New York: Oxford University Press.

7

Inference to the Best Explanation, Bayesianism, and Knowledge

Alexander Bird

1. Introduction

How do scientists reason? How ought they reason? These two questions, the first descriptive and the second normative, are quite distinct. However, if one thinks, as a scientific realist might, that scientists on the whole reason as they ought to then one will expect the two questions to have very similar answers.

That fact bears on debates between Bayesians and explanationists. When we evaluate a scientific hypothesis, we may find that the evidence supports the hypothesis to some degree, but not to such a degree that we would be willing to assert outright that the hypothesis is true. Bayesianism is a theory about how such evaluations should be made. *Standard Bayesianism* I take to be the view that involves the following commitments:

Personalism: A thinker's evaluations in such cases are *credences*, doxastic states of the subject that may be operationalized as dispositions to accept a set of bets on the truth of propositions.

Probabilism: The evaluations, if rational, can be represented as *probabilities*; they should obey the axioms of probability.

Conditionalization: The evaluations are subject to *updating* in the light of new evidence e, so that the new evaluation of a hypothesis, $P_n(h)$, should be equal to the old evaluation of the hypothesis conditional on that evidence $P_o(h|e)$.

The standard Bayesian holds that credences are an analogue to belief simpliciter: traditional epistemology concerns what it is rational to believe outright while Bayesian epistemology concerns the rationality of partial belief. Various arguments (Dutch book arguments, accuracy-dominance arguments) are used to show that a personalist should be a probabilist, i.e. that rational credences are probabilities.[1] If probabilism is

---

[1] See Joyce (1998); Leitgeb and Pettigrew (2010a, 2010b). The argument that Bayesian conditionalization, in addition to Bayesian coherence (the probabilist requirement that a subject's credences satisfy the probability axioms), is required by rationality is thus a diachronic Dutch book argument. Whether a diachronic Dutch book implies irrationality is far from clear (Briggs 2009; Mahtani 2012).

true, then the evaluation function, P, is also a probability distribution. And a consequence of the probability axioms, Bayes's theorem, holds of the evaluations of any propositions, e, h, and their conditional relations:

$$(B) \quad P(h \mid e) = \frac{P(e \mid h)P(h)}{P(e)}.$$

Conditionalization says that we should update those evaluations (probability distributions) in the light of new evidence; so if e is in fact new evidence, then:

$$(S) \quad P_n(h) = P_o(h \mid e) = \frac{P_o(e \mid h)P_o(h)}{P_o(e)},$$

where P_o is the probability distribution before the evidence was acquired and P_n is the new probability distribution in the light of that evidence. In the context of conditionalization, $P_o(h)$ is the prior probability of the hypothesis, $P_o(e)$ is the prior probability of the evidence, also known as the expectedness of the evidence, and $P_o(e|h)$ is the likelihood; $P_n(h)$ is the posterior probability, the probability now given the hypothesis having received the evidence. (S) is known as *strict* conditionalization. As we shall see, strict conditionalization is not the only updating rule that the Bayesian may employ.

Standard Bayesianism is normative, a view that asserts requirements about how reasoning ought to proceed, in addition to a descriptive kernel (personalism). A standard Bayesian is also a subjectivist if she holds that probabilism and conditionalization are the only rational constraints on credences. An objective Bayesian extends standard Bayesianism with further constraints, such as a principle of indifference.[2] Some Bayesians are non-standard in that they do not accept all the features of standard Bayesianism. For example, one might accept that there are counterexamples to conditionalization.

Descriptive explanationism I take to be the view that people, scientists included, often hold the explanatory character of a hypothesis to be relevant to its epistemic evaluation. Inference to the Best Explanation (IBE), understood as a description of an inferential practice central to science, is explanationism *par excellence*: it holds that such subjects come to accept a hypothesis because it provides a better explanation of the evidence than its rivals. Per se, normative and descriptive claims are easy to reconcile. Even if IBE and Bayesianism are entirely different, it might be that IBE describes how we do reason while Bayesianism describes how we ought to reason (but do not).

A scientific realist may well claim that scientists do reason, by and large, as they ought to: IBE is both descriptively and normatively correct. In which case the task of reconciliation becomes pressing. Some anti-realists may escape the issue by denying the descriptive or normative claims of IBE or both. Notable critics of IBE, such as van

---

[2] Other extensions to standard Bayesianism include commitments to Regularity, Countable Additivity, the Principal Principle, and Maximum Entropy.

Fraassen (1989: 160–70) and Salmon (2001), think that they are not compatible. Van Fraassen argues that if explanationist considerations were relevant, then they would encourage a thinker to boost the credence $p(h|e)$ above what is mandated by Bayes's theorem, when h is a better explanation than competitor hypotheses while reducing the corresponding credences for those competitors. Such a thinker would be open to a Dutch book (a set of bets he takes to be fair but which will lose money whatever the outcome). An account of reasoning cannot be normatively correct if it enjoins accepting a Dutch book. So IBE is not a normatively correct account of reasoning.

So the explanationist who holds that IBE is the right answer to both the descriptive and normative questions must reject van Fraassen's claim that IBE would encourage different credences from those that issue from Bayes's theorem. Rather, Bayesianism and IBE are compatible answers to both questions. Since Bayesianism and IBE are prima facie so very different, showing that they are compatible is not straightforward.

In this chapter I develop three challenges for compatibilism between Bayesianism and explanationism. These three challenges relate to the fact that IBE ought on some occasions to be able to give us *knowledge*. To answer the challenges I argue that we should (i) reject personalism, replacing credences by assessments of plausibility (hence departing from standard Bayesianism), and (ii) embrace eliminative abduction.

2. Heuristic Compatibilism

Peter Lipton (2004: 103–20) has argued that there is in fact no incompatibility between Bayesianism and IBE.[3] In Lipton's view IBE is a *heuristic* we employ in lieu of directly applying (S). This heuristic view of compatibilism is examined and advanced in detail also by Ted Poston (2014: 149–81).[4] In this section I articulate how I think this idea should be understood.

A kind of problem may have an optimal method of solution, but that method may not be easy to implement properly. Implementing it properly might be costly in terms of effort. Or that attempt may be prone to error, and so lead to a mistaken answer. A heuristic for solving a problem is a method that may be in principle suboptimal but is more easily implemented. It may well be rational to employ the heuristic if the saving of effort outweighs the loss in reliability. Or the heuristic might be more reliable than the optimal method as poorly implemented by us inadequate thinkers. For example, the recognition heuristic is employed when subjects are asked to say which of two named cities has a greater population: when only one city is recognized, then judge that that city has the larger population. This is easy to implement in the relevant cases; and it turns out also to be more reliable than the process of making inferences from

---

[3] Lipton does not say which form of Bayesianism he has in mind, so I assume that his focus concerns standard Bayesianism in its subjectivist form.

[4] See also Schupbach (Chapter 4, this volume) for an account of explanatory power that has a heuristic value relative to Bayesianism.

relevant knowledge (subjects performed *less* well when they knew something about both cities)—this is the *less is more* effect.

Lipton notes the considerable evidence that we find thinking with probabilities difficult and are liable to error. So it would be rational for us to employ a heuristic in place of Bayesian reasoning. I propose that the heuristic is this:

(H)   If hypothesis A is a lovelier potential explanation of the evidence than B, then judge A to be more likely to be true than B.

('Loveliness' is Lipton's term for the explanatory goodness of a potential explanation; it is what, according to explanationism, we evaluate when evaluating a hypothesis.) If (H) is a rational heuristic for Bayesianism then it should yield much the same results as (S) would. We can show that it does by arguing that elements of the explanationist heuristic correspond to components of the Bayesian updating in (S). The can be done, in Lipton's view, as follows. (I) If propositions h and e are ones that the subject has not given thought to before, then considering both as hypotheses, we make an assessment of their explanatory loveliness, taking into account such features as unification and simplicity.[5] This corresponds to the Bayesian process of assigning values to the priors $P_o(h)$ and $P_o(e)$. (II) We evaluate the explanatory loveliness of h in relation to the evidence e, and we then change the overall loveliness assessment of h accordingly. This corresponds to conditionalization, (S), whereby we move from the prior probabilities, $P_o(h)$ and $P_o(e)$ to the posterior probability, $P_n(h)$. (III) Explanationist considerations determine which evidence is relevant to the hypothesis—the injunction to consider the total evidence, although implicit in Bayesianism, is not one we can actually implement, so we need a heuristic to guide us to the relevant evidence, viz. whether the evidence could be explained by the hypothesis.

To complete the explanation of why IBE may be understood as a heuristic for Bayesianism, we should show that using IBE will tend to get similar results to Bayesianism. For example, we should show that (H) says that we should judge that A is more likely than B when (S) says that $P(A)$ is greater than $P(B)$. First, however, I shall expand on the details of Lipton's proposal a little. Consider the question: how good an explanation is hypothesis h? Now some aspects of the answer will concern h alone, whereas others will concern h's explanatory relationship to the evidence e. For example, the simplicity of h will concern h alone. Newton's law of gravitation plus the laws of motion have a high degree of simplicity and strength, and that fact (other things being equal) makes it a lovelier theory than one with more laws or more complex ones.

---

[5] In some case h itself determines a probability for e. Could it be that two competing hypotheses, h_1 and h_2 both give the same value to the probability of e yet h_1 is a better explanation of e than h_2? If so, then should that fact be reflected, according to the explanationist, in a higher value for $P(e|h_1)$ than for $P(e|h_2)$? If it were, then that would conflict with the Principal Principle, which says, in effect, that $P(e|h)$ must be equal to the probability that h determines for e. (And it would also amount to another instance of van Fraassen's kind of criticism.) I think that the compatibilist must regard the probabilities $P(e|h_1)$ and $P(e|h_2)$ (or, rather, their ratios to $P(e)$) as capturing all that there is to be said about external explanatory goodness. In effect, explanatory goodness has a role to play when h does not entail an easily ascertained value for e.

On the other hand, whether the explanation of e by h is lovely is a matter of the relationship between the two. For example, the hypothesis that an asteroid impact caused the K-T (K-Pg) extinction is a lovely explanation of the fact that the geological K-T layer coincides with an iridium anomaly, because the fact that asteroids have high iridium concentrations directly explains that coincidence. Following Einstein, I call these the *internal* and *external* explanatory virtues of a hypothesis.[6]

Armed with this distinction between the internal virtues of h per se and the external virtues of h's explanation of e, we can articulate a slightly different set of correspondences between IBE and Bayesianism and argue that they are plausible.

First, the internal explanatory virtues of h correspond to $P(h)$, insofar as h is a novel hypothesis. Subjective Bayesianism specifies no constraints on the priors, and so there is no danger of an explanationist account of the internal explanatory virtues of h conflicting with the requirements of the subjective Bayesian. As far as the latter is concerned the priors can be anything—they could come from a ouija board, so they might as well come from explanationist considerations. However, the heuristic view aims at showing that the heuristic (H) (mostly) yields the conclusion that an ideal Bayesian would reach. And that would not be the case if the Bayesian were to choose unusual priors (which the subjective Bayesian will feel entitled to do). Nonetheless, as a matter of descriptive fact, our actual assessment of the probability of priors does correlate with our assessment of their explanatory value. In advance of the evidence we do think that simpler, more unifying explanations are more likely to be true.

Indeed, the very propositions that we regard as prime drivers of updating reflect this. Consider the principle of indifference—one should accord equal probability to propositions regarding which one's evidence is symmetric (i.e. one's evidence bears the same relationship to each outcome; see Jon Williamson's (2010: 2) 'equivocation norm'). For example, nothing that one knows about a coin discriminates between heads and tails: it might be fair; or it might be biased, but if it is biased one knows nothing that says that it would be biased towards heads rather than tails or vice-versa. If so, one should accord the hypothesis that the next toss will yield tails the same probability as the hypothesis that it will land heads. Now think about a sequence of ten tosses that one is about to make. Here are two approaches to thinking about the outcome by applying the principle of indifference. Approach 1: consider hypotheses each of which specifies how the sequence of tosses will turn out—there are 1,024 such hypotheses. Apply the principle of indifference to these, giving each a probability of

---

[6] In his 'Autobiographical Notes' Einstein (1991) contrasted:

> The first point of view...the theory must not contradict empirical facts...[it] is concerned with the confirmation of the theoretical foundation by the available empirical facts. The second point of view is not concerned with the relation of the material of observation but with the premises of the theory itself, with what may briefly but vaguely be characterized as the 'naturalness' or 'logical simplicity' of the premises...The second point of view may briefly be characterized as concerning itself with the 'inner perfection' of a theory, whereas the first point of view refers to the 'external confirmation'.

Quoted by Salmon (1990: 181).

1/1,024. One can then calculate the probability of various proportions of heads/tails in the ten tosses. Approach 2: consider first the hypotheses that assert the proportion of heads/tails—there are eleven such hypotheses. Apply the principle of indifference to these eleven hypotheses of the second type, giving each a probability of 1/11. One can then use these probabilities to calculate probabilities for particular sequences. The two approaches lead to different probabilities for particular sequences: the first option says that a sequence of ten heads has a probability of 0.001 whereas the second option says that that sequence has a probability of 0.09. Likewise the two approaches lead to different probabilities for the proportions of heads/tails. More importantly, Huemer (2009) shows that the choice leads to a difference between scepticism and inductivism: the first choice is sceptical, in that it does not allow the probability assigned to heads on the tenth toss to change, even if one has learned that the first nine tosses have all yielded heads, whereas the second choice does assign a high probability to heads in the light of such evidence. The second set of hypotheses are explanatory hypotheses whereas the first set are not. And that, argue Huemer (2009) and Poston (2014), shows that Bayesianism *needs* explanationism if Bayesianism is to be a satisfactory account of inductive reasoning. Descriptively, we do think that we should consider the hypotheses concerning proportions/probabilities first and allow the probabilities concerning particular sequences to follow as a consequence rather than vice-versa. That strongly suggests that we are indeed moved by explanatory considerations in the assignment of priors to hypotheses. Put another way, we are as a matter of fact inductive thinkers, and since being a Bayesian inductivist requires (or at least motivates) explanationist thinking about priors, then we should expect that we are explanationists about priors.

Secondly, the external explanatory virtues of h relative to e correspond to the *ratio* of the likelihood to the expectedness of the evidence, $P(e|h)/P(e)$. External explanationist considerations do not correspond to the likelihood $P(e|h)$ alone. $P(e|h)$ can be very high even if h does *not* explain e well—if e is itself very probable: famously John Jones's taking a birth-control pill does not provide a good explanation of his not getting pregnant, since it was certain anyway that he would not get pregnant; still $P(e|h)$ is very high. And, in reverse, $P(e|h)$ can be low even if h does explain e well. For example, Holmes suspects Professor Moriarty of having murdered Major Arbuthnot and of having disposed of his body, meticulously erasing evidence of the crime. Holmes then discovers a minute trace of the highly unusual Uzbek tobacco smoked by Arbuthnot in the cuff of Moriarty's coat. Now given the meticulousness that the hypothesis ascribes to Moriarty, finding that tobacco trace is itself highly unexpected, so $P(e|h)$ is low. But Holmes's hypothesis is still an excellent explanation. The latter fact is still consistent with the demands of Bayes's theorem, since $P(e)$ is even lower. I propose therefore that the external explanatory virtue of h in connection to e corresponds to the *ratio* $P(e|h)/P(e)$, i.e. the measure of how much more probable e becomes when conditioned on h.

Bayes's theorem states that the ratio $P(e|h)/P(e)$ is equal to $P(h|e)/P(h)$, the ratio of the probability of the hypothesis conditional on the evidence to the prior probability of the hypothesis. So, given conditionalization, $P(e|h)/P(e)$ is equal to the proportion by which the probability of the hypothesis changes on learning the evidence. $P(h|e)/P(h)$, or some monotonically strictly increasing function thereof (such as its logarithm), is taken by some Bayesians to be a measure of the confirmation of h by e (Keynes 1921; Horwich 1982; Milne 1996). And so the proposal that (external) explanationist considerations correspond to $P(e|h)/P(e)$ and so correlate with this measure of confirmation means that an explanationist account of confirmation corresponds to the Bayesian one.

The heuristic view holds that Bayesian reasoning and explanationist reasoning are distinct, although leading to the same or similar judgements reasonably frequently. We should therefore not expect every component in the Bayesian calculation to have an explanationist counterpart. So a subject forming a judgement on the basis of explanationist considerations forms a judgement that corresponds to a Bayesian value for $P(e|h)/P(e)$ without that being the result of preceding judgements corresponding to $P(e|h)$ and $P(e)$. That is an advantage of the view I propose, since it means that explanatory virtue can support a hypothesis even if we have no clear idea of what $P(e|h)$ and $P(e)$ should reasonably be: Holmes might have only an imprecise idea regarding the probability of the presence of Uzbek tobacco ash in Moriarty's cuff, both the unconditional probability and the probability conditional on Moriarty having carried out a carefully planned murder. Even so, he might be able to be rather more precise about the degree to which the Moriarty hypothesis *increases* the probability of the presence of the ash. For the heuristic view to hold, it should be the case that as a matter of fact, subjects' judgements of the external explanatory virtues of a hypothesis, i.e. when the evidence confirms a hypothesis by explanationist lights, should correlate with their assessments that the hypothesis makes the evidence more likely than it would otherwise be. And that looks correct: as the Holmes example and others show, if a hypothesis explains some evidence, then we think that the evidence proposition is more likely if the hypothesis is true than if we don't make that assumption.

Having drawn correspondences between the explanatory virtues (internal and external) of a hypothesis and components of the Bayesian calculation, to confirm the heuristic view we need to show that our overall assessment of the probability of a hypothesis correlates with something that we can regard as the (arithmetical) 'product' of our judgements regarding the internal explanatory virtue of a hypothesis and its external explanatory virtue. Since our judgements of explanatory virtue are not numerical, it is difficult to show this explicitly. On the other hand, the way we combine those judgements does seem to mirror multiplication rather than addition or averaging. For example, if a hypothesis is known to be false then even unlikely evidence it explains well will not make us think that the hypothesis is even moderately likely—that corresponds to the product of a finite amount and zero, whereas averaging would suggest that a known false hypothesis could become likely on favourable evidence.

To conclude: the challenge for compatibilism between explanationism and Bayesianism was that if the two give different results regarding the credence we should attach to a proposition or regarding the ordering of our preferences among propositions, then we would have a reason for thinking that explanationism is incoherent. If, on the other hand, the two always agree, then explanationism is otiose: Bayesianism gives a precise and well-motivated account of confirmation, whereas explanationism is imprecise and philosophically controversial. The heuristic approach advocated by Lipton and Poston resolves this challenge, by showing that explanationism is a useful substitute for Bayesian reasoning. For various reasons it is unnatural and difficult for human thinkers to implement the probabilistic reasoning required by standard Bayesianism. So we use explanationist considerations as a heuristic—a manner of reasoning that will by and large get the same result as Bayesian reasoning but without its costs and pitfalls. I have added some detail to this picture, showing how aspects of explanationist thinking correspond to elements of Bayes's theorem. The internal explanatory virtues of a hypothesis correspond to $P(h)$; the external explanatory virtues correspond to $P(e|h)/P(e)$; and we combine these virtues in a manner that corresponds to their product. Explanationism does not give precise values for probabilities, but it does allow us to order our preferences among hypotheses. According to the heuristic approach, the ordering that the heuristic (H) recommends is for the most part the same ordering that explicit and accurate use of Bayes's theorem and conditionalization would have given us had we been in a position to use it.

3. Three Challenges for Compatibilism

In this section I raise three challenges to the compatibilism just articulated. These start from the assumption that IBE can lead to knowledge. One might reject that assumption, but if explanationism is a correct descriptive account of some important inferential practices in science, then that rejection amounts to scepticism about those parts of science that employ those practices.

3.1 Knowledge and objectivity

What does this compatibilist argument really show? It shows that if Bayesianism is the correct normative account of scientific reasoning, then explanationism need not be regarded as in conflict with Bayesianism. On the contrary, explanationism describes the heuristic mechanism whereby scientists typically implement Bayesianism. While this ought to satisfy the subjective Bayesian, it ought not satisfy the explanationist. For the latter thinks that there is more to epistemology than is found in subjective Bayesianism. The explanationist thinks that on some occasions IBE can give us knowledge. She holds explanationist considerations to be a guide to the truth. So the explanationist cannot regard standard Bayesianism as a complete normative epistemology. She should say that priors and likelihoods informed by explanationism are better than those not so informed. Explanationist Bayesianism must reject subjective Bayesianism, and a strong form of objective Bayesianism looks more suitable, with constraints on which priors are rationally acceptable.[7] Some priors—those guided by explanationist

---

[7] See Psillos (2004) and Huemer (2009) on explanationism as a rational constraint on priors.

considerations—are better or more acceptable in that they will lead to a posterior probability function such that hypotheses with high probabilities will more often be true than those with low probabilities.

Lipton himself needs to say something like this. Otherwise many of the prima facie problems that he raises for IBE (and addresses) would not be concerns at all. For example, Voltaire's problem asks why it is that we should think that the explanatorily loveliest explanation is the actual (true) explanation. Should we think that the actual world is the loveliest of all possible worlds? But that's irrelevant if *all* the epistemic work is done by Bayes's theorem while explanationist concerns are merely the source of the values of the priors and likelihood, without making any normative contribution. However, concern with Voltaire's problem is required if we want the resulting epistemic attitude to correlate with truth and IBE possibly to lead to knowledge.

So while Lipton's compatibilism can defuse the criticism from subjective Bayesians, the explanationist holds that there is more to be said about the epistemological function of explanationist factors—specifically regarding their connection to truth and knowledge.

3.2 Knowledge, evidence, and credence

If a subject has already acquired evidence e, then $P(e)$ is trivially equal to 1. If b is a proposition known by the subject, then $P(b)$ also equals 1. This is a trivial consequence of the form of Bayes's theorem that takes background knowledge into consideration:

$$(B') \quad P(h \mid e \& b) = \frac{P(e \mid h \& b) P(h \& b)}{P(e \& b)}.$$

In effect Bayesianism treats evidence and background knowledge equivalently. (I take the significance of 'background' to be simply knowledge other than e, which is knowledge in the foreground because when conditionalizing on new evidence e it is the probability of e that changes to 1 whereas the probability of b is already at 1.) In a sense, therefore, Bayesianism endorses Timothy Williamson's equation of evidence and knowledge, E = K (Williamson 1997). (Alternatively, if one accepts E = K for Williamson's independent reasons, then that implies that the propositions that make up b in (B') really are the propositions *known* to the subject rather than, e.g., those believed by the subject.)

Yet the view that $P(e)=1$ and $P(b)=1$ has highly unpalatable consequences for personalism (the view that the probabilities in question are credences, canonically measured by betting quotients). For a credence of 1 in p means that one would be willing to take a bet that would yield a one penny gain if p is true and unending torture if p is false. But few if any propositions are such that reasonable people would take such a bet. Certainly the contingent propositions that we take to be knowledge or to be our evidence are not typically of that sort. That looks to lead to the sceptical conclusion that we do not know those propositions after all and that no such propositions can ever really be our evidence. Furthermore, that conclusion, apart from being sceptical, also means that Bayesian updating (conditionalization) will never occur, thereby rendering it pointless.

3.2.1 JEFFREY CONDITIONALIZATION

This concern might lead one away from standard ('strict') Bayesian conditionalization, (S), to Jeffrey conditionalization (Jeffrey 1983) (so long as one retains personalism—but see below). Jeffrey conditionalization does not conditionalize just on propositions that have probability equal to 1 (as in strict conditionalization), but also on propositions stating 'uncertain evidence'. If a proposition is uncertain, then so is its negation. Consequently, Jeffrey conditionalization conditionalizes both on the uncertain evidence proposition and on its negation, weighted according to one's subjective confidence. So:

$$(J) \quad P_n(h) = \frac{P_o(e \mid h)P_o(h)}{P_o(e)}P_n(e) + \frac{P_o(\neg e \mid h)P_o(h)}{P_o(\neg e)}P_n(\neg e)$$

where $P(e)_n$ and $P(\neg e)_n$ are the weightings, *viz.* the new credences in e and its negation, resulting from some new and uncertain observation, for example.

Strict conditionalization is the special case of Jeffrey conditionalization where $P(e)_n = 1$ (and so $P(\neg e)_n = 0$). So both strict conditionalization and Jeffrey conditionalization involve a process where a proposition, e, changes its credence in a way that is not itself the result of conditionalization, from $P(e)_o$ to $P(e)_n$, where $P(e)_n = 1$ for the strict conditionalization case and $P(e)_n < 1$ for the Jeffrey conditionalization case.

Note that both Jeffrey and strict conditionalization can be seen as special cases of a more general Bayesian updating rule that is a consequence of finite additivity, the third of the Kolmogorov probability axioms:

$$(F) \quad P(a) = P(a \wedge b_1) + \ldots + P(a \wedge b_i),$$

where $\{b_1 \ldots b_i\}$ is a partition, i.e. a set of mutually exclusive and jointly exhaustive propositions. Given the definition of conditional probability, (F) is equivalent to:

$$(F') \quad P(a) = P(a \mid b_1)P(b_1) + \ldots + P(a \mid b_i)P(b_i).$$

Let us say some evidence comes in that (rationally) causes the subject to change some of the $P(b_1) \ldots P(b_i)$, so they are now $P_n(b_1) \ldots P_n(b_i)$. Then the Bayesian will expect the subject to change $P(a)$ accordingly to $P_n(a)$:

$$(F'') \quad P_n(a) = P_o(a \mid b_1)P_n(b_1) + \ldots + P_o(a \mid b_i)P_n(b_i)$$

(where P_o is the old probability distribution P in (F')). The proposition a will itself be a member of another partition, so that $P(c)$ stands to $P(a)$ and $P(c \mid a)$ just as $P(a)$ stands to $P(b_1)$ and $P(a|b_1)$ in (F'). So a change in the $P(b_1)\dots P(b_n)$ has a ripple effect, requiring the updating of the probabilities of many propositions to which those $b_1\dots b_n$ are related, transitively, by equations such as (F").

Now note that Bayes's theorem tells us that $P(a|b_1) = P_o(b_1|a)P_o(a)/P_o(b_1)$, etc. Hence:

$$(\text{F}''')\ \ P_n(a) = \frac{P_o(b_1\,|\,a)P_o(a)}{P_o(b_1)}P_n(b_1)+\dots+\frac{P_o(b_i\,|\,a)P_o(a)}{P_o(b_i)}P_n(b_i).$$

Since (F") is an updating rule so is (F'''). An updating rule identical to Jeffrey conditionalization, (J), is the special case of (F''') where we update using the partition $\{e, \neg e\}$.

So what is distinctive about Jeffrey conditionalization is not the updating rule, which is in use whenever there is a change in the probability distribution. For example, as a result of strict conditionalization, some probabilities will change and (F''') and its kin will be used iteratively to update further propositions. So what is at issue is not the updating rule but instead which changes in the probability distribution should rationally *initiate* updating.

The strict conditionalizer has a clear if incomplete answer to that question. Updating is initiated when a proposition achieves the status of evidence. This answer is incomplete in two respects. First, Bayesianism tells us almost nothing about when a proposition is evidence—the only constraint is that evidence propositions are those propositions that have a probability equal to 1. Secondly, for updating to be a rational requirement, the change of a proposition's status to evidence must itself be rational. If a subject who starts with a rational probability distribution changes the probability he attributes to a proposition p from less than 1 to 1 without any rational cause, then it is not rationally required that the subject update the rest of his probability distribution in line with conditionalization. Another option (and the one that does seem rationally required) is to reverse the change. Put another way, if the subject keeps the rest of his probability distribution fixed (which he ought, given that it was rational to begin with and the subject has undergone no rational change), then he is rationally obliged to change $P(p)$ again, in line with Bayes's theorem, which will set $P(p)$ back to where it started.

Consequently the Bayesian has to outsource some important epistemological questions: Which propositions are evidence propositions? Why is it rational to update when a proposition becomes an evidence proposition?[8] As we shall see, a response can be given that answers both questions. Evidence propositions are precisely those that are known (Williamson 1997); as already discussed, known propositions are indeed accorded probability of 1. And clearly if one's *knowledge* changes then it is rational to change one's probability distribution accordingly. As we shall see, this answer requires dropping personalism, but we will see that is a price well worth paying.

---

[8] See Williamson (2000: 217) on the need for supplementing Bayesianism with an account of evidence.

If we turn now to Jeffrey conditionalization, we see that analogous questions arise, but answering them becomes that much more difficult. If 'evidence' propositions whose changes in probability are ones that rationally initiate updating, then which propositions are they? The answer that evidence propositions are the known propositions is not available. The new probabilities of these propositions are less than 1, whereas known propositions have probability equal to 1. Jeffrey conditionalization allows for the retention as epistemically possible (probability > 0) propositions inconsistent with the evidence, but that makes no sense if the evidence propositions are known: propositions inconsistent with the subject's knowledge are not epistemically possible. That is really just to say that there is no natural place for an account of evidence as knowledge in Jeffrey conditionalization. However, that means that the explanation given above about why it is rational to update on changes in evidence ('because it is rational to update on changes in what one knows') is not available.

So Jeffrey conditionalization needs an account of evidence that allows evidence to be less than knowledge, yet explains why it is rational to update in response to changes in evidence. So far no such account is forthcoming. Jeffrey conditionalizers might aim to get by with something weaker: an account of evidence such that updating on evidence propositions is rationally permissible, even if not rationally required. Which propositions are those then? I suggest that something like the following thought is part of the Jeffrey conditionalizer's outlook. It would be rationally *impermissible* for a change in the probability of p to *initiate* updating, if that probability is fixed by its relationships to the probabilities of other propositions, via (F″) and (F‴) and the like. For a change in $P(p)$ in such a case would have to come about *as a result of* changes in those other probabilities. So the propositions that are permitted to initiate changes in a probability distribution are those whose probabilities are not dependent on the probabilities of other propositions. Einstein's GTR is a proposition whose probability depends on the probabilities of other propositions, concerning the outcomes of various experiments and observations as well as other theories, etc.; so it would not be permissible to update in response to a change in the probability accorded to GTR if that change comes about directly as opposed to as a result of changes in the probabilities accorded to those experiments, observations, and other theories. So it is not surprising that the propositions that Jeffrey and others use as examples for e in (J) are propositions arising from sensory perception: famously the subject who looks at a piece of cloth by candlelight and who is uncertain as to its colour and so gives various probabilities to the propositions that the cloth is green, that it is blue, and that it is violet (Jeffrey 1983: 165).

Nonetheless, even those propositions do not meet the requirement that their probabilities are not dependent on the probabilities of other propositions. As Williamson (2000: 217) points out, those probabilities will depend on what I may know about the reliability of my eyesight, the nature of the lighting conditions, the character of the cloth, etc. It is difficult to see that *any* proposition will meet this independence condition. One might think that reports of sensory appearances do meet the condition

('It seems to me that this cloth is probably green'). However, if these internal reports are fallible, then they will be dependent on my assessment of my own reliability in delivering such reports. On the other hand, if internal reports are infallible then we may ascribe probabilities of 1 to those reports, in which case we do not need Jeffrey conditionalization.

Jeffrey conditionalization looked liked a plausible option to take account of the fact that if personalism is true, then hardly ever, if at all, will a proposition have a probability of 1—certainly not the propositions of interest to science. Bayesian updating, whether strict or Jeffrey, needs an account of the circumstances in which a change in a probability distribution may rationally initiate updating of other propositions, since coherence can always be achieved by reversing the initial change. Such an account will come from outside the Bayesian framework. Strict conditionalization can be supplemented by a natural account of when that is: when a proposition becomes known, that proposition may (and should) initiate updating.[9] No such account is available to the Jeffrey conditionalizer.

3.3 Knowledge and IBE

One important difference between Inference to the Best Explanation and standard Bayesianism is that the former tells us which propositions to infer from others—it includes a rule of detachment—whereas Bayesianism does not. So IBE as typically conceived might be summarized in a rule such as:

> (IBE)   When hypothesis A is a much lovelier potential explanation of the evidence than all its competitors (and it is in itself a good explanation, there is a good quantity of evidence, and alternative explanations have been sought assiduously), then infer that A is the actual explanation of the evidence and so is true.

Bayesianism might be supplemented by a rule of detachment—infer those propositions that when conditioned on the evidence reach a threshold probability (e.g. 0.95). But such rules are highly problematic unless the threshold is equal to 1.[10]

In this section I raise a related problem for compatibilism. Above I took it that the explanationist believes that using the rule (IBE) can lead to knowledge of the inferred proposition. This leads to the following problem in the Bayesian context if this proposition is true:

> In a typical explanationist inference, the evidence is consistent with more than one of the competing hypotheses, and no hypothesis has a prior equal to 1.

---

[9] For completeness, it should be added that updating should also be initiated when a proposition loses its status as knowledge.

[10] E.g. such a rule might permit the inference of each of two propositions individually but not their conjunction. In other cases, such as the proposition asserting that a given lottery ticket will not win, we might think that the proposition, though reaching the threshold, is not one that should be inferred.

If this is correct, then a typical explanationist inference will produce a preferred hypothesis whose probability is less than 1. We saw above that (B′) entails that known propositions have a probability equal to 1. These together entail that a typical explanationist inference will not lead to knowledge. So explanationism seems to condemn us to scepticism. (Note that this problem is distinct from the challenge raised in the preceding subsection, 3.2. The latter arises from personalism. This problem is independent of personalism and arises simply from the probabilism in the form of (B′), even if we take the probabilities to be something other than credences, as this chapter will propose.)

One might attempt to resolve this with the following. Bayes's theorem gives one the value of $P(h|e)$. Conditionalization says that on receipt of evidence e one should alter the probability attached to h to this value. However, if the epistemic circumstances are favourable then the subject gets to know, and the probability of the proposition rises to 1. That is, under such circumstances one can depart from conditionalization. This, however, is just a special case of the boost view of explanationism. While there are counterexamples to Bayesian conditionalization, this departure seems unmotivated.

Jeffrey Dunn (2014) raises the same problem in connection with evidence. Can a proposition that is learned by inductive inference itself be evidence? Let p be the proposition that a new Pope has been elected, and s be the proposition that there is white smoke coming from the Sistine Chapel. My credences are $P(p) = 0.5$, $P(s) = 0.5$, and $P(p|s) = 0.9$. I then see white smoke over the Sistine chapel, so s is now in my evidence set. So, with conditionalization, my credences are now $P(p) = 0.9$, $P(s) = 1$, and $P(p|s) = 0.9$. Can I now use p as evidence (e.g. with regard to some proposition concerning the immediate future of the Church)? According to Bayesianism, I cannot. For if p were now evidence, then it would have to have a probability of 1, whereas its probability is 0.9.

(Dunn concludes that p is not in my evidence, and so this makes trouble for $E = K$, since I can get to *know* that a new Pope has been elected by seeing white smoke. I don't think that this is really the problem. For as remarked above, the Bayesian is committed to $E = K$. If I really do know p, then p now becomes part of my background knowledge and so, thanks to (B′), $P(p) = 1$. So the probability assigned by conditionalization, $P(p) = 0.9$, is inconsistent both with the proposition that p is now in my evidence and with the proposition that I now know p. I return to Dunn's argument below.)

3.4 Three challenges summarized

To summarize, the three challenges to compatibilism between Bayesianism and explanationism are:

1. Mere compatibilism with subjective Bayesianism answers Bayesian objections to IBE but fails to articulate how IBE can lead to the truth and so be a source of knowledge.
2. Bayesianism requires that evidence propositions and other known propositions have probability equal to 1. If that probability is a credence then, since no

one reasonably has credence 1 in most non-logical propositions, very few propositions will ever be known/evidence. The challenge is a consequence of Bayesianism's commitment to personalism.

3. If IBE selects among hypotheses that are all consistent with the evidence, Bayesian compatibilism implies that the inferred hypothesis has a probability of less than 1. But a known proposition cannot have a probability of less than 1. The challenge is a consequence of Bayesianism's commitment to probabilism plus the claim that IBE selects among hypotheses that are all consistent with the evidence.

I mooted that the explanationist might prefer some strong kind of objective Bayesianism. So in the next section I consider whether objective Bayesianism suffices to provide what we want for explanationist Bayesianism. I suggest that it is not enough. Objective Bayesianism will not suffice to answer 2, since objective Bayesians are personalists—the probabilities in question are credences. Below I consider a move beyond objective Bayesianism that rejects personalism and embraces a different conception of probability. This conception associates the probability of a hypothesis with its *plausibility*. I then return to challenge 3, which I answer by retaining probabilism but rejecting the assumption that IBE inevitably selects among hypotheses that are all consistent with the evidence.

4. Probability and Plausibility

IBE requires objectivity if explanatory considerations are a guide to the truth. So we must rule out subjective Bayesianism. Should explanationism instead endorse objective Bayesianism, accepting further constraints on conditionalization? Standard approaches to objective Bayesianism will deploy versions of the principle of indifference, discussed above. Objective Bayesians will often hold that one's prior probabilities should be calibrated according to evidence of frequencies: if all one knows about some possible outcome o is that in the relevant circumstances such outcomes occur with frequency f, then one should accord o the probability f. Nonetheless, these kinds of constraint (indifference, calibration) do not suffice to capture what is needed to provide a form of objective Bayesianism that marries well with IBE. For often scientists are faced with competing hypotheses that are not neatly symmetric, for example: the hypothesis that the K-T extinction was caused by an asteroid collision versus the hypothesis that it was brought about by a period of rapid evolution. Furthermore, it is difficult to bring frequentist considerations to bear on such hypotheses—this is most especially obvious if the competing hypotheses concern the laws of nature. The other principles that objective Bayesians typically appeal to are, like indifference and calibration, principles that have an a priori appeal—Jaynes's (1968) principle of maximum entropy, for example. The problem we faced was that compatibilist explanationism needs explanatory considerations to inform or parallel the use of Bayes's theorem in such a way that there is a correlation between higher posterior probabilities and the truth. Further a priori principles seem too thin to be able to do that—if they did, that would amount to an a priori justification of induction. So objective Bayesianism, as it is typically articulated, turns out to be an unsatisfactory answer to problem 1 after all.

We need therefore to move beyond objective Bayesianism's supplementation of standard Bayesianism with additional a priori principles. We may contemplate a 'super-objective' Bayesianism where good priors and likelihoods are determined by explanatory considerations that cannot be captured as a priori principles but instead have a *de facto* reliability. Super-objective Bayesianism would be a significant departure even from objective Bayesianism. If the relevant constraints on credences are a priori and assuming that credences, being doxastic attitudes, are introspectively knowable to the subject, then the subject is in a position to know a priori that her credences obey all the relevant constraints—the subject can know that she is being rational.[11]

So super-objectivism would undermine one motivation for personalism, that it permits the subject to assess the rationality of her evaluations. At the same time, super-objectivism does not itself solve problem 2, since that problem arises from the commitment to personalism. A natural step therefore is to adopt super-objectivism while renouncing personalism. I agree with Salmon (1990: 182) in holding that 'Plausibility considerations are pervasive in the sciences; they play a significant—indeed, *indispensable*—role.' Accordingly, I propose that our epistemic evaluations of hypotheses are evaluations of plausibility in the light of the evidence—such evaluations I shall call 'plausibilities'. Salmon's objective Bayesianism is an important attempt to link Bayesian confirmation to Kuhn's five criteria of theory choice via plausibility considerations, to which the proposals in this chapter have some similarity, as I shall explain. Salmon, however, regards plausibility arguments as supplying priors within the context of personalism, whereas I follow Williamson (2000) in regarding plausibilities as an alternative to credences. (I also regard plausibility considerations as determining the likelihood and the expectedness—or, directly, their ratio.)

We retain probabilism regarding plausibilities: the plausibility of a hypothesis in the light of the evidence can indeed be modelled by probabilities.[12] Insofar as we can quantify plausibilities, we can represent them as ranging from a minimum, 0, the plausibility of a hypothesis inconsistent with the evidence, to a maximum, 1, the plausibility of a hypothesis entailed by the evidence. Thus represented, the plausibility of a disjunction of mutually exclusive hypotheses is the sum of the plausibilities of the hypotheses individually; in particular, the plausibility of a disjunction of a hypothesis and its negation

---

[11] I would deny that a subject is always able to know her own credences, but the thought that credences are transparent to introspection and that Bayesianism allows us to police our own rationality is a common one (see Savage 1954: 20; Egan and Elga 2005).

[12] This approach builds on Williamson's (2000) notion of *evidential probability*.

is maximal. Thus finite additivity is satisfied.[13] If my evidence allows me to know, for example, that the Earth orbits the Sun, then that hypothesis has maximum plausibility and could not gain additional plausibility from additional evidence. Likewise, if my theory entails that water is frozen at 80°, then my evidence shows that this theory has zero plausibility. Since plausibilities can be given values of 0 and 1, this approach avoids problem 2.

Above I identified two aspects to the explanatory loveliness of a hypothesis: an internal component, which is independent of the evidence, and an external component, which concerns the ability of the hypothesis to explain the evidence in question. The internal component corresponded to the prior probability of the hypothesis, that is to say, the plausibility of the hypothesis independent of its evidential support. (The 'internal/external' terminology here does *not* align with a distinction between factors that are acceptable to an internalist epistemology and those acceptable only to an externalist epistemology.[14]) In science we do find assessments of plausibility of this sort. For example, scientists in the eighteenth century wondered about the force of attraction and repulsion between charged objects. In 1785, Charles-Augustin de Coulomb proposed an inverse square law. Even before the torsion balance experiments that he then carried out, this hypothesis carried a high degree of (evidence-independent) plausibility because of its analogy to the inverse square law of gravitation. This is evidence-independent plausibility in that it is not the product of any process of conditionalization (e.g. on the law of gravitation), not even in an idealized way. Why should we not treat the structural similarity between Coulomb's hypothesis and Newton's law as a piece of evidence (since that similarity is a known fact) in favour of the former? We don't typically think in that way—in the context of the heuristic approach this fact is psychologically most plausibly seen as influencing our prior for Coulomb's hypothesis. It would be difficult to implement Bayesianism with this as evidence. For that would require setting an independent prior for the hypothesis, and where does that come from? And it would require setting a prior for the evidence, the similarity between Coulomb's hypothesis and Newton's law, and it is also unclear where that could come from. This influence on the plausibility of Coulomb's proposed law can come about even if we hold that the law of gravitation is not itself known (and so not evidence). More generally, the gamut of a scientist's experience can influence the assessment of plausibility in ways that cannot simply be

---

[13] A further question asks how plausibilities relate to credences. Clearly there is no neat relationship, in that some propositions will have a plausibility of 1 but a credence of less than 1. On the other hand, if a subjective Bayesian does treat a subject's evidence/knowledge as having credence equal to 1 and has priors coinciding with those of the plausibility theorist, then conditionalization will lead to the same posterior probabilities.

[14] The internal explanatory virtue is what Williamson (2000: 211) calls the 'intrinsic plausibility [of the hypothesis] prior to investigation'. It does not seem strictly correct, though, that plausibility can be a truly *intrinsic* property of a hypothesis. If it were, there would be a plausibility that is independent of the psychological make-up of even an ideal human scientist, including their innate psychology. Such a notion looks as if it would be close to the a priori objectivism we considered but rejected above.

reduced to an evidential relationship (rather as one may correctly recognize the facial similarity of two members of the same family without being able to articulate the basis of that similarity). So a hypothesis can gain plausibility by its connection to elements of the scientific context in a way that is independent of conditionalization on evidence. Analogy with well-confirmed hypotheses is an example of how this occurs. And so we have one answer to the question: where does evidence-independent plausibility come from? (There may be other sources of evidence-independent plausibility.[15]) The remaining question concerns the objectivity of this plausibility.

5. Assessing Objective Plausibility

I have just argued that we make judgements of plausibility of a hypothesis that are independent of its conditionalization on the evidence—in so doing we respond to internal explanatory virtues, and these may be influenced, for example, by analogy with other theories. Similar remarks may be made about external explanatory virtues, which allow us to assess the factor by which the degree of plausibility changes when the evidence is obtained (i.e. the ratio $P(e|h)/P(e)$). These too may be informed by exemplars of successful explanatory relations between theories and evidence.

Indeed, we can draw on a central lesson of the work of Thomas Kuhn (1970), which is that scientists' assessment of any hypothesis is crucially influenced by their prior experience of other successful theories.[16] According to Kuhn, such theories, which he called 'exemplars' or 'paradigms', are models that inform the development and assessment of newer hypotheses. Because Newton's theory of gravitation was so successful, it became a model (in multiple ways) of what a good theory might look like, and so, as we saw, Coulomb's proposed law acquired plausibility because of its similarity to that model.

We can link this point to explanationism with the following proposal from Walker (2012). Scientists acquire their sense of explanatory virtues from their training with and use of exemplars. Let us assume some degree of scientific realism—the theories in question have a high degree of truthlikeness. Let us also assume that theories similar to true or highly truthlike theories are more likely to be truthlike than less similar theories. If these assumptions are correct (and they are themselves at least plausible), then the practice of allowing one's preferences among theories to be influenced by their explanatory virtues will be a truth-conducive practice. Exemplars change over time, and so the explanatory virtues acknowledged by a community will change over time

---

[15] For example, another source might be the epistemological conservatism advocated by Poston (2014). That a proposition is believed by a subject suffices to give it some degree of plausibility. Poston provides arguments for conservatism from an internalist perspective, while I can see externalist, naturalistic reasons in its favour.

[16] We can accept this result of Kuhn's work without taking it to have any anti-realist consequences, as this section argues. See Bird (2000). Salmon (1990: 196–9) does likewise, though concentrates on Kuhn's five shared criteria of theory choice, which Salmon takes to play a role in determining the plausibilities of hypotheses and thereby the priors of those hypotheses.

also—we can see this as a process of the evolution of a scientific community's values in the direction of greater reliability.[17] We should not expect these values and their applications to be the same in all scientific fields. Different fields have different exemplars and so values (or, more frequently, what counts as instantiating a given value) may differ between fields.[18]

So now let us imagine a scientist who considers some candidate hypothesis in advance (or independently) of the evidence. Her assessment of its plausibility is influenced in part by the degree to which it possesses the explanatory virtues acknowledged by the community. In the light of the foregoing, and if the two assumptions are correct, we may expect there to be some correlation between the plausibility assessments and truth. Consequently, probabilities interpreted as (quantified) plausibilities will be objective in the sense relevant to challenge 1: probabilities will relate to truth, in that hypotheses assessed as more plausible will be more likely to be true.

Note that it does not need to be claimed that this correlation is perfect. From the Bayesian perspective the assessment of prior plausibility might be only rough and ready and still good enough. For the phenomenon of the washing-out of priors will mean that imperfections in the prior probabilities will be increasingly insignificant as ever more evidence comes in. The washing-out of priors argument does not help the subjective Bayesian since that position accepts no epistemologically relevant constraints on priors, meaning that priors are acceptable that are so extreme that they cannot be washed out by a reasonable quantity of evidence. If plausibility is a rational constraint on priors, then extreme priors will be excluded (unless warranted in particular cases by extreme plausibility or implausibility).

6. Plausibility, Knowledge, and Eliminativism

Since plausibilities are not credences, a subject who assigns a plausibility of 1 to a proposition is not obliged to accept wild bets on the truth of that proposition. So the second challenge is not a problem for this approach. A plausibility of 1 means that the proposition is maximally plausible *given the evidence and background knowledge*. Rationally assessing a proposition as maximally plausible does not seem unachievable in the way that a rational credence of 1 looks unachievable for most propositions.

However, $P(h) = 1$ does mean that any proposition inconsistent with h (e.g. a competitor hypothesis) must have zero probability, that is to say maximum implausibility in the light of the evidence. This looks to be in tension with IBE, since the latter is typically understood as making inferences on the basis of one explanation being a *better* explanation than the competitor hypotheses. The latter implies that one can make a

---

[17] By 'value' here I mean the features of a theory that the community regard as virtues—as making for explanatory loveliness.
[18] As Salmon (1990: 186) remarks, 'simplicity...[is] a highly relevant characteristic, but one whose applicability varies from one scientific context to another'.

reasonable IBE when the best explanation is just rather *more* plausible than its competitors; the latter thus retain some degree of plausibility. If that is right, then the inferred proposition will not be knowledge, since knowing h requires $P(h)=1$ and $P(h')=0$ for all competitor hypotheses h'.[19]

So knowledge of a proposition inferred on the basis of explanationist considerations requires that the competing hypotheses must be *ruled out* by the evidence and background knowledge: what may be called *eliminative abduction* (Bird 2005). It might be wondered whether explanationist considerations can ever achieve this. Perhaps the most that can be achieved is that a hypothesis is shown to be highly implausible.

This challenge cannot be covered comprehensively here. But I can indicate some of the lines of response I advocate:

- Scientists do think that hypotheses can be shown to be false: Harvey showed that blood could not continuously be produced afresh by the liver; Galileo's observations of the phases of Venus proved that the Sun and Venus did not both orbit the Earth; Pasteur's experiments with swan-necked flasks refuted the hypothesis that spontaneous generation was the cause of the spoiling of organic liquids such as broth or wine;[20] Rutherford's gold-foil experiments disproved the plum-pudding model of the atom; Wu's experiment ruled out the conservation of parity; and so forth. The pre-theoretic judgement of scientists is that in cases such as these it is possible to know that a hypothesis is false.
- That pre-theoretic judgement is challenged by the use of Duhem's contention that use of an experiment to refute a hypothesis will typically depend on the use of auxiliary hypotheses. All one can know, it is said, is that the conjunction of the hypothesis under test *and* the various auxiliary hypotheses is false, not that the test hypothesis itself is false. While such an argument appears to *lead to* scepticism, it is instead the *product* of an antecedent scepticism. For the conclusion holds only if the auxiliary hypotheses are themselves not known. If a scientist does know the auxiliary hypotheses to be true, then she can know that the test hypothesis is false. So a sceptical conclusion about the possibility of refutation can only be drawn if one is already sceptical about the possibility of knowing the truth of auxiliary hypotheses. While this may be problematic for Popper, since he is a sceptic of that sort, there is no such problem for those who are not already sceptical about the possibility of scientific knowledge. The truth of an auxiliary hypothesis may always be disputed. But that fact does not show that

---

[19] Since a subject's known propositions and evidence propositions are the same, and evidence propositions have a plausibility of 1, so also do known propositions. $P(h) = 1$ is consistent with fallibilism, when the probabilities are plausibilities. If the probabilities are credences then fallibilists will want probabilities to be less than 1 even for knowledge.

[20] Pasteur is reported (Vallery-Radot 1926: 109) to have concluded a public lecture at the Sorbonne in 1864 by saying 'Never will the doctrine of spontaneous generation recover from the mortal blow of this simple experiment.'

the hypothesis is not known by any scientist—immunity from challenge is not a condition on knowledge.

- Even if one is not antecedently a sceptic, one might think that sceptical hypotheses are among the competitor hypotheses that would need to be refuted in order for the favoured hypothesis to have a probability of 1, and that the very nature of such hypotheses is that they cannot be refuted. For example, if one is considering competing explanations of the evidence e, is not one of the hypotheses to be considered the hypothesis that an evil demon has made it falsely appear that e is the case? If so, how would one refute that hypothesis? I say that *that* hypothesis need not be considered. For it is inconsistent with and so ruled out by e. An explanation of why things falsely appear as if e is necessarily not an explanation of the evidence e. *Ex hypothesi* our evidence is e not merely the proposition that things appear as if e. If one responds that the former should not be included in our evidence, and only the latter proposition, then that is to adopt an antecedent scepticism that the sort of propositions we take to be our evidence are never known.[21]

- One might nonetheless remain concerned that there are yet other sceptical hypotheses consistent with our total evidence (as non-sceptically conceived). For example, we might propose the following hypothesis for the outcome of Pasteur's swan-necked flask experiments: there is spontaneous generation in the flasks but the evil demon uses special powers to kill off the organisms thus generated before they can spoil the organic infusion. So Pasteur's evidence is just as we take it to be (there is no perceptual illusion caused by the evil demon). To this kind of sceptical hypothesis, the following responses are worth considering:

  o It is not entirely clear that such hypotheses really are consistent with our evidence. The existence of an evil demon with a power of unspecified mechanism is hypothesized. If some specific mechanism were hypothesized, then it might be clear that what we know from physics and chemistry and such like is inconsistent with the hypothesis. Failure to specify the mechanism doesn't remove concerns about inconsistency, for it is far from obvious that there is some possible mechanism consistent with our scientific evidence.

  o Even if such possibilities are consistent with our evidence, we may still be entitled either to ignore them or to give them zero plausibility. For such possibilities are remote possibilities, in the sense that they are true only in remote

---

[21] Of course, the sceptic may ask of some evidence proposition e how we can be sure that e is in our evidence—perhaps it only appears to be evidence? Can we distinguish between genuine and merely apparent evidence? While we may be able in many cases to make that distinction, the important point is that for present purposes it does not need to be answered. Being able to demonstrate that e is in S's evidence is not a condition of its in fact being in S's evidence—the KK principle is rejected. Our current question asks which hypotheses can be ruled out by a subject's evidence, not which hypotheses can be ruled out by what a subject knows to be her evidence.

possible worlds.[22] And it may be sufficient for eliminative knowledge of the kind being proposed here that all competing hypotheses true in close worlds are eliminated by the evidence. This might be endorsed by those who take knowledge to be tantamount to satisfying a requirement of safety (roughly: S knows that p iff S could not easily be mistaken in believing that p). Some contextualists will endorse a similar proposal, that in the scientific context it is proper to ignore (not consider) such hypotheses. (Since the probabilities are not credences but plausibilities, a contingent but remote possibility can have a probability equal to zero.)

In addition to being able to rule out hypotheses, the subject needs also to know that the true hypothesis is among the competing hypotheses she considers. While a subject can be wrong about this, I have argued elsewhere that such knowledge is certainly possible (Bird 2005): in a popular example, the detective can know that (at least) one of the fifteen passengers on the train committed the crime, but examples also abound in both daily life and in science.

Returning to Dunn's case, if the subject knows that a new Pope has been elected when she sees white smoke above the Sistine Chapel, then it must also be the case that the subject knows that if there is white smoke above the Sistine Chapel then a new Pope has been elected. (This could well be a case of eliminative abduction: another explanation for the smoke is that there is an accidental fire. But if so, the smoke would not be white and there would be fire alarms. So that alternative hypothesis is ruled out by the evidence.) If the subject does know that conditional proposition, then her conditional probability $P(p|s) = 1$, not 0.9 as in Dunn's example. In which case, $P(p) = 1$, and Dunn's problem disappears; there is no Bayesian problem with p being evidence. On the other hand, if $P(p|s) = 0.9$ then the subject does not know that if there is white smoke then there is a new Pope, in which case acquiring the evidence that there is white smoke cannot lead to knowledge that there is a new Pope (and so that proposition is not an evidence proposition). Either way, E = K is safe. More importantly, in the first case, where $P(p|s) = 1$ and so $P(p) = 1$, the inferred proposition is known (and so is evidence and has probability = 1).

7. Conclusion

The heuristic approach to compatibilism between Bayesianism and IBE can see off the challenge from the subjective Bayesian. Explanationist considerations do not supplement Bayesian considerations by giving a boost to the probability of a hypothesis when conditionalized on the evidence. Rather we can see explanationist factors as

---

[22] The condition that the possibilities in question are remote, not close is important. One cannot ignore an implausible hypothesis that is true in a close world. Police may suspect a known burglar who is suddenly rich of having pulled off a heist. But they should not ignore the possibility of a lottery win, since such a possibility (if the burglar bought a ticket) is a close possibility even though it is an improbable one.

corresponding to the various elements of Bayes's theorem, in particular the prior $P(h)$ and the ratio of the likelihood to the expectation of the evidence, $P(e|h)/P(e)$. Still, that cannot be all that explanationist considerations achieve, for IBE reasonably aspires to give us knowledge, whereas subjective Bayesianism does not. This may motivate a move from subjective to objective Bayesianism. But that position faces the problem that no scientific proposition can reasonably be accorded credence of 1, in which case no proposition can ever be known or be evidence. So we need to move beyond the personalism of both subjective and objective Bayesianism: we should not understand the probabilities with which we are concerned as credences but rather as something else. That something else must allow it to be possible to have evidence and knowledge and to accord the relevant propositions probability 1. The proposal here is that this is achieved by identifying the probabilities with assessments of plausibility in the light of the evidence (thus drawing on Williamson's evidential probability).

Consideration of how a hypothesis acquires plausibility on explanationist grounds (e.g. explanationist values that feed into plausibility are generated by assessments of similarity to successful, highly truthlike theories) shows that it can be objective and be correlated with the truth. IBE does face an additional problem in that if it is to lead to knowledge, explanationist considerations must in fact be able to rule out competing hypotheses, i.e. reduce their plausibility to 0. While showing that takes us beyond this chapter, I have argued that so doing does not face any obvious and insurmountable obstacle.[23]

References

Bird, A. 2000. *Thomas Kuhn*. Chesham: Acumen.
Bird, A. 2005. Abductive knowledge and Holmesian inference. In T. S. Gendler and J. Hawthorne (Eds), *Oxford Studies in Epistemology*, pp. 1–31. Oxford: Oxford University Press.
Briggs, R. 2009. Distorted reflection. *Philosophical Review* 118: 59–85.
Dunn, J. 2014. Inferential evidence. *American Philosophical Quarterly* 51: 201–13.
Egan, A. and A. Elga. 2005. I can't believe I'm stupid. *Philosophical Perspectives* 19: 77–93.
Einstein, A. 1991. *Autobiographical Notes*. La Salle, IL: Open Court.
Horwich, P. 1982. *Probability and Evidence*. Cambridge: Cambridge University Press.
Huemer, M. 2009. Explanationist aid for the theory of inductive logic. *British Journal for the Philosophy of Science* 60: 345–75.
Jaynes, E. 1968. Prior probabilities. *Institute of Electrical and Electronic Engineers Transactions on Systems Science and Cybernetics* SSC-4: 227–41.
Jeffrey, R. 1983. *The Logic of Decision* (2nd ed.). Chicago, IL: University of Chicago Press.
Joyce, J. M. 1998. A nonpragmatic vindication of probabilism. *Philosophy of Science* 65: 575–603.

[23] My thanks to Kevin McCain, Richard Pettigrew, Ted Poston, and Timothy Williamson for their instructive discussion, comments, and advice. I am also grateful to the Warden and Fellows of All Souls College, Oxford, for a Visiting Fellowship that enabled this chapter to be written.

Keynes, J. M. 1921. *A Treatise on Probability*. London: Macmillan.

Kuhn, T. S. 1970. *The Structure of Scientific Revolutions* (2nd ed.). Chicago, IL: University of Chicago Press.

Leitgeb, H. and R. Pettigrew. 2010a. An objective justification of Bayesianism I: Measuring inaccuracy. *Philosophy of Science* 77: 201–35.

Leitgeb, H. and R. Pettigrew. 2010b. An objective justification of Bayesianism II: The consequences of minimizing inaccuracy. *Philosophy of Science* 77: 236–72.

Lipton, P. 2004. *Inference to the Best Explanation* (2nd ed.). London: Routledge.

Mahtani, A. 2012. Diachronic Dutch book arguments. *Philosophical Review* 121: 443–50.

Milne, P. 1996. log [P(h/eb)/P(h/b)] is the one true measure of confirmation. *Philosophy of Science* 63: 21–6.

Poston, T. 2014. *Reason and Explanation*. Basingstoke: Palgrave Macmillan.

Psillos, S. 2004. Inference to the best explanation and Bayesianism. In F. Stadler (Ed.), *Induction and Deduction in the Sciences*, pp. 83–91. Dordrecht: Kluwer.

Salmon, W. C. 1990. Rationality and objectivity in science or Tom Kuhn meets Tom Bayes. In C. W. Savage (Ed.), *Scientific Theories*, Volume 14 of *Minnesota Studies in the Philosophy of Science*, pp. 175–204. Minneapolis: University of Minnesota Press.

Salmon, W. C. 2001. Explanation and confirmation: A Bayesian critique of inference to the best explanation. In G. Hon and S. S. Rakover (Eds), *Explanation: Theoretical Approaches and Applications*, pp. 61–91. Dordrecht: Kluwer.

Savage, L. J. 1954. *The Foundations of Statistics*. New York: Dover.

Vallery-Radot, R. 1926. *The Life of Pasteur*. Garden City, NY: Garden City Publishing Company. Translated by R. L. Devonshire.

van Fraassen, B. 1989. *Laws and Symmetry*. Oxford: Oxford University Press.

Walker, D. 2012. A Kuhnian defence of inference to the best explanation. *Studies in History and Philosophy of Science* 43: 64–73.

Williamson, J. 2010. *In Defence of Objective Bayesianism*. Oxford: Oxford University Press.

Williamson, T. 1997. Knowledge as evidence. *Mind* 106: 1–25.

Williamson, T. 2000. *Knowledge and Its Limits*. Oxford: Oxford University Press.

8

The Evidential Impact of Explanatory Considerations

Kevin McCain and Ted Poston

Explanationism is an attractive family of theories of epistemic justification.[1] In broad terms, explanationism is the idea that what a person is justified in believing depends on their explanatory position. At the core, explanationists hold that the fact *p* would explain a body of evidence if *p* were true is itself evidence that *p* is true. In slogan form: explanatoriness is evidentially relevant. Although explanationism has been out of the limelight for quite a while, there has been a resurgence of interest in these views. We hope that this resurgence is in part due to our recent work defending explanationist views, but even if not, we will be happy if such views continue to garner the attention that they deserve.[2]

Despite the plausibility of explanationism, not all of the recent interest in it has been complimentary. Recently, William Roche and Elliott Sober (2013 and 2014) have argued that "explanatoriness is evidentially irrelevant" (2013: 659).[3] R&S's argument against the evidential relevance of explanatory considerations begins with what they call the "Screening-Off Thesis" (SOT):

Let H be some hypothesis, O be some observation, and E be the proposition that H would explain O if H and O were true. Then O screens-off E from H: $\Pr(H|O \,\&\, E) = \Pr(H|O)$.   (2014: 193)[4]

R&S contend that *SOT is true if and only if "explanatoriness is evidentially irrelevant."* We refer to this claim as (IRRELEVANCE). Putting these two together yields R&S's conclusion that *"explanatoriness is evidentially irrelevant."*

---

[1] Hereafter, we will drop the qualifier "epistemic," but any use of "justification" or its cognates should be understood to mean epistemic justification unless specified.

[2] See, for example, McCain (2014), (2016), and Poston (2014).

[3] Henceforth, we refer to Roche and Sober as "R&S."

[4] The relevant notion of probability in SOT and throughout this discussion is epistemic probability. This sort of probability is often referred to as "rational credence." It is because of this that we will often transition back and forth from speaking of probabilities to speaking of credences. Our hope is that where we do so the discussion is made clearer by our using either the term "probability" or "credence."

R&S's conclusion is surprising and far-reaching. As we noted above, explanationism is an attractive view of justification. After all, it seems clear that a proposition's being part of the best explanation of a body of data is itself a reason to think that the proposition is true. However, R&S's conclusion, if correct, would spell serious trouble for explanationism. What is more, R&S's argument does not simply threaten explanationism as a theory of justification, it threatens the viability of inference to the best explanation. If explanatoriness is evidentially irrelevant then we cannot be justified in inferring the truth of a proposition on the basis of its being part of the best explanation of a set of facts. This implies that Charles Darwin in *The Origin of the Species* actually gives a bad argument when he defends the use of explanatory reasoning in science:

It can hardly be supposed that a false theory would explain, in so satisfactory a manner as does the theory of natural selection, the several large classes of facts above specified. It has recently been objected that this is an unsafe method of arguing; but it is a method used in judging of the common events of life, and has often been used by the greatest natural philosophers.   (1962 [1872]: 476)

Similarly, if IRRELEVANCE is true, Antoine Lavoisier provides poor support for his oxygen theory of calcification and combustion when he appeals to its explanatory virtues:

I have deduced all the explanations from a simple principle, that pure or vital air is composed of a principle particular to it, which forms its base, and which I have named *the oxygen principle*, combined with the matter of fire and heat. Once this principle was admitted, the main difficulties of chemistry appeared to dissipate and vanish, and all the phenomena were explained with an astonishing simplicity.[5]

If R&S are correct then the explanatory considerations that Darwin and Lavoisier track in their momentous scientific work provide no evidence that their theories are true. Furthermore, the widespread and common use of inference to the best explanation from motorcycle repair to administrative planning to medical diagnosis is misguided.

Fortunately, R&S are incorrect, and their argument fails. In a recent article (2014), we argued that R&S overlook an important dimension of evidential support when making their case that IRRELEVANCE is true, viz., the resilience of a probability function. Resilience is essentially how volatile a probability function is with respect to new evidence; a probability function with low volatility is more resilient than a function with high volatility. We maintained that IRRELEVANCE is false because there are clear cases where explanatory considerations increase the resilience of a probability function. Additionally, we argued that there are numerous cases where SOT fails to hold.

R&S (2014) argue that we were mistaken on both accounts. The arguments R&S offer are not persuasive, but they do significantly clarify the disagreement. We pick up on this

---

[5] Lavoisier (1862), Vol II: 623—quotes and translation here from Thagard (1978).

improved dialectical situation to further defend our position that explanatoriness is evidentially relevant. The upshot of our discussion is that both SOT and IRRELEVANCE are false because explanatory considerations may be captured in logical and mathematical relations encoded in a probability function. Thus, both inference to the best explanation and explanationism are safe from attack.

1. Explanatoriness Matters even if SOT is Granted

We provided a case to illustrate that IRRELEVANCE is false (2014: 149). The case shows how explanatoriness can make an evidential difference by making a probability function more resilient even if SOT is granted. In this case we focused on two subjects, Sally and Tom. Both are aware that there are 1,000 "x-spheres" in an opaque urn. Sally and Tom both observe the same random drawing (with replacement) of ten x-spheres. An equal number of blue x-spheres and red x-spheres are drawn. Sally and Tom have the same updated credences after observing the outcome of the experiment. Namely, they both assign a probability of 0.5 to the next x-sphere randomly drawn being blue. However, there is an important difference between Sally and Tom. Sally, but not Tom, has an explanation of why the experiment turned out the way it did. As we explained, "Sally knows that blue and red x-spheres must be stored in exactly equal numbers because the atomic structure of x-spheres is such that if there are more (or less) blue x-spheres than red, the atoms of all of the x-spheres will spontaneously decay resulting in an enormous explosion" (2014: 149). We pointed out that the explanatory difference between Sally and Tom makes Sally's credence more resilient to future misleading information. That is, Sally's credence remains 0.5 that the next x-sphere randomly drawn will be blue even given an improbable run of ten blue x-spheres, while Tom's credence significantly changes in response to this new (misleading) information. Thus, we concluded that even if it is granted that SOT is true, and so Sally's having an explanation is screened off from raising her credences in various outcomes, the explanatory difference between Sally and Tom makes an evidential difference. Thus, IRRELEVANCE is false.

R&S (2014) agree that Sally's credence is more resilient than Tom's in our case, but they deny that the reason for this difference is that Sally has an explanation and Tom does not. They highlight four features of our case:

(i) Sally and Tom have a credence of 0.5 in proposition H (that the x-sphere drawn on the next random draw will be blue), (ii) Sally's credence is more resilient than Tom's, (iii) Sally but not Tom knows that if blue and red x-spheres are stored in unequal numbers, then there will be an enormous explosion, and (iv) Sally but not Tom has an explanation of why the probability of the blue x-sphere on a random drawing from the urn is 0.5.   (196)

R&S write, "It is true that Sally but not Tom has an explanation for why the probability of a blue x-sphere on a random drawing from the urn is 0.5, but this difference between

Sally and Tom is doing no work" (196). Rather, on their view, what does all the work is (iii). They explain, "Our point is that (ii) is true because (iii) is true, and (iv) does nothing to make (ii) true once (iii) is taken into account" (197). This is to say that (iv) is screened off once (iii) is taken into account. On their view, explanatoriness is evidentially irrelevant in this case because Sally's knowledge concerning the atomic structure of x-spheres makes (ii) true.

This move fails. In our example, the property of *Sally having the explanation* is one and the same property as the property of *her knowing the relevant fact about x-spheres*. In order to see this, consider an argument parallel to R&S's that denies that water has the property of *extinguishing fire*. Water is H_2O, and H_2O has special chemical properties that make it an excellent chemical to extinguish fire. Because of the high degree of hydrogen bonding between water molecules, H_2O has the second highest heat capacity of all known substances. In virtue of its high heat capacity, its transition from a liquid to a gas requires a significant amount of energy, which enables it to rapidly quench flames. Once we account for these facts about H_2O the fact that water is present is screened off. Thus, water is irrelevant to extinguishing fire because H_2O is doing all the work.

The response to the water/H_2O argument is obvious: Water = H_2O. The properties of H_2O in virtue of which it makes an excellent fire extinguisher are the properties of water. They are one and the same. The same consideration applies to our x-sphere case. We might put our point thusly: to the extent that water has the property of extinguishing fires, explanatoriness in our x-sphere case is evidentially relevant. Water extinguishes fire in virtue of its chemical structure. Explanatoriness is evidentially relevant in virtue of it specifying certain relations between H and E that get encoded in a probability function.

It may be replied that the parallel argument breaks down because water is identical to H_2O, but Sally's having the explanation is not identical to her knowing the relevant facts about x-spheres. R&S may think that Sally can know that blue and red x-spheres are stored in equal numbers to prevent an enormous explosion without having an explanation of why the probability of a blue x-sphere on a random drawing from the urn is 0.5. This claim, however, is contentious, and R&S do not provide support for it. The claim is clearly false given a causal account of explanation. If knowledge of causal relations constitutes knowledge of explanatory relations, as causal accounts of explanation hold, then knowledge of causes is necessary and sufficient for explanatory knowledge. In which case, (iii) and (iv) are describing the same state of affairs.

Our initial (2014) reply to R&S left open the nature of the explanatory relation, but as causal accounts of explanation make clear, if explanations are constituted by such causal facts then conditionalization on causal facts will screen off explanatoriness. But this is just the sense in which water is screened off from H_2O. In other words, it is not really screened off at all because those facts have already been taken into account. Conditionalizing separately on Sally's knowledge of the causal relation (as expressed in (iii)) and her knowledge of the explanation (as expressed in (iv)) is counting the same facts twice.

R&S do acknowledge that explanatoriness may be evidentially relevant if it is indicative of some fact, I, which specifies a probabilistic relation between H and E. I might be the proposition that H entails E, for example. In such a case R&S allow that if explanatory considerations in some way indicate to us that I is true, then such considerations may be evidentially relevant. However, they point out that Bayesian confirmation theory assumes agents are logically omniscient, and such omniscient agents would know logical and probabilistic facts like I. Thus, they maintain "explana-toriness has no confirmational significance, once purely logical and mathematical facts are taken into account" (195).

We have two replies to this move. First, this response assumes that explanatory considerations are entirely separate from probabilistic relations between a hypothesis and evidence. If knowing the explanatory facts amounts to knowing some probabilistic relations between a hypothesis and evidence, then this response fails. It is exactly this sort of thing that we argue occurs in our x-sphere case—knowledge of the causal facts *just is* knowledge of the explanatory facts. In terms of the previous paragraph this means that in some cases knowledge of I just is knowledge of the explanatory considerations. So, there is no way to take I into account without taking explanatory considerations into account. Thus, we have good reason to think that IRRELEVANCE is false.

Second, there is a larger issue about what counts as "logical and mathematical" facts. Is it a logical or mathematical fact that water extinguishes fire? If a person had com-plete knowledge of chemical theory could they deduce that water extinguishes fire in a normal environment (or, that the probability of its doing so was high)? Arguably, yes. But if these kinds of facts are facts a logically omniscient agent is aware of, then clearly once they are taken into account explanatoriness is evidentially irrelevant. Again, this is because the logical/mathematical facts encompass the explanatory facts. So, explan-atoriness appears evidentially irrelevant, but this is simply because those facts have already been counted. Thus, if this is how one should understand R&S's SOT thesis, then it amounts to the prohibition of counting the same facts twice. This should not concern any explanationist.

2. Why SOT Does Not Hold in all Cases

In addition to arguing that explanatoriness is evidentially relevant even in cases where it is granted that SOT holds, we argued that SOT is false in a number of cases. Here is the particular case we mentioned in our earlier response to R&S:

The ability of Newton's theory to explain the orbits of the planets is evidence that Newton's theory is true, even if we lack observational evidence regarding the nonepistemic, objective chance that Newton's theory is true. Similarly, the discovery that Einstein's theory of general relativity explained the precession of the perihelion of Mercury increased the probability of Einstein's theory. So, in these cases Pr(H|O&E)>Pr(H|O). (146)

R&S maintain that we are mistaken on this point. They claim:

Suppose you know that a theory H (e.g. Newton's) logically implies O (given the background information codified in Pr) and so you realize that $Pr(O|H)=1$. You then work out the value of $Pr(H|O)$ by first obtaining values for $Pr(H)$ and $Pr(O)$. It follows (if neither H nor O has a probability of 1 or 0) that $Pr(H|O)$ is greater than $Pr(H)$. You then learn that O and as a result increase your credence in H. Suppose you later learn E. M&P's view [that is, our view] entails that upon learning E you should *further* increase your credence in H. It seems clear, however, that you should not do this. (195)

R&S maintain that even in cases like those depicted by us explanatory considerations are not playing an evidential role. Rather, they maintain that all of the evidential work is being done by logical and mathematical truths. Thus, R&S maintain that we failed to show that there are cases where SOT does not hold.

One thing that explanationists sympathetic to our position could immediately question is how the values for $Pr(H)$ and $Pr(O)$ are obtained. The explanationist could plausibly argue that setting such prior probabilities will require an appeal to explanatory considerations. Plausibly, if these priors are set without appealing to explanatory considerations, the door to inductive skepticism will be left wide open. Huemer (2009) and Poston (2014) both argue that the threat of inductive skepticism can only be avoided by way of explanatory reasoning. That is to say, without appealing to explanatory considerations we cannot establish a probability distribution that will allow for inductive confirmation. We mention this because it is a reasonable point for explanationists to press against R&S, but we will not pursue this line of argument here. Instead, we will point out the more glaring problem with R&S's argument.

The mistake that R&S are making here is similar to the mistake we pointed out in the first section. R&S are assuming that a particular kind of knowledge is distinct from having an explanation—they are assuming that coming to know that H entails O is not explanatory knowledge. This assumption, at least in the cases they are discussing, is mistaken. It is exceedingly plausible that knowing that H entails O (especially when H is a purported natural law such as we have in Newton's theory) is the same as having an explanation of O, assuming that H is true. After all, this is part of the Deductive-Nomological (D-N) model of explanation.[6] The D-N model does not work in all cases (e.g., flagpoles and shadows, hexed salt, and so on), but it does work in the sort of case that R&S describe where a natural law predicts/entails an observation. This is arguably a paradigm case of the sort that Hempel and Oppenheim had in mind when formulating this theory of explanation.

Apparently, R&S assume that an explanation must in all cases be something beyond knowing that H entails O. As they say in their original (2013) discussion of this topic, "even if proposition (E) has entailments about the logical and probabilistic relations of O and H, there is more to explanation than this... reasonable candidates for that

---

[6] Hempel and Oppenheim (1948).

something more will vindicate our screening-off thesis" (663). It seems that R&S see their screening-off argument as similar to van Fraassen's (1989) Dutch book argument against inference to the best explanation. Both arguments rely on the assumption that in order to be evidentially relevant explanatory considerations must be something over and above logical and probabilistic relations. But, this is a mistake. While there are clearly cases where more is desired by an explanation, there are contexts where information about logical and probabilistic relations is explanatory. For instance, explanations in pure mathematics will plausibly consist entirely of information concerning logical and probabilistic relations. Insofar as explanations in pure mathematics are genuine explanations information about logical and probabilistic relations will be sufficient for explanation in at least some cases.[7]

It is worth briefly summing up a few points here. We agree with R&S that it would be a mistake to increase your credence in Newton's theory (H) upon learning E (H explains some observation, O) in the case as they describe it. In their discussion of the case R&S assume that you have already increased your credence in H on account of learning O. They assume that you possess the correct value for $Pr(O|H)$ and $Pr(H)$; so by Bayesian learning you update $Pr(H)$ by $Pr(H|O)$. You then learn E, which amounts to realizing that the probabilistic relation encoded in the value of $Pr(O|H)$ is an explanatory relation. In this case it is clear that you should not further increase your credence upon learning E because you have already taken account of E when you include your knowledge of $Pr(O|H)$. The evidential relevance of explanatory considerations does not license double counting. Thus, it seems that the case of Newton's theory is plausibly a case in which SOT does not apply. Similar considerations apply to the case we described concerning Einstein's theory of relativity as well as purely mathematical explanations in general.

Perhaps R&S think that in the cases of Newton's theory, Relativity, and pure mathematical explanations one does not actually possess an explanation unless she has a particular psychological state of understanding, an "aha!" moment. It may be that they think that this psychological state is required for explanation, and yet at the same time when a person has the "aha!" moment it adds nothing to confirmation.

While R&S may be correct that such a state does not further increase the probability of the relevant hypothesis (though such an experience may increase resiliency and so be evidentially relevant), it is a mistake to think that such a psychological state is required for possessing an explanation. It is one thing to have the information, and hence the evidential confirmation provided by an explanation, and it is another thing to have this particular psychological state.[8] So, attempting to argue that information about logical and probabilistic relations fails to be explanatory because it may not include an "aha!" moment is not promising. There are contexts

---

[7] For more on mathematical explanations see Hafner and Mancosu (2005), Kitcher (1989), and Steiner (1978).

[8] For more on the relation between explanation and the psychological sense of understanding see de Regt (2004), Grimm (2009), and Trout (2002), (2005).

where such information is explanatory whether or not it is accompanied by a particular feeling of understanding.

R&S mistakenly think that knowing that E (if H and O were true, H would explain O) is something beyond knowing that H and O are true and that H entails O because explanation is not a purely logical fact (2013: 664). This assumption coupled with some basic considerations from Bayesian confirmation theory lead R&S to wrongly claim that explanation is evidentially irrelevant. Bayesian confirmation theory assumes that agents are logically and mathematically omniscient. R&S play along with this but note that it can give the wrong results in some cases. They give the following case. Consider some proposition I, which says that O logically implies H. I can be evidence for H. Suppose you know that O is true, but you don't know I. You then learn I. Your credence in H should change (assuming you were not already certain that H is true). Yet on the assumption of logical omniscience, I is not evidence. I is screened off from H since $Pr(H|O)=Pr(H|O\&I)$. R&S hold that in this case the screening-off test "isn't a good test for the confirmation relevance of purely logical facts" (2013: 664). But they think it is fair game for explanatory facts because explanatory facts are not logical facts.

The crucial question then is whether explanatory facts are, at least at times, indicative of facts about full or partial entailment (probabilistic facts are often thought of as facts about partial entailment). If E is so indicative, then the screening-off test is not a good test for confirmational relevance. In our case of Sally and Tom described above E is indicative of a probabilistic fact because the explanation specifies that the probability of random selection must be 0.5. Thus, the screening-off test is not a good test for the confirmational relevance of E in our x-sphere case. Moreover, often explanations work by specifying some previously unknown fact—e.g. the fact that space-time is curved. R&S (2014) discuss this in some detail. They acknowledge that there can be cases where E increases the probability of some hypothesis conditional on O: $Pr(H|O\&E) > Pr(H|O)$, especially where E is indicative of a logical fact like I. But they claim "explanatoriness has no confirmational significance, once purely logical and mathematical facts are taken into account" (2014: 195). This brings us back to our earlier criticism. To the extent that water has the property of extinguishing fire, explanatoriness has the property of being evidentially relevant. On our view R&S's screening-off test amounts to a prohibition against counting the same facts twice. While good methodological advice, it does nothing to show that explanatoriness is evidentially irrelevant.

In light of the considerations adduced here it is reasonable to conclude that both SOT and IRRELEVANCE are false. Explanatoriness is evidentially relevant.[9]

[9] We are grateful to Kenny Boyce, Nevin Climenhaga, Brad Monton, Mark Newman, and Bill Roche for helpful comments on, and discussion of, earlier drafts.

References

Darwin, C. *The Origin of Species*, 6th Edition. New York: Collier, 1962 [1872].

de Regt, H.W. "Discussion Note: Making Sense of Understanding." *Philosophy of Science* 71 (2004): 98–109.

Grimm, S. "Reliability and the Sense of Understanding." In H.W. de Regt, S. Leonelli, and K. Eigner (eds), *Scientific Understanding: Philosophical Perspectives*. Pittsburgh, PA: University of Pittsburgh Press, 2009: 83–99.

Hafner, J. and P. Mancosu. "The Varieties of Mathematical Explanation." In P. Mancosu, K. Frovin Jørgensen, and S.A. Pedersen. (eds), *Visualization, Explanation and Reasoning Styles in Mathematics*. Berlin: Springer, 2005: 215–50.

Hempel, C.G. and P. Oppenheim. "Studies in the Logic of Explanation." *Philosophy of Science* 15 (1948): 135–75.

Huemer, M. "Explanationist Aid for the Theory of Inductive Logic." *British Journal for the Philosophy of Science* 60 (2009): 345–75.

Kitcher, P. "Explanatory Unification and the Causal Structure of the World." In P. Kitcher and W. Salmon (eds), *Scientific Explanation*. Minneapolis: University of Minnesota Press, 1989: 410–505.

Lavoisier, A. *Oeuvres de Lavoisier*. Paris: Imprimerie Imperiale, 1862–93.

McCain, K. *Evidentialism and Epistemic Justification*. New York: Routledge, 2014.

McCain, K. *The Nature of Scientific Knowledge: An Explanatory Approach*. Switzerland: Springer, 2016.

McCain, K. and T. Poston. "Why Explanatoriness Is Evidentially Relevant." *Thought* 3 (2014): 145–53.

Poston, T. *Reason and Explanation: A Defense of Explanatory Coherentism*. New York: Palgrave-Macmillan, 2014.

Roche, W. and E. Sober. "Explanatoriness Is Evidentially Irrelevant, or Inference to the Best Explanation Meets Bayesian Confirmation Theory." *Analysis* 73 (2013): 659–68.

Roche, W. and E. Sober. "Explanatoriness and Evidence: A Reply to McCain and Poston." *Thought* 3 (2014): 193–9.

Steiner, M. "Mathematical Explanation." *Philosophical Studies* 34 (1978): 135–51.

Thagard, P. "The Best Explanation: Criteria for Theory Choice." *Journal of Philosophy* 75 (1978): 76–92.

Trout, J.D. "Scientific Explanation and the Sense of Understanding." *Philosophy of Science* 69 (2002): 212–33.

Trout, J.D. "Paying the Price for a Theory of Explanation: de Regt's Discussion of Trout." *Philosophy of Science* 71 (2005): 198–208.

van Fraassen, B. *Laws and Symmetry*. Oxford: Oxford University Press, 1989.

PART III

Justifying Inference to the Best Explanation

9

Inference to the Best Explanation and Epistemic Circularity

J. Adam Carter and Duncan Pritchard

1. The Epistemic Status of IBE: a Puzzle

Inference to the best explanation—or, IBE—tells us to infer from the available evidence to the hypothesis which would, if correct, best explain that evidence. As Peter Lipton (2001, 184) puts it, the core idea driving IBE is that explanatory considerations are a guide to inference. IBE is widely regarded as a hallmark of scientific methodology.[1] Moreover, IBE features in the background of our everyday (i.e., non-scientific) inquiries in such a way that—as Igor Douven (2011, §1.2) notes—is so 'routine and automatic that it easily goes unnoticed'.[2] Given our widespread dependence on IBE, one might naturally ask what justifies our employment of it.

One response that might be proffered in this regard is to contend that IBE is basic to our epistemic practices in such a way that it does not require a justification.[3] However, as David Enoch & Joshua Schechter (2008) note, it would surely be 'philosophically unsatisfying' if we could not do so. As they put the worry:

There are many different possible belief-forming methods that could be employed as basic. Some, such as MP [*modus ponens*], IBE, and relying on perception, we presumably are justified in employing. Others, such as Affirming the Consequent, Inference to the Third Worst Explanation, and relying on wishful thinking, we presumably would be unjustified in employing. It is highly implausible that it is merely a brute fact that we are justified in employing certain methods as basic and not others. It is much more plausible that there is a principled distinction between the two classes . . . relevant to justification, one that presents [MP, IBE, etc.] in a rationally positive light. (Enoch & Schechter 2008, 557–8)

---

[1] See, for example, Boyd (1981, 1984), Lipton (1991, 2004), and Psillos (1999). Moreover, IBE is widely taken to underwrite certain popular arguments for scientific realism on the basis of the success of science.

[2] See, for instance, Adler (1994) for an argument to the effect that IBE routinely features in our justification for accepting testimony. See here also Fricker (Chapter 17).

[3] Another potential response would be to appeal to the ubiquity of inductive knowledge and then reason that inductive knowledge requires IBE.

The thinking here is that it is not enough to observe that we employ IBE as if it were basic—that is unlikely to be in doubt—since we still need a rationale for why we are right to so employ it, and that question takes us right back to the need to justify our use of IBE.[4]

Paul Boghossian (2001) takes things a step further by suggesting that if we can't justify our basic inference rules, then this speaks against their objectivity. Boghossian's reasoning is that 'if there are objective facts about which epistemic principles are true, there should be humanly accessible circumstances under which those facts can be known' (2001, 3), which of course entails that one is able to have justified beliefs about them.[5] On Boghossian's line, then, it would not only be intellectually unsatisfying were we to lack any way to justify IBE, it would also be evidence against the objectivity of IBE.[6]

Let's assume for the sake of argument that it is incumbent upon us to at least give *some* kind of account of how IBE is itself justified—i.e., how IBE has a kind of positive epistemic status that is not shared equally by other, epistemically problematic, belief-forming methods, such as 'Inference to the Third Worst Explanation'. How would this be done? Here, Enoch & Schechter (2008) claim that a puzzle arises. We will first outline this puzzle and then show how there is, in fact, a more challenging way to formulate it. The remainder of the chapter will engage with this more challenging version of the puzzle.

As Enoch & Schechter see it, the puzzle that arises once we ask how our reliance on IBE is to be justified is most effectively framed as a choice between two *prima facie* non-starters. One option is to use IBE in the service of its own justification.[7] Another option is to use some other belief-forming method to justify IBE. Neither option is workable, they tell us, the former because it is objectionably circular, and the latter because one cannot justify a basic rule by appeal to other belief-forming methods. The puzzle goes as follows.

The Justificatory Puzzle for IBE

1.   We cannot justify our use of IBE with a justification that relies upon IBE (or otherwise assumes its privileged epistemic status), since such a justification would be objectionably circular.

2.   We cannot justify our use of IBE by appealing to other belief-forming methods, since IBE is a basic rule.

3.   Thus, there is nothing in virtue of which we are justified in using IBE.

---

[4] Cf., Fumerton (Chapter 5) for a view on which IBE is not interestingly different from induction.
[5] Boghossian (2001, 3) formalizes this strand of reasoning, which he engages with, as follows:

  1. Assume that there are objective facts about which epistemic principles are true.
  2. If there are objective facts about which epistemic principles are true, these facts should be knowable: it ought to be possible to arrive at justified beliefs about them.
  3. It is not possible to know which epistemic principles are objectively true, therefore
  4. there are no objective facts about which epistemic principles are true.

[6] Boghossian (2006) revisits this line of reasoning. See Wright (2001) for a reply to Boghossian (2001).
[7] Boyd (e.g., 1984) has been charged with defending IBE in such a way. See Fine (1984) for a criticism.

Enoch & Schechter attempt to resolve this puzzle in a way that draws inspiration from Hans Reichenbach's (1949) pragmatist response to the problem of induction. While we think their positive proposal—though interesting—is not ultimately compelling, this point won't concern us here.[8] Rather, we think that there is a more difficult version of the puzzle than the one they presented.

To appreciate the more difficult puzzle, consider premise (2). In virtue of what is an inference rule *basic*? Either an answer to this question excludes (i.e., by definition) the very *possibility* of vindicating that rule's epistemic status with reference to another fundamental inference rule, or it doesn't. Enoch & Schechter regard the property of basicness as an

intuitive one. The belief-forming methods that are basic for a thinker are those methods that are the most fundamental in how the thinker reasons. All other belief-forming methods employed by the thinker are derivative. This characterization of basicness is not fully precise, and it may be somewhat indeterminate which methods are employed as basic by a thinker. But we find it plausible that MP, IBE, reliance on memory and perception, and reliance on normative and modal intuitions (or close relatives of these belief-forming methods) are basic for most adult human thinkers. (Enoch & Schechter 2008, 551)

Notice that one can agree that IBE is basic in the intuitive sense described without being committed to the further (and considerably stronger) epistemic claim that, in virtue of being basic, a belief-forming process could not *possibly* be justified by some other (perhaps also basic) belief-forming method. After all, there is nothing about a belief-forming method's being fundamental to how one reasons that entails that it has this further property. For all Enoch & Schechter tell us about basicness, two basic belief-forming methods, A and B, could potentially be used to justify one another, even if no derivative belief-forming method could be used to justify A or B. Premise (2) of the puzzle is true only if there is no such possibility.[9]

Enoch & Schechter's formulation of the puzzle is challenging to the extent that the premises are plausible and the conclusion problematic. But premise (2), on closer inspection, looks like a premise that we should accept only if we are already committed to an account of basicness which is stronger than what we must accept by accepting what Enoch & Schechter themselves tell us is involved in a belief-forming method's being basic. The fact that one might reasonably accept the account of basicness they provide and still reject (2) vitiates the force of the puzzle. Or, at least, the version of it they've offered.

---

[8] According to Enoch & Schechter (2008), it is plausible that employing IBE (or a close relative) is needed for successfully engaging in what they call 'the explanatory project', the project of understanding and explaining the world around us. The explanatory project is itself a *rationally required project*; we are, on their view, justified in employing any belief-forming method needed for successfully engaging in a rationally required project.

[9] Put another way, consider Wright's (2001, 4) question, 'Is some substantial form of justification in principle possible for a range of basic beliefs that we have—and if so, what is it?' Wright takes this to be a meaningful question. If basicness excluded such a possibility, this question would be nonsensical.

Here is the crux of the matter. Can the force of the justificatory puzzle be regained by replacing (2) with a weaker premise, one that can be accepted by those who—like Boghossian (2001), for instance—do *not* regard the project of justifying basic inference rules as already excluded by the very definition of basicness?

We think that it can. To appreciate how, let's take as a starting point that the possibility is not foreclosed *ex ante* that we might justify our use of IBE by appealing to some other (basic) belief-forming method(s). One conversational context in which justification for IBE might be requested is a context where our interlocutor subscribes to a very different set of inference rules than we do. Here a thought experiment will be useful. Following Boghossian (2006), let's define an individual's epistemic system or framework as a set of epistemic principles or rules to which the individual subscribes. Imagine now a scenario in which your epistemic framework (for convenience, call this framework 'Western Science') is challenged by an individual who embraces a very different epistemic framework (call this 'Mysticism').

Suppose you attempt to justify the wider system Western Science to your interlocutor, the Mystic, by attempting to justify, one at a time, each of the rules that constitute the framework 'Western Science'. You begin by attempting to justify IBE. Let's add to this story that you take it for granted that justifying IBE by appealing to IBE is objectionably circular (though we'll say more on this point below). Accordingly, you attempt to justify IBE by appealing to another basic belief-forming method (e.g., perception, memory, *modus ponens*, etc.). At this point, your mystic interlocutor reminds you that these other basic inference rules are, no less than IBE, a part of the wider system 'Western Science'. By applying a rule that belongs to Western Science in the service of justifying another rule that's a part of Western Science, you are employing your epistemic framework in support of itself.

Consider at this point Michael Williams' characterization of an argumentative strategy that has been employed in the service of motivating epistemic relativism:

> what about the claims embodied in the framework itself: are they justified? In answering this question, we inevitably apply our own epistemic framework. So, assuming that our framework is coherent and does not undermine itself, the best we can hope for is a justification that is epistemically circular, *employing our epistemic framework in support of itself*. Since this procedure can be followed by anyone, whatever his epistemic framework, all such frameworks, provided they are coherent, are equally defensible (or indefensible).   (Williams 2007, 3–4, our italics)

As Williams sees it, the epistemic relativist proposes what is (for our purposes at least) an idea with important relevance—*viz.*, that there is something objectionably epistemically circular about employing our epistemic framework in support of itself. The relevance, specifically, is that that is precisely what one does when one attempts to justify IBE by appealing to some *other* inference rule which is no less a part of one's wider epistemic system than IBE.[10]

---

[10] For a critical treatment of Williams' (2007) treatment of epistemic relativism, see Pritchard (2010; cf. Pritchard 2009). Note that those, such as Fumerton (Chapter 5) who think IBE is justified only if it's an

The crucial point here is that if employing one's own epistemic framework in support of itself is objectionably epistemically circular (as Williams' epistemic relativist insists that it is), then it looks as though attempting to justify IBE by appealing to some other inference rule which is part of the same epistemic system that IBE belongs to is going to be no more promising as a justification of IBE than would be the objectionably circular strategy of justifying IBE by appealing to IBE itself.[11] With this point in hand, we can now recast Enoch & Schechter's Justificatory Puzzle for IBE in a way that is more challenging:

The Justificatory Puzzle for IBE*
4.   We cannot justify our use of IBE with a justification that relies upon IBE (or otherwise assumes its privileged epistemic status), since such a justification would be objectionably circular.
5.   It is objectionably epistemically circular to employ one's epistemic framework in support of itself.
6.   Justifying our use of IBE by appealing to another inference rule that belongs to the same epistemic framework as IBE is to use one's epistemic framework in support of itself.
7.   We cannot justify our use of IBE with a justification that relies upon another inference rule that belongs to the same epistemic framework as IBE. (From 5, 6)
8.   The only two ways of justifying our use of IBE would involve either relying upon IBE or another inference rule that belongs to the same epistemic framework as IBE.
9.   Thus, there is no way of justifying our use of IBE. (From 4, 7, 8)

Note that the force of the modified version of the puzzle (unlike Enoch & Schechter's original version) does not depend on the contentious claim (featuring in premise (2) of the original puzzle) that basic belief-forming methods are, as such, not possibly justifiable by appeal to other basic belief-forming processes. Put another way: the modified version of the puzzle cannot be dismissed as one which trades on a tendentious notion of what it is for a belief-forming method to be basic. This is thus a stronger version of the puzzle, and so poses a trickier challenge when it comes to our ability to justify our use of IBE.

2.   Rule Circularity: a Distinction

Recall that premise (1) of Enoch & Schechter's original puzzle states that we cannot justify our use of IBE in a way that relies upon IBE (or otherwise assumes its privileged epistemic status), since such a justification would be objectionably circular. As it turns

instance of enumerative induction disagree; however, Williams in response could recast this problem to Fumerton. For discussion on justifying IBE via enumerative induction, and whether and to what extent we regard this as objectionable, see §4.

[11]   A point of clarification. By 'one's own epistemic framework', we mean specifically the epistemic framework which one subscribes to, as a function of which epistemic principles one embraces.

out, there is already an established way of thinking about this particular variety of epistemic circularity, one which occurs when justification for an inference rule proceeds by following that very same rule. Stathis Psillos (1999, 82) and Boghossian (2001) call arguments of this kind *rule-circular* (as distinct from what Psillos, following Richard Braithwaite (1953), calls 'premise circular').[12]

Notice, however, that the kind of circularity Williams' epistemic relativist objects to when denying the legitimacy of employing one's epistemic system in its own defence—e.g., when one attempts to justify IBE by employing some *other* inference rule that is part of the same epistemic system as IBE—is *not* a matter of justifying a rule (within a system) by following that very same rule. By regarding it as objectionably epistemically circular to employ one's own epistemic system in one's own defence, Williams' epistemic relativist objects to the justifying of one rule within an epistemic system by employing another rule within the same epistemic system.[13]

We can distinguish then between two kinds of rule circularity which feature in the modified puzzle outlined at the end of §1, *narrow* and *wide*. An argument is *narrow rule-circular* when one uses a particular inference rule in the service of justifying that very *same* inference rule (i.e., by taking at least one step in accordance with this rule). This is the kind of rule circularity Enoch & Schechter took for granted to be objectionable in (1) of their original puzzle, and which also features in (4) of the modified puzzle. However, with reference to the kind of reasoning Williams attributes to the epistemic relativist, we can describe *wide rule-circular* arguments as ones which employ an epistemic framework in support of itself in the following way: when, for some epistemic framework F, one employs one F-inference rule in support of another F-inference rule.[14]

With reference to this distinction between narrow and wide rule-circular arguments, we can now restate the crux of the modified justificatory puzzle for IBE very simply: arguments which attempt to justify IBE will either be narrow rule-circular (because they rely on IBE to justify IBE) or wide rule-circular (because they employ an epistemic system in support of itself, by relying on some other rule that's part of the same system as IBE to justify IBE). If both of these varieties of rule circularity are sufficient to make a piece of reasoning have a defective justificatory structure, then it really does look like IBE cannot be justified in a satisfactory way.

Here, in summary form, is the line that we shall advance against this argument:

  (i) All narrow rule-circular arguments have a defective justificatory structure.
  (ii) But only *some* wide rule-circular arguments have a defective justificatory structure.

---

[12] We will say more about 'premise-circular' arguments below, in distinguishing between some varieties of premise-circularity as noted by Pryor (2004).

[13] For related discussion on this issue, see Poston (2014, §6.2.2)

[14] See Poston (2011, 413–15) for a related suggestion to the effect that the explanatory virtues are plastic in the sense that they aren't specific, fixed rules.

The upshot of (i) is that we should simply grant premise (4) of the puzzle and concede that narrow rule-circular arguments have a defective justificatory structure. Such arguments are not *merely*, as Boghossian (2001) and Psillos (1999) have argued, *dialectically ineffective*—i.e., they don't merely lack rational force against one who antecedently doubts the conclusion—but rather are defective *period*.[15,16] The upshot of (ii) is that we should reject (7), and do so because we reject (5). Wide rule-circular arguments do not necessarily have a defective justificatory structure. Whether a given wide rule-circular argument has a defective justificatory structure depends importantly on just *how* the conclusion depends on the support offered for it.

Our rationale for both (i) and (ii) draws from a wider picture, one which can be found in the recent literature on perceptual warrant, concerning how conclusions may (or may not) depend for support on their premises.

3. Perceptual Warrant: an Analogy

In this section, a very plausible general position about *premise circularity* will be outlined, one that has been developed in most detail in the perceptual warrant literature, particularly by Jim Pryor (e.g., 2004).[17] *Modulo* some small refinements, we will be arguing that this position can be extended to *rule*-circular arguments, of both the narrow and wide varieties articulated in §2. A straightforward rationale then emerges for why narrow rule-circular arguments have a defective justificatory structure, but only *some* wide rule-circular arguments have a defective justificatory structure. This rationale implies that we should accept (4) in the modified justificatory puzzle for IBE but reject (5). More generally, this means that while we cannot provide a satisfactory justification for IBE in the way that Boghossian and Psillos think, there are nonetheless some other potential ways to do so.

To this end, consider a general question with regard to which *dogmatists* (e.g., Pryor 2000, 2004) about perceptual warrant and their traditional opponents, *conservatives* (e.g., Wright 2003, 2004, 2007), are divided: *under what conditions does the dependence of an argument's conclusion on one (or more of) its premises undermine the justificatory structure of an argument?*[18] A natural frame of reference here is Pryor's

---

[15] Cf., McCain (2016) for a qualified defence of such a rule-circular defence of IBE.

[16] For a defence of rule-circular arguments for induction, see van Cleve (1984) and Papineau (1993).

[17] As Moretti & Piazza (2013, §1) note, most epistemologists who weigh in on this debate use the term 'warrant' though 'they all seem to use the term "warrant" to refer to some kind of epistemic justification' and in doing so 'broadly identify the epistemic property capable of being transmitted with *propositional* justification'. We shall use these terms interchangeably in the present discussion as nothing really hangs (for our purposes) on this terminological difference.

[18] Alston (1986) offers the following view of what would be both necessary *and* sufficient for a belief *p* to confer warrant to another belief *q*. (A) S is justified in believing the premises, *p*. (B) *p* and *q* are logically related in such a way that if *p* is true, that is a good reason for supposing that *q* is at least likely to be true. (C) S knows, or is justified in believing that the logical relation between *p* and *q* is as specified in (B). (D) S infers *q* from *p* because of her belief specified in (C). Boghossian rightly worries that Alston's criteria are too demanding. These criteria require, as a necessary condition that one must know or justifiably believe,

(2004) discussion of two kinds of conclusion-premise dependence that he calls *Type 4 dependence* and *Type 5 dependence*, the latter of which he thinks is significantly more problematic than the former:

> *Type 4 Dependence*:   the conclusion is such that evidence against it would (to at least some degree) undermine the kind of justification you purport to have for the premises.[19]
>
> *Type 5 Dependence*:   having justification to believe the conclusion is among the conditions that make you have the justification you purport to have for the premise. (Pryor 2004, 359–60)

Pryor offers the following piece of reasoning as a straightforward instance of Type 5 dependence:

10.   I intend to walk to Lot 15 and drive home.
11.   So I will walk to Lot 15 and drive home.
12.   So my car will still be in Lot 15 when I get there.

The strand of reasoning from (10) to (12) is Type 5 because justification for (10) *relies* on your already having justification to believe (12) and so it can't make (12) any more credible for you. Crucially, Pryor takes it that it's possible for an argument to exhibit Type 4 dependence while failing to exhibit Type 5 dependence.[20]

Here is one such example case, which he appeals to in making this point. Suppose you watch a cat stalk a mouse, and so your visual experiences justify you in believing:

13.   The cat sees the mouse.

Suppose further that you reason:

14.   If the cat sees the mouse, then there are some cases of seeing.
15.   So there are some cases of seeing.

that the premises and conclusion are logically related in such a way that if the premises are true, that is a good reason for supposing the conclusion is likely to be true. But this criteria quickly leads one into the trap of Carroll's (1895) regress. As Boghossian puts it:

> at some point it must be possible to use a rule in reasoning in order to arrive at a *justified* conclusion, without this use needing to be supported by some knowledge about the rule that one is relying on. It must be possible to simply move between thoughts in a way that generates justified beliefs, without this movement being grounded in the thinker's justified belief about the rule used in the reasoning.   (Boghossian 2001, 27)

[19] Note that Type 4 dependence is not the claim that the conclusion is such that evidence against it would (to at least some degree) be evidence (directly) against one of the premises. Rather, the idea is that the conclusion is such that evidence against it would (to at least some degree) undermine the kind of *justification* you purport to have for the premises, and this is a claim about the relation between the conclusion and what you take to *justify* your premises.

[20] Neta (2007, 17) criticizes this case by remarking that 'for the example above to do the argumentative work that Pryor wants it to do, we need to know why we should believe that what makes me propositionally justified in believing the conclusion is not precisely the same thing that makes me propositionally justified in believing the premises'.

Pryor's (2004, 361) assessment is that (13–15) exhibit Type 4 dependence but *not* Type 5 because, firstly, evidence against (15) would undermine the justification one has for believing (13). Secondly, though, Pryor says, 'I don't think you need antecedent justification to believe (15), before your experiences can give you justification to believe (13). I also think it's plausible that your perceptual justification to believe (13) contributes to the credibility of (15)' (2004, 361).

In fact, this is precisely what Pryor, as a dogmatist about perceptual warrant, takes to be going on in G. E. Moore's (1939) famous proof of the external world—i.e., Type 4 but not Type 5 dependence. In contrast, a conservative like Crispin Wright regards Moore's proof as failing to transmit warrant, and so exhibiting Type 5. But diagnosing Moore's proof needn't concern us here. Our goal will rather be to first highlight a key *difference* between what dogmatists and conservatives say about arguments more generally which exhibit Type 4 and Type 5 dependence and, secondly, to note a point of *agreement* between dogmatists and conservatives about these two kinds of reasoning. Indeed, all that will matter for our purposes going forward is a point that Wright and Pryor can agree upon.

Pryor (2004, §§4–5) contends that arguments that exhibit merely Type 4 dependence are *not* epistemologically objectionable—that is, warrant for believing the premise *can* transmit from the premise to the conclusion. According to the dogmatist, however, Type 4 arguments are nonetheless *dialectically ineffective* against one who antecedently doubts the conclusion.[21] That is, dogmatists grant that while Type 5 dependence 'ruins an argument' (Pryor 2004, 360), Type 4 arguments are such that—while there's nothing wrong with their justificatory structure as such (i.e., justification *does* transmit from premises to conclusion)—they are not effectual in bringing one who *already* doubts the conclusion to rational conviction of the conclusion on the basis of the premises. Conservatives such as Wright, by contrast, deny that warrant transmits in *both* Type 5 and Type 4 cases; on this view, *both* forms of argument exhibit a defective justificatory structure.

The issue of who's right in this debate won't matter for our present purposes. What matters is rather that we can conveniently extract from this debate two entirely general theses about warrant transmission that both dogmatists and conservatives, despite their other differences, can accept:

- *Type 5 arguments* fail to transmit warrant.
- *Type 4 arguments* are (at least) dialectically ineffective.

Given that these general theses about Type 4 and Type 5 arguments are common ground between the two sides, they provide a helpful reference point from which to answer the question of interest to us, which is whether the kind of narrow rule-circular and wide rule-circular arguments at issue in the modified justificatory puzzle for IBE are themselves of a defective justificatory structure.

---

[21] For a similar move, with respect to the variety of epistemic circularity that arises in bootstrapping arguments, see Markie (2005).

We'll do this in two steps, which will be the focus of the next section. First, we'll construct 'rule-circular' analogues to Type 4 and Type 5 dependence.[22] That way, we'll have a principled basis for explaining why, for a given rule-circular argument, it would either fail to transmit warrant (if corresponding with Type 5) or be at least dialectically ineffective (if corresponding with Type 4).

Next, we'll conclude by showing how narrow rule-circular arguments (i.e., attempts to justify IBE by reasoning in accordance with IBE) plausibly feature Type 5 rule dependence (and so fail to transmit warrant), while at least some but not all wide rule-circular arguments exhibit Type 4 rule dependence (and so some but not all wide rule-circular arguments are dialectically ineffective).

4. The Modified Justificatory Puzzle for IBE, Redux

Arguments that feature Type 4 and Type 5 dependence are, at least as they are presented by Pryor (e.g. 2004, §4), framed in terms of conclusion-*premise* dependence relations. However, we can very naturally think of Type 4 and Type 5 arguments as a genus of which Type 4 and Type 5 *rule-circular* arguments are a species. To bring this idea into sharp relief, let's consider Type 4 dependence first. According to Pryor, Type 4 dependence occurs when the conclusion is such that evidence against it would (to at least some degree) undermine the kind of justification you purport to have for the premises. Here is a plausible way of thinking about a rule-circular twist on this idea: let's say an argument exhibits what we can call *Type 4 rule dependence* when the conclusion is such that evidence against it would (to at least some degree) be evidence against the legitimacy of employing one (or more) rules one reasons in accordance with in moving from premises to conclusion.[23] Moreover, we can now say that arguments which exhibit Type 4 rule dependence are (like arguments which feature Type 4 dependence, more generally) at least dialectically ineffective.

We now need a rule-circular twist on Type 5 dependence. This is a bit trickier to model, but a plausible candidate goes as follows. An argument exhibits *Type 5 rule dependence* if the legitimacy of reasoning in accordance with the rule—i.e., the conclusion of the piece of reasoning—being justified just is (or is partly constitutive of) the legitimacy one purports to have for moving from premises to conclusion.[24]

[22] See Carter (2016) for a more general strategy for making this kind of move.
[23] For example: suppose you have two closely connected inference rules, I and I^*, such that inferring in accordance with I will usually not violate I^*, and vice versa. Suppose further that you know this, and so are aware of the significant overlap. Now, suppose you are reasoning attempting to prove I^* by reasoning in accordance with I. In such a circumstance, evidence against I^* will at least to some extent be (defeasible) evidence against the legitimacy of moving from premise to conclusion via rule I.
[24] The term 'legitimacy' is used here so as to make Type 4 rule circularity as closely analogous as we can to Pryor's Type 4 (premise) circularity—*viz.*, just as the kind of justification you purport to have for believing a premise can be undermined, so analogously the, and we are using this term inclusively, *legitimacy* you have for following a rule can be undermined. Here is a paradigmatic case: if, for example, you have excellent reason to believe that *modus ponens* is false, then this counts against the legitimacy of reasoning in accordance with *modus ponens*.

The presence of Type 5 rule dependence will ruin an argument just as any Type 5 dependence does.

We are in a position to submit the following general theses (modelled from points on which Wright and Pryor agree) about the conditions under which rule-circular arguments are objectionable.

> *Type 4 Rule Dependence*:   the conclusion is such that evidence against it would (to at least some degree) be evidence against the legitimacy of employing one (or more) rules one reasons in accordance with in moving from premises to conclusion. (Corollary: arguments exhibiting Type 4 rule dependence, like Type 4 arguments more generally, are at least dialectically ineffective.)

> *Type 5 Rule Dependence*:   the legitimacy of reasoning in accordance with the rule being justified (i.e., the conclusion of the piece of reasoning) just is (or is partly constitutive of) the legitimacy one purports to have for moving from premises to conclusion. (Corollary: arguments exhibiting Type 5 rule dependence, like Type 5 arguments more generally, are of a defective justificatory structure: they fail to transmit warrant.)[25]

These are all the tools we need to now revisit our original puzzle. Let's begin by diagnosing the claim that features in premise (4) of the modified justificatory puzzle for IBE. This premise, recall, states:

> 4.   We cannot justify our use of IBE with a justification that relies upon IBE (or otherwise assumes its privileged epistemic status), since such a justification would be objectionably circular.

We are now armed with a principled way to evaluate whether an attempt to justify IBE that exhibited these features would be (as the premise states) 'objectionably circular' in a way that matters for whether one can satisfactorily justify IBE.

First, let's consider: what would an argument look like which used IBE to justify IBE? Here's an example: suppose you attempt to justify IBE by pointing to the fact that IBE, if correct, would best explain some body of evidence (e.g., success of certain scientific theories). Is the conclusion of this reasoning (i.e., that IBE is an epistemically justified rule) such that evidence *against* it would (to at least some degree) be evidence against the legitimacy of employing one (or more) rules in accordance with which one reasons in moving from premises to conclusion? Absolutely. Evidence against IBE would be at the same time evidence against the legitimacy of employing one of the rules one reasons with here—namely, IBE. Thus, we have a principled explanation for

---

[25] A paradigmatic instance of this kind of reasoning would be as follows: where some piece of reasoning attempts to justify *modus ponens*. And, further, this piece of reasoning proceeds to do this by taking at least one step in accordance with *modus ponens*. In such a case, note that the legitimacy of reasoning in accordance with the rule being justified (i.e., in this case, *modus ponens*) just is (or is partly constitutive of) the legitimacy one purports to have for moving from premises to conclusion. We discuss cases like this in more detail further in this section.

why this form of reasoning (like any Type 4 argument) is at least dialectically ineffective. The next relevant question is this: is the legitimacy of reasoning in accordance with IBE (the conclusion of the argument) also whatever legitimacy one would have for moving from premises to conclusion? It's hard to see how it would not be. After all, the legitimacy of reasoning in accordance with the rule that argument attempts to justify (when reasoning via IBE to IBE) *just is* the legitimacy one has for reasoning to the conclusion in the way one does in such a case, by IBE. What this suggests, though, is that *narrow circular* arguments are in fact in worse shape than Boghossian and Psillos thought. Such arguments are best understood as not merely dialectically ineffective, but moreover as having, *qua* a Type 5 argument, a defective justificatory structure. A more general point here is that we have a principled rationale for why an attempt to justify IBE via a narrow rule-circular argument really is objectionably circular (as per premise (4) of the modified puzzle). It's objectionably circular because it is not merely dialectically ineffective, but it also has a defective justificatory structure.

What is the epistemic status of wide rule-circular arguments? These, recall, are when one attempts to use one's epistemic system in its own service (an activity Williams' epistemic relativist objects to) by employing one basic rule within the same system to justify another.[26] For example, this will be the case when a thinker attempts to justify IBE by relying on perhaps one or more other basic rules which also form part of that thinker's broader epistemic system, such as perception, memory, *modus ponens*, and so on. Are such arguments, like narrow rule-circular arguments, objectionably circular? As with the case of narrow rule-circular arguments, we simply need to ask whether such arguments feature merely Type 4 rule dependence or also Type 5 rule dependence.

Let's consider such a case. For example, suppose one attempts to justify IBE by reasoning in accordance with *modus ponens* (i.e., by taking at least one step in accordance with *modus ponens*), which is also a basic inference rule within the wider system that this thinker accepts (again, let's just call this 'Western Science'). Because the reasoning described is an instance of employing one's epistemic system in support of itself, it is widely rule-circular. In assessing to what extent the alleged circularity is objectionable, we can quickly point out that the argument does *not* feature Type 5 rule dependence. After all, the legitimacy of reasoning in accordance with IBE (i.e., the rule which the thinker is attempting to justify with the argument) is independent of whatever legitimacy one has for moving from premises to conclusion by *modus ponens*. The wide rule-circular argument described is thus not a Type 5 argument.[27]

---

[26] An interesting issue, in connection with wide rule circularity (and the epistemic status of arguments which feature it) is the matter of the epistemic status of the more general methodology of reflective equilibrium as a method of justifying epistemic principles. We are open to an interpretation of wide rule circularity according to which it is a form of reflective equilibrium; it is beyond the scope of the present chapter, however, to assess more generally the epistemic status of the methodology of reflective equilibrium (under its various guises).

[27] At this point, it is worth considering the following line of objection: any defence of a basic inference method, such as IBE, will have to make use of IBE somewhere as either a step in the argument or as a support for the premise. Does that mean that all arguments for IBE are doomed to Type 5 rule dependence?

But, interestingly, the argument also does not feature Type 4 rule dependence either. Consider: would the reasons one has for antecedently doubting (IBE) also constitute a reason to doubt the legitimacy of reasoning via *modus ponens*? It's not obvious at all that it does. To see why, consider, for example, Bas van Fraassen's (1989) criticism of certain probabilistic versions of IBE. The reasons for van Fraassen's criticisms have to do with considerations about Bayes' Theorem and Dutch Book arguments. These are simply no considerations that count against *modus ponens*. They are, rather, orthogonal to *modus ponens*.

What this means is that at least *some* kinds of wide rule-circular arguments are not themselves epistemically objectionable in a way that features either Type 5 or Type 4 rule dependence, and so would *not* have a defective justificatory structure. Moreover, such arguments needn't be dialectically ineffective either. Note that we have not attempted to defend the idea that appealing to *modus ponens* suffices to justify IBE. That would depend on whether a particular instance of doing so is sound. Our point, rather, is that if one were to do so, such an argument—even though widely circular in the sense discussed in §2—is not such that we can give any principled reason for why it is epistemically objectionable. This much is already reason to reject premise (5) of the modified justificatory puzzle for IBE. Not *all* wide rule-circular arguments are problematic.

We suggested at the outset that while all narrow rule-circular arguments are epistemically objectionable, *some* kinds of wide rule-circular justifications for IBE are considerably less objectionable than others. We've already seen how at least one kind of wide rule-circular justification for IBE is not obviously objectionable (in that it exhibits neither Type 4 nor Type 5 rule dependence). We conclude by noting why *some* wide rule-circular arguments are in a worse position than others (and, thus, why not all wide rule-circular arguments are equally epistemically objectionable/unobjectionable).

Consider now the following example: suppose you attempt to justify IBE by using simple enumerative induction, a move which has been made by, among others, Alexander Bird (1998), Richard Fumerton (1980, Chapter 5 in this volume), Philip Kitcher (2001), and Douven (2002).[28] As Douven writes:

The common idea of these attempts is that every newly recorded successful application of abduction—like the discovery of Neptune, whose existence had been postulated on explanatory grounds...adds further support to the hypothesis that abduction is a reliable rule of inference, in the way in which every newly observed black raven adds some support to the hypothesis that all ravens are black.   (Douven 2002, §3.2)

The right response here, we think, is to deny the supposition that any defence of IBE must make use of IBE somewhere. A position according to which it would is what Fumerton (Chapter 5) calls 'extreme explanationism'. We should accept the supposition that any defence of IBE must make use of IBE somewhere only if embracing an overly inclusive conception of what counts as reasoning in accordance with IBE. For example, we see no reason to think that all instances of reasoning in accordance with *modus ponens* are thereby reasoning with IBE, even if reasoning in accordance with *modus ponens* is *compatible* with reasoning in accordance with IBE. Thanks to the editors for requesting elucidation on this point.

[28] See Douven (§3.2) for discussion.

In short, each time inferring to the best explanation seems to work, we have more evidence to believe it's true—and given its track record of success, we've thus got substantial evidence in its favour. IBE and enumerative induction are separate inference rules, so attempting to justify IBE by enumerative induction is not to simply justify IBE *via* IBE (as, for instance, Boyd (e.g., 1984) is inclined to).[29] But since IBE and enumerative induction are both rules that feature within the same framework, such an argument is wide rule-circular.

Whereas, as we've seen, a wide rule-circular argument which attempts to justify IBE by using *modus ponens* exhibits neither Type 4 nor Type 5 dependence (at least, not in virtue of taking a step in accordance with *modus ponens*), we suggest now that attempting to justify IBE by using induction in the fashion sketched above does exhibit at least Type 4 rule dependence, and possibly also Type 5.[30]

In order to bring this point into focus, let's ask: in a situation where a thinker attempts to justify IBE by enumerative induction, would reason to doubt IBE be at the same time reason to doubt the legitimacy of reasoning in accordance with enumerative induction? This is a difficult question, but we submit that, plausibly, it would. After

---

[29] This point is disputed by some, such as Fumerton (1980; Chapter 5 in this volume). Fumerton, in suggesting that IBE is best understood as just a form of induction, attempts to show how Peirce's (1903) description of abductive reasoning can be redescribed as inductive reasoning. Peirce's case involved the discovery of fossilized remains of fish skeletons in rocks that are far from any body of water. Fumerton (Chapter 5, 68) remarks:

> It strikes us that this needs an explanation...Perhaps we have a half-way decent inductive argument that when we look hard enough for causal explanations we eventually find them. In the case of our desert rock with remains of fish, it probably wouldn't have been that hard for ancient people to have reached the conclusion that the water once covered the land where the remains were found. They might have made a further epistemic leap by also inferring that the most common way that water ends up covering land that is usually dry is through flooding—and when the dry land is a long way from any water the flood must have been impressive indeed. Eureka!—we have an explanation of the prevalence of flood myths in ancient cultures.

If Fumerton is right that IBE is a kind of induction, then one obvious result on our model is that inductive attempts to justify IBE will be Type 5 as well as Type 4, and so be of a defective justificatory structure. However, this might be an academic point. For one thing, Fumerton' has shown how a case of IBE can be redescribed as a case of multiple inductions. Harman (1965) has appealed to similar considerations of redescription to suggest a move in the opposite direction: that induction is a form of abductive reasoning. We are agnostic about whether cases of potential redescription recommend pulling toward a reduction in one direction rather than another. Our point is that unless we have good reason to think that potential redescription cases favour Fumerton's direction of reduction, we've no *pro tanto* reason to align ourselves with his reduction. And, again, even if we did, the result on our model would simply be a Type 5 diagnosis in all cases of justifying IBE by induction. Thanks to the editors for requesting we engage with Fumerton on this point.

[30] Note that some arguments that reason in accordance with *modus ponens* will also rely on some other inference rule in order to establish IBE. For example, suppose an argument for IBE that proceeded via enumerative induction claimed that certain facts entail that IBE is reliable. Notice that such an argument relies on *modus ponens* and induction. The point we advance in this section implies that *if* such an argument exhibits Type 4 or Type 5 dependence, it won't be *in virtue* of taking a step in accordance with *modus ponens*.

all, suppose a thinker doubts IBE on the grounds G, where G is some very general belief the thinker has about the relation between explanation and inference. It's not hard to imagine that some such general doubt about the relation between explanation and inference could at least to some degree count against the legitimacy of reasoning by induction—*viz.*, by reasoning from observed frequencies. After all, explanatory considerations plausibly play at least an implicit role in inductive inferences. It seems, then, that such reasoning exhibits Type 4 rule dependence and will accordingly be dialectically ineffective. Given that induction and IBE are closely connected—Gilbert Harman (1965), for example, thought the former was really just a species of the latter—this should not be entirely surprising.[31]

While justifying IBE by induction wouldn't obviously exhibit Type 5 rule dependence, we can imagine some wide rule-circular arguments which plausibly would do so: for example, using *modus tollens* to justify *modus ponens*, or some forms of induction to justify some very closely related forms of induction. So there is at least the possibility that Type 5 rule dependence is present here too, at least until it is demonstrated otherwise.

The moral of the story is that while all narrow rule-circular arguments have a defective justificatory structure, only *some* wide rule-circular arguments have a defective justificatory structure. Moreover, in the case of wide rule-circular arguments, whether such arguments have a defective justificatory structure (and furthermore, whether such arguments are dialectically ineffective) depends importantly on which particular basic inference rules are being justified, and by which other rules. In slogan form: employing our own epistemic system in support of itself is less objectionable in some cases than in others, and taking a cue from the perceptual warrant debate provides a rationale for explaining why.

5. Conclusion

The aim of this chapter has been to show, first, how the task of justifying IBE has been for various reasons regarded as philosophically problematic. Second, we've argued that Enoch & Schechter's (2008) version of the puzzle can be reformulated in a way that is considerably more challenging to address. Third, in the service of addressing this more challenging version, we've drawn some connections from an already established debate in the literature on perceptual warrant in order to give a principled response to the puzzle. Our response is broadly pessimistic in the case of narrow rule circularity, but considerably less so in the case of wide rule circularity. More importantly, though, we hope to have moved the debate forward by showing

---

[31] Likewise, as the editors have noted, some philosophers, including McCain (2014) and Poston (2014), suggest that explanatory reasoning is required in order to make the projection from observed cases to unobserved cases justified.

how some distinctions in the literature on perceptual warrant offer very useful applications with respect to the issue of how, and why, basic inference rules such as IBE might be satisfactorily justified.[32]

References

Adler, J. (1994). 'Testimony, Trust, Knowing', *Journal of Philosophy* 91, 264–75.

Alston, W. (1986). 'Epistemic Circularity', *Philosophy and Phenomenological Research* 47, 1–30.

Bird, A. (1998). *Philosophy of Science*, London: UCL Press.

Boghossian, P. (2001). 'How Are Objective Epistemic Reasons Possible?' *Philosophical Studies* 106, 1–40.

Boghossian, P. (2006). *Fear of Knowledge*, Oxford: Clarendon Press.

Boyd, R. (1981). 'Scientific Realism and Naturalistic Epistemology', in *PSA* (Vol. II), eds. P. D. Asquith & R. Giere, 613–62, East Lansing, MI: Philosophy of Science Association.

Boyd, R. (1984). 'The Current Status of Scientific Realism', in *Scientific Realism*, ed. J. Leplin, 41–82, Berkeley: University of California Press.

Braithwaite, R. (1953). *Scientific Explanation*, Cambridge: Cambridge University Press.

Carroll, L. (1895). 'What the Tortoise Said to Achilles', *Mind* 4, 278–80.

Carter, J. A. (2016). *Metaepistemology and Relativism*, London: Palgrave Macmillan.

Douven, I. (2002). 'Testing Inference to the Best Explanation', *Synthese* 130, 355–77.

Douven, I. (2011). 'Abduction', in *Stanford Encyclopedia of Philosophy*, ed. E. Zalta, <http://plato.stanford.edu/archives/spr2011/entries/abduction/>.

Enoch, D. & Schechter, J. (2008). 'How Are Basic Belief-Forming Methods Justified?' *Philosophy and Phenomenological Research* 76, 547–79.

Fine, A. (1984). 'The Natural Ontological Attitude', in *Scientific Realism*, ed. J. Leplin, 83–107, Berkeley: University of California Press.

Fumerton, R. A. (1980). 'Induction and Reasoning to the Best Explanation', *Philosophy of Science* 47 (4), 589–600.

Harman, G. (1965). 'The Inference to the Best Explanation', *Philosophical Review* 74, 88–95.

Kitcher, P. (2001). 'Real Realism: The Galilean Strategy', *Philosophical Review* 110, 151–97.

Lipton, P. (1991). *Inference to the Best Explanation*, London: Routledge.

Lipton, P. (2001). 'Is Explanation a Guide to Inference? A Reply to Wesley C. Salmon', in *Explanation*, eds. G. Hon & S. Rakover, 93–120, Dordrecht: Springer.

Lipton, P. (2004). *Inference to the Best Explanation*, 2nd edn, London: Routledge.

Markie, P. (2005). 'Easy Knowledge', *Philosophy and Phenomenological Research* 70, 406–16.

McCain, K. (2014). *Evidentialism and Epistemic Justification*, London: Routledge.

McCain, K. (2016). *The Nature of Scientific Knowledge: An Epistemological Approach*, Cham, Switzerland: Springer.

Moore, G. E. (1939). 'Proof of an External World', *Proceedings of the British Academy* 25, 273–300.

---

[32] Thanks to Kevin McCain and Ted Poston for very helpful comments on an earlier version of this chapter.

Moretti, L. & Piazza, T. (2013). 'Transmission of Justification and Warrant', in *Stanford Encyclopedia of Philosophy*, ed. E. Zalta, <http://plato.stanford.edu/archives/win2013/entries/transmission-justification-warrant>.

Neta, R. (2007). 'Fixing the Transmission: The New Mooreans', in *Themes From G. E. Moore: New Essays in Epistemology and Ethics*, eds. S. Nuccetelli & G. Seay, 62–83, Oxford: Clarendon Press.

Papineau, D. (1993). *Philosophical Naturalism*, Oxford: Blackwell.

Peirce, C. S. (1903). *The Essential Peirce: Selected Philosophical Writings*, Vol. 2 (1893–1913), Bloomington: Indiana University Press.

Poston, T. (2011). 'Explanationist Plasticity and the Problem of the Criterion', *Philosophical Papers* 40 (3), 395–419.

Poston, T. (2014). *Reason and Explanation: A Defense of Explanatory Coherentism*, London: Palgrave Macmillan.

Pritchard, D. H. (2009). 'Defusing Epistemic Relativism', *Synthese* 166, 397–412.

Pritchard, D. H. (2010). 'Epistemic Relativism, Epistemic Incommensurability and Wittgensteinian Epistemology', in *Blackwell Companion to Relativism*, ed. S. Hales, 266–85, Oxford: Blackwell.

Pryor, J. (2000). 'The Skeptic and the Dogmatist', *Noûs* 34, 517–49.

Pryor, J. (2004). 'What's Wrong with Moore's Argument?' *Philosophical Issues* 14, 349–78.

Psillos, S. (1999). *Scientific Realism: How Science Tracks Truth*, New York: Routledge.

Reichenbach, H. (1949). *The Theory of Probability*, 2nd edn, trans. E. H. Hutten & M. Reichenbach, Berkeley: University of California Press.

van Cleve, J. (1984). 'Reliability, Justification, and the Problem of Induction', *Midwest Studies in Philosophy* 9, 555–67.

van Fraassen, B. (1989). *Laws and Symmetry*, Oxford: Clarendon Press.

Williams, M. (2007). 'Why (Wittgensteinian) Contextualism Is Not Relativism', *Episteme* 4, 93–114.

Wright, C. (2001). 'On Basic Logical Knowledge; Reflections on Paul Boghossian's "How Are Objective Epistemic Reasons Possible"', *Philosophical Studies* 106, 41–85.

Wright, C. (2003). 'Some Reflections on the Acquisition of Warrant by Inference', in *New Essays on Semantic Externalism, Skepticism and Self-Knowledge*, ed. S. Nuccetelli, 57–78, Cambridge, MA: MIT Press.

Wright, C. (2004). 'Warrant for Nothing (and Foundations for Free)?' *Aristotelian Society Supplementary Volume* 78, 167–212.

Wright, C. (2007). 'The Perils of Dogmatism', in *Themes From G. E. Moore: New Essays in Epistemology and Ethics*, eds. S. Nuccetelli & G. Seay, 25–48, Oxford: Clarendon Press.

10

In Defense of Rationalism about Abductive Inference

Ali Hasan

1. Rationalism, Abductivism, and Rationalism about Abduction

According to rationalism, as I shall understand it here, some of our beliefs and inferences are justifiable purely *a priori*; these beliefs and inferences do not depend for their justification on any of the sources traditionally classified as *a posteriori*: testimony, sensory perception, or introspection. On the traditional rationalist view I favor, our *a priori* justification depends on some sort of grasp or awareness of abstracta.[1]

According to the abductivist response to external world skepticism—"abductivism" for short—belief in the external world is justified by explanatory considerations: the hypothesis of an external world of spatial objects, one that is in line with our ordinary beliefs in at least some central respects, is the best explanation of our experiential regularities.[2]

Rationalism about abduction is the view that, at a fundamental level, explanatory or abductive inference is justifiable *a priori*. Rationalism about abduction to the external world is the view that explanatory or abductive inference from our experiential regularities to the external world is justifiable *a priori*. One might be a rationalist about abduction and yet be a skeptic about our external world beliefs, i.e., hold that some abductive reasoning is justifiable a priori, but deny that our external world beliefs can be justified, perhaps because we have no good reason to prefer the external or real world hypothesis over the hypothesis that we are brains in vats or victims of a Cartesian evil demon. I'll focus on the rationalist about abduction who is not a skeptic, and in fact accepts an abductivist response to external world skepticism.

---

[1] This view remains very much out of favor, though there are signs of a resurgence. See, for example, BonJour (1998: 180–6). For recent views of *a priori* knowledge as involving a direct awareness of abstracta, see Bengson (2015) and Chudnoff (2013).

[2] See, for example, BonJour (1999), Vogel (1990) and (2008), Huemer (2016), and Hasan (Forthcoming).

Rationalism about abduction is not a popular view. As Beebe notes, "practically everyone who works on abductive inference believes that such inferences are justified empirically and that the theoretical virtues [like simplicity and explanatory power] are broadly empirical and contingent marks of truth" (2009: 625). A large part of this rejection of rationalism about abduction derives from the concern that, at least when it comes to non-deductive inferences, it just seems implausible that we could have an *a priori* reason or justification for thinking that the premises make the conclusion probable.[3] Sometimes the concern seems to be that any such principle will be synthetic rather than analytic, and so those who reject the possibility that synthetic judgments can be justified *a priori* will reject the possibility of *a priori* justification of non-deductive inferences. Of course, traditional rationalists are not likely to be moved by this sort of objection. Sometimes the concern is that the connection between the truth of the premises and the truth of the conclusion in such inferences is bound to be contingent, and so those who reject the possibility of contingent *a priori* will reject the possibility of *a priori* justification of non-deductive inferences. Some might worry that even when we set these arguments aside and leave open the possibility, in principle, of an *a priori* justification of non-deductive forms of inference, we really don't seem to have the sort of awareness or grasp of, or insight into, the inductive or probabilistic connections required, let alone have enough to bridge the epistemic gap between our experiential evidence and our external world beliefs.[4]

My main focus here is to examine these reasons against the existence of *a priori* justification of inductive or probabilistic relations. I'm inclined to think that this rejection or restriction of the *a priori* is unjustified. I argue that the case against the existence of such an insight is much weaker than contemporary epistemologists seem to think, and that there are good reasons to be optimistic about our ability to have such an insight.

2. Motivating Rationalism about Inferential Justification

I have characterized rationalism as the view that some of our beliefs *and inferences* are justifiable *a priori*. It is intuitive and relatively uncontroversial that, in order for a subject to have justification for believing that Q inferentially, on the basis of another belief that P, (i) the subject must have justification for P, and (ii) the subject must have

---

[3] See Douven (2011: sec. 3.2):

> Hardly anyone nowadays would want to subscribe to a conception of truth that posits a necessary connection between explanatory force and truth—for instance, because it stipulates explanatory superiority to be necessary for truth. As a result, a priori defenses of abduction seem out of the question. Indeed, all defenses that have been given so far are of an empirical nature in that they appeal to data that supposedly support the claim that (in some form) abduction is a reliable rule of inference.

See also Sober (1988), Boyd (1999), and Psillos (1999).
[4] See Ramsey (1926) and Fumerton (1995).

justification for the inference from P to Q. The latter condition has been understood in different ways, but a very natural interpretation is that it involves the following requirement: *in order for the inference from P to Q to be justified, the subject must have justification for believing that there is a suitable relation between P and Q, one that is potentially relevant to the truth or probability of Q, such as a relation of entailment or probability.* Call this the *principle of inferential justification* (PIJ). We can motivate PIJ by considering a person who infers Q from P, where P does in fact logically entail or makes probable Q, but where the entailment or probability is far too complex for the person to grasp or even understand. A subject S might be disposed to form the belief that Q as a result of believing that P, perhaps because someone has, unbeknownst to S, implanted a chip in S's brain that leads S to believe Q if and when S believes P. Surely, the belief that Q is not justified if the person who reaches the conclusion couldn't see how P entails or makes probable Q.[5]

Why is PIJ so intuitive? It is often associated with access internalism about epistemic justification, and the requirement can be seen as a natural application of the core intuition behind the view. According to access internalism, in order for the subject to have justification for believing something, the subject must be aware of or have access to something that makes a difference to the subject's perspective on the truth, or probable truth, of the proposition believed. Without some such justification, the internalist claims, there is no epistemically relevant difference to the subject's perspective on the truth or probability of the inferred belief; the subject seems to just be flying blind. The access internalist thus hopes to secure two intuitively attractive and historically prominent requirements for epistemic justification: internal access to good reasons, and a robust connection to truth or probability. Applying this to inferential justification: in order for a subject to have justification for believing Q by inference from P, the subject must have a good reason to think that the truth of P is relevant to the truth or probability of Q.

A belief's justification is inferential if it depends on one's justification for other beliefs. My justification for Q depends on my justification for P, and the latter might depend in turn on my justification for various other propositions. The same might be said of *inferences*: I might be justified in inferring one belief from another, but only because I am justified in believing something about the connection between the propositions believed. But foundationalists have long argued that, on pain of vicious regress, if any of our beliefs are justified then there must be some beliefs that are "basic" or "foundational": they are justified but do not depend for their justification on any other beliefs. Foundationalists who accept PIJ have also argued that, on pain of vicious regress, inferential justification requires, at some fundamental level, a *non-doxastic* grasp or awareness of the relevant inferential connection.[6] In the case of deductive inferences, one must have, or at least be capable of having, a *non-doxastic* grasp of the

[5] See Hasan and Fumerton (2014) for more detailed discussion.

[6] For the inferential internalist, this seems to be the lesson of Lewis Carroll's regress. See Fumerton (1995: ch. 7) and BonJour (2005).

entailment between the premises and conclusion (or between the premises and conclusions of sub-arguments) so that the regress of justification comes to a halt. The same holds for non-deductive inferences. Perhaps our justification for abduction depends on enumerative induction or (more plausibly, in my view) enumerative induction depends on abduction, but it is implausible to think that either are reducible to deduction. What is required to halt the regress of justification for such inferences is a *non-doxastic* grasp of a relation, other than entailment, that is relevant to the truth or probability of the inferred belief. It seems implausible to construe our awareness of these relations of entailment or probability as *empirical* or *a posteriori* sources of justification; our reasoning or inference must be wholly or partly *a priori*. True to a form of awareness or access internalism, this traditional rationalism about deductive and non-deductive inference requires a positive grasp of broadly logical and probabilistic relations, and not merely that these logical and probabilistic relations do in fact hold. To deny that we have any such grasp thus seems to lead to a radical skepticism, undermining our justification for any inferences or inferentially justified beliefs.

I have put it in more general terms here, but this sort of motivation for rationalism has been discussed in detail elsewhere.[7] One thing I want to stress, however, is that on its own such an argument does not show that we have *a priori* justification. At best, it shows that *if* we have any justification for certain beliefs and inferences, then the justification of some fundamental deductive and non-deductive inferences must be (wholly or partly) *a priori*. Rationalism about abduction may be the most defensible *non-skeptical* view, but even granting that for sake of argument, why should we think we do indeed have the sort of *a priori* justification the rationalist claims we need in order to avoid radical skepticism? Indeed, some have argued that we just don't have the sort of justification the rationalist needs.[8] As I've indicated, my main task here is to respond to some of these concerns and motivate the claim that we do have an *a priori* justification for non-deductive inferences, including explanatory ones.

I want to briefly discuss two likely concerns related to the argument of this section before moving on. First, coherentists about epistemic justification might object that the argument assumes that coherentism about epistemic justification is false, for it does not consider the possibility that one's justification for believing that some propositions make other propositions probable could take the form of beliefs that are justified by virtue of their *coherence* with the subject's other beliefs. A vicious regress may result if we think of justificatory relations as linear or one-directional, but the coherentist rejects this linear conception of justification for a holistic one. I have two responses to this concern.

First, coherentism cannot accept PIJ, for the problem is not just that there is some sort of regress that ensues, but that satisfying that requirement leads to a hierarchical

---

[7] See BonJour (1998) and (2005). For some criticism, see Casullo (2003). For attempts to respond to Casullo and develop the argument further, see Thurow (2009) and Hasan (2014).

[8] See notes 3 and 4.

regress of beliefs of ever increasing complexity. In order for S to be justified in believing that *P makes probable Q*, S must have some reason *R* to think this, and, given PIJ, must have justification for believing that *R makes probable that (P makes probable that Q)*. In order for the latter belief to be justified, S must have some reason *T* to believe this, and have justification for believing that *T makes probable that (R makes probable that (P makes probable that Q))*. And so on.[9] The problem is not just that there is some regress here (which coherentists might not be too worried about), but that it is just implausible that we are able to form such complex beliefs even at a few iterations, let alone be able to have an infinite hierarchy of them.

There is a second, related worry with coherentism, which is that although it was largely motivated by the internal access requirement and the alleged failure of foundationalism to satisfy this requirement,[10] it turns out that coherentism itself is, ironically, incapable of satisfying it.[11] Suppose for simplicity that the subject has only three beliefs that cohere with each other: *P*, *Q*, and *R*. (This is implausible, but it will be easier to state the problem with this simplification in place, and once the problem is clear it will become clear that adding further beliefs won't help.) For the coherentist, access to one's reasons for believing that *P* will have to take the form of (i) access to the other beliefs that support or cohere with *P*, namely, *Q* and *R*; and (ii) access to the fact that *Q* and *R* support or cohere with *P*. Now, for the coherentist, any such access cannot take the form of a direct, non-doxastic awareness of one's beliefs or of the fact that they support or cohere with *P*; for the coherentist, any such access must take the form of other, *higher-order beliefs* to the effect (i) that one believes *Q* and *R* and (ii) that *Q* and *R* support or cohere with *P*. But these higher-order beliefs were beliefs that, by hypothesis, the subject did not already have, and so the subject does not have the relevant access. It won't help to add these higher-order beliefs to the subject's belief set, for then access to *these* higher-order beliefs will require still further beliefs that the subject did not already have. The result is that internal access to any of one's reasons for belief turns out to be impossible. (Note that the problem remains even if the coherentist accepts a weak form of internalism that requires only access to cohering beliefs and not to their coherence.) The coherentist might drop the internal access requirement, as some recent coherentists seem to have done, and perhaps find some other way to motivate coherentism rather than accept externalist foundationalism.[12] But for those who want to hold onto access internalism, a direct, non-doxastic grasp of relations between propositions is needed.

[9] See Bergmann (2006: ch. 1 and 2) for a similar argument that such a requirement leads to an infinite regress of beliefs of ever increasing complexity. Bergmann's argument for this is one horn of a dilemma against access internalism. The other horn of the dilemma attacks versions of access internalism that require only a non-doxastic grasp of the connection between one's grounds and the beliefs they allegedly justify. For a defense of the latter form of access internalism, see Hasan (2011).

[10] See BonJour (1985). [11] See Moser (1989) and BonJour (2003) for similar arguments.

[12] See Poston (2014) for an attempt to defend coherentism without accepting the internal access or awareness requirement.

The second concern might come from someone sympathetic to phenomenal conservatism rather than coherentism. I am interested in defending the claim that we can be directly aware or have a direct grasp of relations between propositions, including non-deductive or probabilistic relations. Indeed, I am inclined to think that, in Bertrand Russell's terms, we are *acquainted* with such relations, that we have an awareness of such relations that is "factive." But why not hold that what is required is a non-factive "seeming" that one proposition entails or makes probable the other? That would, according to the phenomenal conservative, be a non-doxastic state capable of providing justification and avoiding any regress. Perhaps those who are inclined to appeal to such seemings are also able to respond to Ramsey's concerns about our access to probabilities as I do below, and I welcome their attempts to develop similar responses. One worry with this suggestion, however, is that if all we have are non-factive seemings then there is no necessary connection between epistemic justification and objective probability. For those who accept a conception of epistemic justification as involving reasons that are *both* internally accessible *and* objectively truth-conducive, a mere seeming that some probability relation obtains won't do for justified inference.

3. Interpretations of Epistemic Probability

One of the major challenges facing a rationalist account of abduction is to provide an interpretation of probability compatible with our knowing statements of probability *a priori*. I will be concerned in this section with how to understand probability. To get a sense of the challenge, consider the rationalist who favors the abductivist response to skepticism regarding the external world. Suppose I accept this sort of rationalism. I grant that it is possible that I am the victim of a new evil demon world, but I claim that, in light of my experiential evidence and what I can draw from it, I am justified in believing that this world is *not* a demon world. I might naturally claim: it is *possible* but *improbable* that I am an evil demon victim, and this is because (i) I have such-and-such experiences, and (ii) my having these experiences is best explained by, and hence makes highly probable, or at least more probable than not, that the real world hypothesis (RWH) holds—i.e., that my beliefs are more or less correct with respect to the spatial features of objects in my local environment. And the latter probability claim is, according to the rationalist, knowable or justifiable *a priori*, and it holds true even *if* it turns out, unbeknownst to me, that I am actually in an evil demon world or that RWH is false in some other way.

But how, then, should we understand the claim of probability? First, suppose we understand it as some claim about *actual frequency*, the frequency of beliefs of a certain kind that are *actually* true in this world. The claim then is roughly that most, or at least more than half, of the actual (past, present, and future) cases of perceptual belief (beliefs based on sensory or perceptual experience) turn out to be true. In that case the probability claim would be false in a demon world, since the external world beliefs are all false in that world. We need some understanding of probability according to which such experiences make the RWH probable *even in a demon world*, and the actual frequency interpretation does not allow for that. The same holds for ordinary and scientific inferences that assume the falsity of external world skepticism: if justified belief requires that our observations or empirical evidence make scientific hypotheses probable in the actual frequency sense, then all sorts of scientific hypotheses that we apparently have excellent evidence to accept would be unjustified if our evidence turns out to be misleading and such hypotheses are never or very rarely true in this world.

It's worth noting that this is not a problem about our *access* to such probabilities; the problem, rather, is just that the actual frequency interpretation of probability, if it is intended as an interpretation of *epistemic* probability, is vulnerable to counterexamples, for there are intuitively epistemically justified beliefs that are not probable in this sense. Necessary falsehoods provide clear cases: it is possible for one to have a justified belief in mathematics or logic that turns out to have a probability of 0 on the actual frequency interpretation. Any who hold that epistemically justified beliefs must be probable in some sense, and this includes externalists of various sorts, must have some alternative way of thinking about this probability, something other than the actual frequency interpretation. It is for this reason that various contemporary epistemological theories that wish to retain some sort of "truth-connection" turn to a notion of probability that employs subjunctive conditionals, holding, for example, that only those beliefs that would usually be true under such-and-such (possibly false) conditions are epistemically justified.

The rationalist about abduction has the *additional* problem that actual frequencies don't seem to be the sort of thing that is even in principle accessible to us *a priori*. These actual frequencies are clearly *contingent*, not *necessary*, and so cannot be justified *a priori*. The same holds if we understand probability in terms of truth in nearby possible worlds, or in terms of subjunctive conditionals regarding what *would* hold under certain conditions: these too are contingent, and so cannot be justified *a priori*.[13]

The rationalist might prefer a *modal frequency* interpretation of probability. On this interpretation, the relevant probabilities are limiting frequencies across all possible worlds rather than just in the actual world or similar, "nearby" possible worlds. Worlds where induction and abduction fail systematically are, as BonJour puts it, "quite rare and unlikely within the total class of possible worlds, which would make the claim that the actual world is not such a world itself highly likely to be true" (1998: 209). BonJour

---

[13] This is to assume that we cannot come to know contingent truths *a priori*, and it may turn out that there are some interesting counterexamples to this. (Turri 2011 provides one of the best counterexamples.) While I cannot discuss these examples here, they do not suggest that the rationalist should embrace these subjunctive analyses of epistemic probability. Even if such cases of contingent *a priori* knowledge or justification are genuine, it's arguable that we can arrive at them only via a grasp of a *necessary* truth that these contingent truths are *probable*. This will make better sense after we have discussed the Keynesian view of probability.

is aware that this talk of the "rarity" of certain sorts of worlds in the class of all possible worlds involves a controversial assumption:

This way of putting the matter assumes in effect that it is possible to make sense of the relative size of classes of possible worlds, even though both those classes and the total set of possible worlds are presumably infinite. But I have no space to go into the issues surrounding this assumption and must be content here with saying that its intuitive credentials in other cases (e.g., the claim that there are twice as many positive integers as even integers) seem to me strong enough to make it reasonable to construe the difficulties as problems to be solved and not as insuperable objections.   (1998: 209, n. 24)

However, as Beebe (2009) points out, this interpretation yields frequencies that are undefined in the absence of some ordering of the members of these infinite sets or classes, and unless we can see how this ordering can be determined *a priori*, we lack a workable interpretation of probability that can serve the rationalist's purposes. Beebe explains:

The critical difficulty is that there is no privileged or natural ordering of worlds to which we can appeal. Yet if his [BonJour's] claims about limiting frequencies are not relativized to any particular sequence(s) of worlds, the frequencies in question will be undefined and hence his claims about them cannot be true.   (Beebe 2009: 630)

Consider a simple example for illustration: what is the frequency of even numbers in the set of all natural numbers? If the set were *finite*—if, for example, we consider the ratio of even numbers that are less than a thousand to all numbers that are less than a thousand, then there is a clear way to determine this frequency or ratio. Simply divide the number of even numbers less than a thousand by the number of natural numbers less than a thousand. The problem of course is that the set of evens and the set of naturals both have *infinite* members, and so determining the frequency of evens is a tricky matter: it seems that we can make sense of frequency in this case only if there is already some way that the members are ordered, something we didn't need to worry about in the finite case. Under one ordering, the natural ordering of "greater than" (0, 1, 2, 3 . . .) the limiting frequency of even numbers would be one half, since the frequency of even numbers approaches one half as the series approaches infinity. But other orderings are possible. Even numbers might be more spaced out than odd ones, yielding much lower frequencies (e.g., 1, 3, 5, 7, 2, 9, 11, 13, 15, 4 . . . which yields a limiting frequency of 1/5). It should be easy to see the problem as it occurs for sets of possible worlds: if the set of counter-inductive worlds, or worlds where abductive forms of inference fail regularly, or worlds controlled by an evil demon are infinite in number, then it seems that their frequencies are undefined without some sort of ordering. If we are offered no indication of how to impose a particular ordering *a priori*, then we lack an adequate rationalist account of our justification for claiming that such worlds are rare or improbable.

How, then, could we understand the epistemic probabilities so that they can be known *a priori*? What the rationalist requires is some notion of probability that holds *necessarily*. The actual and subjunctive frequency interpretations won't do since they are only contingent, and the modal frequency interpretation won't do because it is undefined in the absence of a particular ordering, an ordering that would have to be justified *a priori*.

On the *Keynesian* interpretation, epistemic probabilities are "internal relations" between propositions in the sense that they are relations that depend on the propositions themselves and nothing else. Entailment is an internal relation between propositions in this sense: if an entailment relation holds between two propositions, then it holds of necessity; that the relation holds depends only on what the propositions are, and not in any way on other factors such as the state of the external world (or nearby possible worlds) or the state of the believer. On the Keynesian view, epistemic or logical probability is also an internal relation between propositions—indeed, the case of entailment might be understood as the limit of the "making probable" relation between propositions, a case where one proposition makes another probable to the highest possible degree. Fumerton (1995: ch. 7) has argued that internalists who want to avoid skepticism should accept a Keynesian theory of probability and hold that the probability relations between our ultimate evidence and our inferred, external world beliefs hold necessarily.[14]

In an excellent discussion of the Keynesian theory, and despite finding much to recommend it, Bertrand Russell (1948) ultimately rejects the Keynesian theory of probability. His central dissatisfaction with it seems to be that it fails to secure a connection to the actual truth of our beliefs, and only some sophisticated version of a frequency view will do. Our evidence might make a proposition probable in the Keynesian sense, but this yields no necessary connection to the truth. So it looks like we lose the "truth-connection" that is so dear to many epistemologists. However, as Fumerton argues (1995: 200–3), the Keynesian can respond that there is indeed a necessary connection: Necessarily, if our evidence E makes a proposition P probable and E holds, then P is *probably* true. It is possible that our inferred beliefs are never true, but they may yet remain *probable*, and a necessary connection between our evidence and their *probable* truth remains.

While I accept a Keynesian, or logical interpretation of probability, my aim is not to defend the specific view that Keynes himself held. Indeed, I am inclined to accept certain claims that I am not sure Keynes would accept, though it would take more research to determine whether or not this is so. First, while Keynes, and Ramsey in his interpretation of Keynes, seems to assume that there are probability relations between every pair of propositions, I see no clear reason to accept this; and even if there are (and here

---

[14] In many cases, when we say that P makes probable Q we might mean that this probability relation holds *given some empirical background information*. That my car is not starting this morning may make probable that the battery is dead, but that only holds given a background of empirical information about cars and batteries. For the traditional rationalist, in line with the argument of the previous section, we must have access to some probability relations between propositions that do not depend on further background information. So, like Keynes, we are interested in whether probability relations hold between propositions independently of any further background information.

at least Keynes would agree), there is no clear reason to think that we have access to all these relations. Second, in addition to having access to some probabilistic relations between propositions, I think we can have *a priori* access to the fact that certain propositions are highly likely or unlikely, and not just to relations between propositions. I will say more about both of these points below.

4. Ramsey's Objection to Keynesian Probability

Let us turn now turn to the skeptical concerns regarding our *a priori* access to Keynesian probabilities.

But let us now return to a more fundamental criticism of Mr. Keynes' views, which is the obvious one that there really do not seem to be any such things as the probability relations he describes. He supposes that, at any rate in certain cases, they can be perceived; but speaking for myself I feel confident that this is not true. I do not perceive them, and if I am to be persuaded that they exist it must be by argument; moreover I shrewdly suspect that others do not perceive them either, because they are able to come to so very little agreement as to which of them relates any two given propositions.   (Ramsey 1926: 161)

Ramsey thus doubts that there are any such things as Keynesian probability relations, and there seem to be two main sources of this doubt. The first is a phenomenological one: speaking for himself, he feels confident that he does not perceive them. The second concern is that there is so little agreement about what probability relations hold between any two given propositions that we should be skeptical that any subjects have access to probability relations. I want to argue that there is in fact plenty of agreement on some simple cases, and while there is a sort of philosophical disagreement that persists, this is not surprising and is not a problem unique to the Keynesian view.[15]

These two concerns are related, but let us begin with the second. It may help to distinguish two ways to understand the nature of this disagreement. We might understand the disagreement to be about, on the one hand, the claim that some proposition makes another probable, and on the other hand, the claim that in some cases I perceive or am directly aware of an internal probability relation between propositions. Suppose it is the latter. I admit that there is little agreement on this matter. But this is no surprise, and while disagreements between philosophers may be epistemically significant and require us to lower our confidence or even revise our stance in some cases, this is not a problem peculiar to the Keynesian view. To see the point, it may help to compare a parallel distinction in the debate on the nature of perception between the claim that we perceive certain ordinary objects, and the claim that what is involved in our perception

---

[15] One might be inclined to think that the widespread agreement regarding some cases is due to shared background knowledge. However, I am not arguing that we have access to Keynesian probabilities *merely* because there is widespread agreement on simple cases. My main task in this section is to respond to objections to the claim that we have such insights, including the objection from lack of agreement. I do want to motivate the idea that we have such insights, but that depends on phenomenological and theoretical considerations, and not merely on the existence of significant agreement on simple cases.

of these objects is a direct awareness of a sense datum, a way of appearing, an abstract sensible profile or universal, an external world object, or the view that there really is no genuine "awareness" at all, and perhaps still other views. There is substantial agreement that we do perceive all sorts of objects, but very little philosophical agreement as to what sort of awareness perception involves, of what sorts of entities, belonging to what ontological categories. The fact that there is substantial disagreement in such debates may raise serious questions about the epistemic significance of disagreement in philosophy, but these epistemological and meta-philosophical problems are not specific to particular theories of perception. The same is true of the claim that such irreducible internal relations do or do not exist, or that we can or cannot perceive such relations between propositions.

It is worth adding at this point that using the term "perception" to describe our access to relations of probability can be misleading, even though proponents of Keynesian probability, including Keynes himself, have used that term. The problem is that the term suggests a relation that is similar to that of sensory perception, which many hold to involve or require a causal relation of some sort between the subject and the object, whether or not that causal relation is taken to be part of the very analysis of perception. Abstracta, such as universals, propositions, and relations of entailment or probability between propositions, are not causes and do not have causal powers, so it seems that we cannot stand in the very same sort of perceptual relation to such entities. This is not to deny that there is a sense in which they may be said to "influence" our cognition or what we think in some way, but the abstracta influence our states by partly constituting them rather than in the standard causal way, much as on classical views regarding our access to phenomenal or experiential qualities, these qualities partly constitute our states of direct awareness or "acquaintance" rather than causing them.

Ramsey is confident that he does not "perceive" probability relations. Nor does his position seem to depend on the use of the latter term; he probably would have also denied that he is ever directly aware of or acquainted with, or has any direct access to, such probability relations. And it is possible that many who reject the Keynesian view do so because they find themselves agreeing with Ramsey's phenomenological report. Indeed, while Fumerton has argued that the internalist who wishes to escape radical skepticism should embrace Keynesian probability relations, he too denies that he is aware of such relations:

The Keynesian strategy described above for avoiding skepticism is, I suspect, dialectically impregnable... For me, it has only one drawback. I cannot quite bring myself to believe that I am phenomenologically acquainted with this internal relation of making probable bridging the problematic gaps... And in the end, I strongly suspect that the probability relation that philosophers *do seek* in order to avoid skepticism concerning inferentially justified beliefs is an illusion.   (1995: 218)

Like Fumerton, and apparently Ramsey too, I think phenomenological appeals can carry significant weight. Indeed, I take them to be indispensable in philosophy. But there are ways we can push back. First, we should admit that phenomenological investigations can be difficult and that our phenomenological claims are susceptible to the influence of our beliefs and judgments. (This of course goes both ways: it is possible that my own confidence is simply the product of bias, or is a mere projection of what I already accept.) While I do take phenomenological appeals to be indispensable, I'm inclined to think that here and with other fundamental epistemological issues, we must evaluate theories on both phenomenological and broadly theoretical grounds.

This is important, and worth further discussion. Some might be inclined to begin with beliefs and inferences that are justified from the point of view of common sense, and then tailor one's fundamental epistemological principles so as to deliver the commonsensical results we want. This "epistemological commonsensism" utterly fails to take skepticism about common sense seriously; it just assumes that skepticism must be false. Some phenomenology is needed to ensure we are not, in the interest of avoiding skepticism, just making things up. If we hope to avoid skepticism without simply assuming it is false—to avoid skepticism while taking the skeptical challenge seriously to some extent—we need some sort of access to what our experiences are like, and also some sort of access to connections between our experiences and what we hope to infer from them. And we must do some phenomenology if we want to check on our ability to have these sorts of access. But, on the other hand, I do not want to deify phenomenological investigation, or pretend that by directing the inner eye in a certain direction I can tell conclusively and without any doubt that I have or do not have access to just what is needed.

To be fair to Ramsey, while he does say that he does not perceive Keynesian probability relations, his claim about the lack of agreement seems directed at the level of examples of probability claims and not at philosophical claims about whether and how we might have access to them. But part of the reason Ramsey seems to deny that there are such examples may have to do with the particular examples he focuses on, and on his assessment of those examples. I will end up agreeing to some extent with Ramsey that many examples of an apparent intuitive grasp of probabilities are less straightforward than Keynes' view suggests, but deny that this means we should abandon the Keynesian view.

5. *A priori* Probabilities, and the Connection to Abduction

Now, speaking for myself, I find it plausible that I am aware of relations of making probable between some propositions. There do seem to be some examples where I have stable intuitions about probability claims, and where I also seem to grasp the relations in question *a priori*. The latter part of this phenomenological appeal—that it not only seems to me that the relations hold but that I also seem to grasp the relations *a priori*, simply by considering the relevant propositional contents or properties involved—is important, for it suggests, even if it does not prove, that the intuitions are not due to some prior or background beliefs. As with all other arguments that depend at least in part on phenomenological appeals, I cannot literally show others what, it seems to me, I am aware of. But what I can do is present and discuss some central examples which are very intuitive, and which, if supplemented with a discussion of how one might be misled to deny that there are any such examples, at least can open one's mind to the possibility that one is in fact aware of such relations.

Let us begin with examples relevant to *deductive* inference, for that may open us up to the possibility of awareness of relations relevant to non-deductive inference. It is intuitively and phenomenologically plausible that we can grasp relations of entailment between propositions. (Recall that, for proponents of the Keynesian or logical conception of probability, entailment is the upper limit of the making-probable relation.) To offer just a few simple examples, I can infer from the fact that some figure is a circle that it has no corners, and from the fact that some figure is a triangle (has three angles) that it is also a trilateral (has three sides). In these cases, what justifies me in the inference is just that I grasp a certain logical relation between these facts: I grasp a logical relation of incompatibility or exclusion between something's being a circle and its having corners; I grasp a logical relation or inclusion or entailment between something's being triangular and its being trilateral. Phenomenologically, the grasp does not seem to be derivative or indirect: it not only seems that circles have no corners, and triangles are trilateral, but I grasp *why* this must be so, and it is in virtue of this that I am justified in inferring from the proposition that something is a circle that it does not have corners, or from the proposition that it is a triangle that it is a trilateral. We can also reflect on our own experiences and that of our students in logic classes. We are able to grasp entailment relations between particular premises and conclusions in some very simple cases with ease, and do so without relying on any principle of inference—indeed, logic students can often just "see" the connection between the relevant propositions prior to knowing anything about such principles.

Turning now to examples of relations of probability between propositions: Consider some simple examples from Euclidean projective or perspectival geometry, the sort of geometry that is involved in the technique artists use to determine how to represent objects in a drawing or painting. *Given* that certain shapes or figures in an approximately Euclidean three-dimensional space have two-dimensional projections, we can tell *a priori* what sorts of projections they are likely to have. A straight line segment will rarely project a point, but is highly likely to project a line of varying lengths. The projection of two straight lines the ends of which meet at an angle (an L-shaped vertex) is likely to be two straight lines whose ends meet at an angle, though the angle and size of the lines in the projection can change; in a small subset of cases (i.e., when both lines share a plane with the line of sight), the two connected lines will project a single line. Two straight detached lines (imagine two chopsticks or pencils) are likely to project two straight lines, but very unlikely to project lines that meet at their ends to form a corner. A circle or disc is unlikely to project a circle or line, and much more likely to project an ellipse. A square (like the square top of a table) is likely to project quadrilateral shapes, and relatively unlikely to project a line or a square. Spheres are highly likely to project circles, and cylinders less likely to do so. Thus, given that such-and-such Euclidean objects have projections, the projections are likely *a priori* to take such-and-such forms, and unlikely to take such-and-such other forms.[16]

Here is another example, involving color, given by Huemer (2016): suppose that a computer is programmed to place a random color at each position in a 1000×1000 grid. While it is *possible* that this produce a completely red grid, say, or the image of Donald Trump, or some other such meaningful or identifiable image, these results are all *highly improbable*. A random assignment of colors to each position or pixel in a 1000×1000 grid makes it highly probable that the result for the entire grid is just meaningless static.

Do these examples involve a direct awareness of probability relations between the relevant propositions? Perhaps not. At a first approximation, it seems that our grasp of the relevant probability relations are applications of the *principle of indifference* to the effect that when we have competing possibilities and no evidence favoring any one over another, we should assign an equal probability to all of them. If we wish to avoid notorious paradoxes in our employment of the principle of indifference, we will need to constrain its application in some way—to atomic states of affairs, perhaps, or to some "natural" partitioning of possibilities. However, perhaps we need not solve all paradoxes or difficulties regarding the application of the principle of indifference in order to be justified in its restricted application in some clear cases.[17] In the projective geometry and color examples just discussed, in the absence of any evidence to favor one possible assignment of one sort of projection rather than another, or one assignment of colors to positions or pixels in the grid over another, we can see *a priori* that they are equally likely. Given that such-and-such a shape has two-dimensional projections, and given no evidence to favor the claim that it is projected toward one particular direction rather than any other, it is equally probable that the shape is projected toward one region of the surrounding space rather than another. This is the fundamental insight, which is not an insight into a general and unrestricted principle of indifference that is taken to apply to all cases and is then applied to the particular case; it is a grasp that certain specific possibilities are equally likely given a specific hypothesis. Just as one does not need to grasp a general logical principle to see that an entailment holds between certain propositions, one need not grasp a general principle of indifference to see that these possibilities are equally likely given a particular hypothesis. One can then also see, *a priori*, that line segments are likely to project line segments of varying lengths, that two detached lines are unlikely to project two lines with ends that meet at a point to form an L-vertex, that circles or discs are likely to project ellipses, and so on. The fundamental insight in the color example is that given a random assignment

---

[16] For a more detailed discussion of such examples and their relevance to the abductivist response to skepticism, see Hasan (Forthcoming).

[17] Huemer (2016) motivates this as well.

of colors to each pixel or position, each possible distribution of colors over the entire grid is equally likely. One can then also see, *a priori*, that random assignments of color to each pixel are highly likely to yield meaningless static images. Whatever our concerns about a general and unrestricted principle of indifference, we seem quite able to grasp that in these cases the possibilities are equally probable, and be able to infer from these that some types of possibilities are highly probable, others highly improbable, and some much more probable than others.

I can know *a priori* that certain propositions are highly probable or highly improbable, and not just that some propositions make others probable or improbable. This follows trivially from the view that we have insight into certain probabilistic relations between propositions: if I see that P makes it highly probable that Q, then I can also see that $P \mathbin{\&} \sim Q$ is highly improbable, and that $\sim P \lor Q$ is highly probable. For example, I know that it is highly probable that the projection of a circle or disc is an ellipse, highly improbable that a 1000 x 1000 grid of pixels with randomly assigned colors produces a uniformly red grid, and so on. But it is not necessarily because I grasp relations between propositions that I grasp probabilities regarding single propositions. I grasp *a priori* that any particular experience I have has a very low *prior* probability, for I grasp that it is but one of indefinitely many *a priori* possible experiences. I also grasp *a priori* that any very specific hypothesis about the nature of this world has a very low prior probability. For example, I grasp *a priori* that it is highly improbable that there are currently exactly 10^{25} stars in this world.

Examples of this sort are, I believe, crucial for rationalism about abduction or inference to the best explanation, for they suggest that we can see, *a priori*, that certain hypotheses make some sorts of experiences or observations highly probable, and others highly improbable, and we can compare different hypotheses on this basis. For example, as just discussed, certain sorts of two-dimensional figures are more likely to result from certain objects as opposed to others; and certain images on a color grid are more likely given certain mechanisms or processes of color assignment as compared to others. But how are we to get from here to a rationalist justification of inference to the best explanation? The basic idea is that the proponent of a Keynesian or logical conception of probability can use this *a priori* grasp of probabilities and Bayes' Theorem (which is widely accepted and can also be justified *a priori*) to guide one's selection of hypotheses. According to Bayes' Theorem, for any hypothesis H and evidence E:

$$P(H \mid E) = \frac{P(H) \times P(E \mid H)}{P(E)}.$$

Suppose that we are trying to determine the posterior probability of H, $P(H|E)$, where E is the total evidence. What should we take into consideration? (i) Well, the probability of E given H, $P(E|H)$ is intuitively relevant; the higher this probability, all else equal, the higher the posterior probability of H. The intuitive idea here, roughly, is that the better a hypothesis accounts for the evidence, the more likely it is that the hypothesis is true; so, other things being equal, if one hypothesis accounts better for some evidence than another, the more likely it is than the other. For example, if my monitor displays an image of Donald Trump, I will take it to be extremely improbable that the image was generated by a random assignment of colors to the pixels, for the probability of such an image is much too low given that hypothesis. Or consider the following more general case: if there is an identifiable pattern or strong correlation of some sort, the probability that this pattern or correlation has some explanation (other than chance or coincidence) is probable.[18] We commonly say when we observe persistent or extensive correlations that "there must be an explanation," but the "must" here is not because we think a lawful explanation is *necessary*, but because we take it to be so probable that we would be very surprised if there were no explanation, and more surprised the stronger the correlation.

But the prior probabilities $P(E)$ and $P(H)$ are relevant too. (ii) The more improbable, *a priori*, that the evidence is, the more probable is the hypothesis that accounts for it. The intuitive idea here is that an explanation that accounts for a surprising bit of evidence is better, all else equal, than one that accounts for something not very surprising. (iii) Finally, other things being equal, the more probable is the *a priori* probability of a hypothesis, the more probable its posteriori probability. If, for example, one hypothesis is *a priori* less probable compared to another because it is much more specific or complex, then, other things being equal, its posterior probability will be lower than the other.

I have been motivating the idea that we can be aware of some of the probabilities that enter into these assessments—not that we can arrive at very precise probabilities, but that we can have at least a rough sense of the probabilities, or be able to determine that some propositions are more probable than others, or make certain other propositions more probable than others. Our *a priori* grasp of probabilities and Bayes' Theorem can thus be used to guide our selection of hypotheses, or to determine that certain hypotheses provide better explanations than others do.

I must remind the reader that my task here is not to offer anything approaching a rigorous defense of abduction or inference to the best explanation. Rather, I am attempting to motivate the idea that we can grasp the truth of some probability claims that seem to depend essentially only on the propositions (or theories) themselves, and not on the way the actual world, nearby worlds, or space of possible worlds is like. I have attempted to offer probability claims of this sort, claims regarding which there is likely to be widespread agreement. Many will disagree that we can know such principles *a priori*, or that we can be aware of these probabilities, but again, that there is this sort of disagreement is not surprising in philosophy, and does not seem to be a problem specific to rationalism about abduction or probabilistic inference.

---

[18] BonJour (1998: ch. 7) relies on a similar principle in his attempt to provide an *a priori* justification of induction, and takes it to be evident *a priori*. It is important that the correlation is strong in the relevant sense. It may be that whenever I'm typing, some butterfly in China is flying, but that would not make a causal explanation probable. For one, it may be that even when I am not typing some butterfly in China is flying. Something like a strong statistical co-variation in observed cases is needed. Moreover, it would not necessarily follow that there is a causal explanation, but only that such an explanation is *probable*.

Let us consider one more kind of example: simple cases of enumerative induction. That the first fifty balls I have drawn from this urn are white makes highly probable that all or almost all the balls in the urn are white. Moreover, the intuition that it is highly probable that all or almost all the balls in the urn are white given that the first fifty draws were white remains even if, as a matter of fact, most balls in the urn are black. Should I later discover that the urn contained 100 black balls and fifty white ones, I would still have said: yes, but given my evidence that was a very improbable result! Some may be inclined to think that we have a direct grasp of the probability relation between the relevant propositions in such cases. While I do believe that this justification is *a priori*, I'm inclined to think it is derivative. When I ask *why* it is that I should expect more white balls, part of the answer is that there must be some explanation (more precisely, it is highly probable that there is some explanation) why I kept getting white balls on the first fifty draws, and an explanation that takes this sample to be representative is a better explanation, for it does not require that there is some additional change or interfering condition responsible for the sample's being unrepresentative (and so it has a higher *a priori* probability). The full story is more complicated, of course, but the justification of induction thus seems to me to depend on a prior grasp of the sorts of probabilities than enter into a comparison of hypotheses, in line with Bayes' Theorem.[19] While these apparently simple examples turn out to be more complex, the Keynesian seems to me to have the resources to provide an intuitive and potentially compelling justification of induction, one that does not simply take a principle of enumerative induction to be basic or foundational, but still regards it as *a priori*, depending on explanatory considerations.

Ramsey admits that there is significant agreement in some cases: "we all agree that the probability of a coin coming up heads is 1/2, but we can none of us say exactly what is the evidence which forms the other term of the probability relation about which we are then judging" (1926: 162). It might be tempting to claim that the other proposition in the relation is that the coin is symmetrical, but though symmetry is relevant, the justification is more complex than that, involving some simplifying assumption that heads and tails are the only two possibilities (disregarding the possibility that the coin lands on its edge) and an application of the principle of indifference to these two possibilities. Still, it's not clear why something like this cannot be used to justify the assignment of probability 1/2 in this case. As a matter of fact, we rely on much more to justify this probability assignment than something like a principle of indifference, including our past experiences of coin flips and background knowledge of our physical world, but it may yet be the case that we would be justified, to some degree, in assigning a probability of 1/2 in the absence of this empirical knowledge. Moreover, even if this example is a problematic case for the Keynesian, the other examples discussed above seem relatively straightforward and enjoy widespread agreement.

---

[19] See BonJour (1998: ch. 7), for an a priori justification of induction along these lines, though one that does not make explicit reference to Bayes' Theorem.

Aside from questioning the simplicity or straightforwardness of examples that enjoy significant agreement, the other main concern that Ramsey raises is that when "we take the simplest possible pairs of propositions such as 'This is red' and 'That is blue' or 'This is red' and 'That is red', whose logical relations should surely be easiest to see, no one, I think, pretends to be sure what is the probability relation which connects them" (1926: 162). First, there is no reason, as far as I can see, that the Keynesian should insist that there be a probability relation between every pair of logically independent propositions. Second, even where such a probability relation exists, there is no reason to think that we must have direct access to it, even when the propositions related are very simple. It may help to note that in some cases of entailment even between very simple propositions, we have extreme difficulty determining whether there is an entailment without a complex proof, if at all. The field of mathematics is filled with examples of rather simple propositions entailed by other simple propositions, but where the entailment is notoriously difficult to prove. But why, then, should we think any different in the case of probability relations, if these are internal relations between propositions?

6. Conclusion

Rationalism about abduction is dialectically attractive for those who want to avoid radical skepticism, but many share Ramsey's concern that we just don't have access to the Keynesian probability relations the rationalist requires. My main task here is the modest one of rescuing rationalism about non-deductive inference and non-deductive principles, including abduction, from the latter objection. In certain respects, Ramsey is right: there aren't all these internal probability relations between all pairs of propositions, or if there are, we aren't aware of them. Moreover, I agree with Ramsey that many examples that enjoy significant agreement are less simple or straightforward upon reflection. However, I have argued that these worries don't affect the view that we can have *a priori* awareness of probabilities of the Keynesian or logical sort, probabilities that are relevant to abductive inference. These are not the only objections Ramsey and others have raised to the Keynesian view, and much work remains to be done to make this interpretation of epistemic probability rigorous and workable.[20] But if I am right, then at the very least such arguments from phenomenology and disagreement are rather weak, and whatever other serious obstacles remain in the way, these are not among them.

Acknowledgments

Some of these ideas were presented at the 2015 Orange Beach Epistemology Workshop, and I am grateful to the participants for their feedback. Thanks especially to Ted Poston and Kevin McCain, who also offered extremely helpful comments on a draft.

[20] See Russell (1948) for an excellent discussion of Keynes' work and formal as well as other problems facing his interpretation of probability.

References

Beebe, James (2009). "The Abductivist Reply to Skepticism." *Philosophy and Phenomenological Research* 79 (3): 605–36.

Bengson, John (2015). "Grasping the Third Realm." *Oxford Studies in Epistemology* 5.

Bergmann, Michael (2006). *Justification without Awareness: A Defense of Epistemic Externalism.* New York: Oxford University Press.

BonJour, Laurence (1985). *The Structure of Empirical Knowledge.* Cambridge, MA: Harvard University Press.

BonJour, Laurence (1998). *In Defense of Pure Reason.* Cambridge: Cambridge University Press.

BonJour, Laurence (1999). "Foundationalism and the External World." *Philosophical Perspectives* 13 (s13): 229–49.

BonJour, Laurence (2003). "A Version of Internalist Foundationalism." In L. BonJour and E. Sosa (eds), *Epistemic Justification: Internalism vs. Externalism, Foundations vs. Virtues,* 5–96. Malden, MA: Blackwell.

BonJour, Laurence (2005). "In Defense of the a priori." In M. Steup and E. Sosa (eds), *Contemporary Debates in Epistemology,* 177–84. Malden, MA: Blackwell.

Boyd, Richard (1999). "Kinds as the 'Workmanship of Men': Realism, Constructivism, and Natural Kinds." In J. Nida-Rümelin (ed.), *Rationalität, Realismus, Revision: Proceedings of the Third International Congress, Gesellschaft für Analytische Philosophie.* Berlin: de Gruyter.

Casullo, Albert (2003). *A Priori Justification.* Oxford: Oxford University Press.

Chudnoff, Elijah (2013). "Awareness of Abstract Objects." *Noûs* 47 (4): 706–26.

Douven, Igor (2011). "Abduction." In E. Zalta (ed.), *The Stanford Encyclopedia of Philosophy,* Spring, <http://plato.stanford.edu/archives/spr2011/entries/abduction/>, (accessed December 20, 2016).

Fumerton, Richard (1995). *Metaepistemology and Skepticism.* Lanham, MD: Rowman and Littlefield.

Hasan, Ali (2011). "Classical Foundationalism and Bergmann's Dilemma for Internalism." *Journal of Philosophical Research* 36: 391–410.

Hasan, Ali (2014). "Review of Albert Casullo, *Essays on A Priori Knowledge and Justification.*" *Notre Dame Philosophical Reviews,* January, <http://ndpr.nd.edu/news/45701-essays-on-a-priori-knowledge-and-justification/>.

Hasan, Ali (Forthcoming). "Skepticism and Spatial Objects." *International Journal for the Study of Skepticism.*

Hasan, Ali and Richard Fumerton (2014). "Foundationalist Theories of Epistemic Justification." In E. Zalta (ed.), *The Stanford Encyclopedia of Philosophy.*

Huemer, Michael (2016). "Serious Theories and Skeptical Theories: Why You Are Probably Not a Brain in a Vat." *Philosophical Studies* 173 (4): 1031–51.

Moser, Paul K. (1989). *Knowledge and Evidence.* Cambridge: Cambridge University Press.

Poston, Ted (2014). *Reason and Explanation: A Defense of Explanatory Coherentism.* Basingstoke: Palgrave Macmillan.

Psillos, Stathis (1999). *Scientific Realism: How Science Tracks Truth.* London: Routledge.

Ramsey, F. P. (1926). "Truth and Probability." In F. P. Ramsey, *The Foundations of Mathematics and Other Logical Essays,* 156–98, ed. R. B. Braithwaite. London: Kegan, Paul, Trench, Trubner and Co.; New York: Harcourt, Brace and Company.

Russell, Bertrand (1948). *Human Knowledge and Its Limits*. New York: Simon and Schuster.

Sober, Elliott (1988). *Reconstructing the Past: Parsimony, Evolution, and Inference*. Cambridge, MA: MIT Press.

Thurow, Joshua C. (2009). "The a priori Defended: A Defense of the Generality Argument." *Philosophical Studies* 146 (2): 273–89.

Turri, John (2011). "Contingent A Priori Knowledge." *Philosophy and Phenomenological Research* 83 (2): 327–44.

Vogel, Jonathan (1990). "Cartesian Skepticism and Inference to the Best Explanation." *Journal of Philosophy* 87 (11): 658–66.

Vogel, Jonathan (2008). "Internalist Responses to Skepticism." In John Greco (ed.), *The Oxford Handbook of Skepticism*. Oxford: Oxford University Press.

PART IV

Inference to the Best Explanation and Skepticism

11

Does Skepticism Presuppose Explanationism?

James R. Beebe

1. The Explanationist Response to Skepticism

In the contemporary epistemological literature, skeptical challenges most often employ skeptical hypotheses that depict situations that are much like what we take our normal circumstances to be in certain respects but in which we fail to have knowledge.[1] These challenges are typically represented as arguments of the following form, where "*p*" is some proposition about the external world that we ordinarily take ourselves to know and "*SK*" is a skeptical hypothesis:

(1.1)   If I know that *p*, then I know that not-*SK*.
(1.2)   I do not know that not-*SK*.
(1.3)   Therefore, I do not know that *p*.

Premise (1.2) is usually supported by considerations that purport to show that one's current evidence somehow fails to favor *p* over *SK*.

A common response to radical skeptical challenges to our knowledge of the external world has been to maintain that, while both commonsense and skeptical explanations of our sensory experiences are in some sense equally consistent with our sensory experience, there are explanatory reasons (e.g., simplicity, coherence, explanatory power, conservatism) for favoring commonsense explanations over skeptical ones. Bertrand Russell (1912, pp. 22–3), for example, writes:

There is no logical impossibility in the supposition that the whole of life is a dream, in which we ourselves create all the objects that come before us. But although this is not logically impossible, there is no reason whatever to suppose that it is true; and it is, in fact, a less simple hypothesis, viewed as a means of accounting for the facts of our own life, than the common-sense hypothesis that there really are objects independent of us, whose action on us causes our sensations.

[1] Thanks to Kevin McCain, Ted Poston, David Sackris, and audience members at the University at Buffalo for helpful comments and feedback on an earlier draft of this chapter.

More recently, this kind of response to skepticism has been defended by William Lycan (1988), Paul Moser (1989), Jonathan Vogel (1990b, 2004), Laurence BonJour (1998, 1999, 2003), and Kevin McCain (2014, ch. 6, Forthcoming).[2] Explanationist responses to skepticism are intended to apply not only to simple, ordinary propositions about medium-sized bits of the external world (e.g., "I have hands") but also to wide-ranging commonsense propositions that concern large-scale features of the world (e.g., "I am not a brain-in-a-vat" or "our sensory experiences are in general caused by objects having roughly the characteristics we commonsensically take them to have").

Despite the degree of visibility this class of response has enjoyed, it has always been viewed with skepticism by the epistemological community because of concerns about the epistemic bona fides of explanatory reasoning. The most common concern about appeals to explanatory virtues has been forcefully articulated by Bas van Fraassen (1980, p. 87):

Judgements of simplicity and explanatory power are the intuitive and natural vehicle for expressing our epistemic appraisal. What can an empiricist make of these other virtues which go so clearly beyond the ones he considers preeminent? There are specifically human concerns, a function of our interests and pleasures, which make some theories more valuable and appealing to us than others. Values of this sort, however ... cannot rationally guide our epistemic attitudes and decisions. For example, if it matters more to us to have one sort of question answered rather than another, that is no reason to think that a theory which answers more of the first sort of question is more likely to be true.

As Lycan (2002, p. 426) notes, in the face of van Fraassen's challenge, the explanationist needs to show that the explanatory virtues are not "merely practical bonbons of no specifically epistemic, truth-conducing value" but are instead "genuine reasons for accepting a theory as more likely to be true than is a competitor that lacks them."

In this chapter, I argue that skeptical challenges that employ skeptical hypotheses presuppose central explanationist tenets and thus that the force of compelling skeptical challenges can be understood only if the explanatory features of skeptical hypotheses are more than "merely practical bonbons" of no epistemic significance. I contend that careful consideration of the explanatory features of skeptical hypotheses paves the way for a better understanding of skeptical challenges in general and a better understanding of dreaming skepticism in particular—a type of challenge that epistemologists have never been able to explain with any satisfaction. I conclude that an appreciation of these facts should raise one's estimation of the strength of explanationist responses to skepticism.

---

[2] Cf. Beebe (2009) for a fairly comprehensive review of this class of responses to skepticism.

2. The Quasi-Logical Approach to Skeptical Challenges

The canonical approach to skeptical challenges rests upon the following assumptions:

SH1. In order for a skeptical hypothesis, *SK*, to raise a significant skeptical challenge to *S*'s putative knowledge that *p*, *p* and *SK* must be incompatible.

SH2. In order for a skeptical hypothesis, *SK*, to raise a significant skeptical challenge to *S*'s putative knowledge that *p*, it must be logically or metaphysically possible for *p* to be false.

SH3. In order for a skeptical hypothesis, *SK*, to raise a significant skeptical challenge to *S*'s putative knowledge that *p*, it must be logically or metaphysically possible for *SK* to be true.[3]

Call the conjunction of SH1 through SH3 "the quasi-logical approach to skeptical challenges." I will argue that each component of the quasi-logical approach is mistaken and that their failures point the way toward a more satisfying explanationist alternative.

Commitment to the quasi-logical approach can be seen in the two most common epistemic principles that are used to articulate radical skeptical challenges. Premise (1.1) in the skeptical argument above is usually taken to be an abbreviated substitution instance of the following epistemic closure principle:

ECP1. If *S* knows that *p*, and *S* knows that *p* entails *q*, then *S* knows (or is in a position to know) that *q*.

Closure principles connect our knowledge (or lack of knowledge) of the falsity of skeptical hypotheses—as represented in (1.2)—with our knowledge (or lack of knowledge) of ordinary propositions—i.e., (1.3). They do this via relations of known entailment obtaining between having hands and not being brains in vats or (alternatively) between being brains in vats and not having hands.

Skeptical arguments that employ underdetermination principles also assume an incompatibility between propositions we think we know and skeptical hypotheses that seek to undermine our knowledge. Consider, for example, the following underdetermination principle from Anthony Brueckner (1994, p. 830):

UP. If *S*'s evidence for believing *p* does not favor *p* over some hypothesis *SK* which *S* knows to be incompatible with *p*, then *S*'s evidence does not justify *S* in believing *p*.

If there is no incompatibility between a putatively known proposition and a skeptical hypothesis, UP is unable to explain how one might be unjustified in rationally preferring the one over the other. Thus, common epistemic principles like ECP1 and UP reflect

---

[3] My discussion of SH1 through SH3 here builds upon Beebe (2010).

the quasi-logical approach to skeptical challenges insofar as they explicitly incorporate conditions like SH1. Because discussions of ECP1 and UP also always focus on contingent propositions, they presuppose SH2 and SH3 as well.

Brueckner's (1985, 1994) influential reconstructions of what he calls "the canonical Cartesian skeptical argument" reflect an explicit commitment to SH1, SH2, and SH3:

[T]he skeptic's reasoning requires the notion of a *counterpossibility* to ϕ. If ψ is a logically possible proposition which is incompatible with ϕ (which logically implies $-\phi$), then ψ is a counterpossibility to ϕ. One counterpossibility principle which... the skeptic's reasoning might use is that if I know that ϕ, then I know that no counterpossibility to ϕ obtains.

(Brueckner 1985, pp. 89–90, emphasis in original)

Without multiplying examples beyond necessity, it should be clear to anyone familiar with the contemporary literature on radical skepticism that commitment to the quasi-logical approach can be seen most anywhere that skepticism is discussed.[4]

Despite the ubiquity of SH1, it can be easily shown to be false. G. E. Moore (1959, p. 245) vividly illustrated the fact that dreaming skeptical hypotheses need not be incompatible with what subjects believe with the following anecdote:

But, on the other hand, from the hypothesis that I am dreaming, it certainly would not follow that I am *not* standing up; for it is certainly logically possible that a man should be fast asleep and dreaming, while he is standing up and not lying down. It is therefore logically possible that I should both be standing up and at the same time dreaming that I am; just as the story, about a well-known Duke of Devonshire, that he once dreamt that he was speaking in the House of Lords and, when he woke up, found that he *was* speaking in the House of Lords, is certainly logically possible.

Thus, since dreaming skeptical hypotheses do not satisfy SH1 and yet clearly underwrite significant skeptical challenges to our beliefs, SH1 must be false. Furthermore, because ECP1 and UP are based upon the same notion of logical incompatibility between putatively known propositions and skeptical hypotheses, they fall short in their ability to explain the full range of skeptical challenges.

There is thus an inconsistency in contemporary discussions of skepticism. On the one hand, principles like SH1 are explicitly endorsed when epistemologists attempt to

[4] Cf. also, e.g., DeRose (1999), Pritchard (2002a, 2002b, 2005), and Greco (2008). Dretske's (1970, p. 1015) seminal formulation of the relevant alternatives approach to knowledge and skepticism also clearly presupposes SH1:

Suppose we assert that x is A. Consider some predicate, "B," which is incompatible with A, such that nothing can be both A and B. It then follows from the fact that x is A that x is not B. Furthermore, if we conjoin B with any other predicate, Q, it follows from the fact that x is A that x is not-(B and Q). I shall call this type of consequence a *contrast consequence*, and I am interested in a particular subset of these; for I believe the most telling skeptical objections to our ordinary knowledge claims exploit a particular set of these contrast consequences.

In spelling out what a contrast consequence or relevant alternative is, Dretske (1981, p. 371) writes, "Call the *Contrasting Set* (CS) the class of situations that are necessarily eliminated by what is known to be the case. That is, if S knows that P, then Q is in the CS (of P) if and only if, given P, necessarily not-Q."

articulate the structure and force of skeptical challenges. On the other hand, it is commonly acknowledged that these principles cannot be true. Thus, not only is it incorrect to argue that epistemic principles like ECP1 and UP (which are formulated as if SH1 were true) form the backbone of skeptical arguments, an exclusive focus on these principles makes it impossible for epistemologists to explain the force of dreaming skepticism—a point to which I will return in Section 4.

Conditions SH2 and SH3 can also be shown to be false by considering the case of belief in the existence of God, understood to be a necessary being. If SH2 were correct and God exists, it would not be possible to raise any skeptical challenges to someone's belief in God's existence simply because that being necessarily exists. For the same reason, if SH3 were true and God exists, no skeptical hypothesis that hypothesized the non-existence of God would be able to get off the ground. This kind of reasoning seems to involve a category mistake. The ontological status of a being is one thing, while the epistemic status of one's belief in such a being's existence is something else entirely. The necessary truth of a belief cannot by itself confer incontrovertible, skeptic-proof epistemic justification upon it. Whether a skeptical challenge based upon the Freudian notion that religious belief is a manifestation of wish fulfillment or the hypothesis that religious belief stems from an evolutionarily adaptive but epistemically substandard belief-forming mechanism can present a compelling epistemological challenge to religious belief seems independent of the alethic modalities of the propositions in question. SH2 and SH3, then, are false.[5] They unduly restrict the set of permissible skeptical hypotheses and the set of putatively known propositions that can be challenged to propositions that are contingent.

I conclude that the quasi-logical approach to skeptical challenges is thoroughly mistaken. Incompatibility between putatively known propositions and skeptical hypotheses is not necessary for the latter to pose a skeptical challenge to the former, and the alethic modalities of the propositions involved are irrelevant to the question of whether such a challenge can be successfully lodged.

3. The Explanationist Approach to Skeptical Challenges

To introduce a better way of understanding radical skeptical challenges, consider Fred Dretske's (1970, p. 1016) famous zebra case, in which you believe that the animal standing before you in the pen at the zoo is a zebra. The following propositions are all incompatible with what you believe:

(2.1)   The animal in the pen is not a zebra.
(2.2)   The animal in the pen is a lion.
(2.3)   The animal in the pen is a mule.
(2.4)   The animal in the pen is a mule cleverly disguised to look like a zebra.

---

[5] For further discussion of these conditions, cf. Beebe (2011).

Proposition (2.1) is the contradictory of the ordinary proposition you believe, while (2.2) and (2.3) are contraries of that proposition. None of these three propositions by themselves has what it takes to underwrite a compelling skeptical challenge to your belief. Only (2.4) does.

The reason why (2.4) counts as a skeptical hypothesis is that it satisfies the following explanatory constraints:

SH4.   In order for a skeptical hypothesis *SK* to raise a significant skeptical challenge to *S*'s putative knowledge that *p*, *SK* must explain *S*'s evidence for *p*.[6]

SH5.   In order for a skeptical hypothesis *SK* to raise a significant skeptical challenge to *S*'s putative knowledge that *p*, *SK* must explain how *S* could believe that *p* on the basis of *S*'s evidence and yet not know that *p*.

(2.1), (2.2), and (2.3) do not attempt to explain (or explain away) your evidence for believing the animal in the pen is a zebra. They are merely incompatible with the proposition you believe. Thus, what makes the difference between a genuine skeptical hypothesis and a merely conflicting proposition are the explanatory relations that obtain between your evidence—e.g., your perceptual evidence that the animal standing before you in the pen at the zoo is a zebra and your background beliefs about how zoos operate and how common it is for zookeepers to perpetrate hoaxes on zoo visitors—and the skeptical hypothesis in question.

Brain-in-a-vat skeptical hypotheses satisfy SH4 and SH5, as does each of the alternatives that Vogel (1990a, pp. 16, 20, 21) sketches in his well-known discussion of "car theft-type cases":

(3.1)   My car is now parked on Avenue A.
(3.2)   My car has been stolen and driven away from where it was parked.
(3.3)   George Bush is the current president of the United States.
(3.4)   George Bush has had a fatal heart attack in the last five minutes.
(3.5)   I can get a good hamburger at a luncheonette several blocks from here.
(3.6)   A fire has just broken out at the luncheonette several blocks from here.
(3.7)   The San Francisco Bay is spanned by the Golden Gate Bridge.
(3.8)   The Golden Gate Bridge was just demolished by a falling meteorite.

The even numbered members of this set do not merely conflict with their odd-numbered counterparts. They explain (albeit briefly) how someone could have evidence for believing the odd-numbered statements without them being true.

It has long been recognized that hypotheses must perform a certain kind of explanatory work in order to count as skeptical hypotheses and that it is the performance of this work that enables them to challenge ordinary knowledge claims. Keith DeRose (1999, p. 1), for example, writes: "Well, skeptical arguments come in many varieties, but some of the most powerful of them proceed by means of *skeptical hypotheses*. Hypotheses explain.

---

[6] As is common in the philosophical literature on explanation, I will use "explains" to mean "would explain if true" rather than "actually explains."

What does a skeptical hypothesis explain? It explains how you might be going wrong about the very things you think you know." However, despite the fact that philosophers are aware that skeptical hypotheses are able to challenge our ordinary beliefs because of the kind of explanations they provide, standard accounts of the structure and force of skeptical arguments fail to reflect this fact, as evidenced by the fact that the conflict between skeptical hypotheses and ordinary beliefs is always explicated in terms of the logical rather than the explanatory relations they bear to one another.[7]

The foregoing reflections also reveal why the necessary condition articulated in SH1 cannot be turned into a sufficient condition for skeptical hypotheses to pose significant skeptical challenges. Logical incompatibility between a putatively known proposition and a skeptical hypothesis is simply not sufficient for the former to pose a challenge to the latter.[8] It is also important to note that ordinary propositions (e.g., "I have hands") and large-scale commonsense propositions about the external world (e.g., "I am not a brain-in-a-vat") both satisfy SH4. Their truth would explain why things appear to us as they do, yet are clearly not skeptical hypotheses. Thus, the explanatory requirement in SH4 cannot be made a sufficient condition for presenting a skeptical challenge. Satisfying SH5 is also required.

Furthermore, satisfying both SH4 and SH5 is not sufficient for posing an effective skeptical challenge. An additional constraint on skeptical hypotheses is that they should be understood as explanations that compete with our more ordinary beliefs about the external world. In other words:

SH6. In order for a skeptical hypothesis *SK* to raise a significant skeptical challenge to *S*'s putative knowledge that *p*, the explanation *SK* provides must compete with the available commonsense explanations of *S*'s belief on which *S* knows that *p*.

Whether one hypothesis competes with another and the degree to which it does so depends in part upon how strong the explanations of the relevant phenomena are that they provide. In general, to be a competing explanation, a minimum threshold of explanatory merit is required.[9] I contend that the only legitimate constraints on skeptical hypotheses are the explanatory ones represented in SH4 through SH6.

---

[7] I do not mean to deny that logical relations can figure as important components of explanations. I merely want to claim that it is their status as explanatory rather than logical relations that matters in the present context.

[8] Cohen (1998, n. 11, italics in original) makes a similar point when he explains the motivation for thinking that one's evidence does not justify the belief that a skeptical hypothesis is false:

> I appeal to the fact that if *SK* were true, it would *explain* the truth of *E* [i.e., one's evidence]. This is because the mere fact that *E* would be true if *SK* were true does not seem to be enough. Let *SK* be the bare hypothesis (not-*P* & *E*). The mere fact that if *SK* were true, *E* would be true, does not seem to count against saying that *E* justifies not-*SK*. We need some skeptical hypothesis that would explain the truth of *E*. That is the reason for specifying the hypothesis that I am a brain-in-a-vat.

[9] There are many conceptions of what a good explanation consists in, and I will not need to take a stand on the explanation debate for present purposes. However, I endorse the view, defended by Poston (2014, ch. 4), that explanation is a primitive relation between propositions that resists analysis into any non-trivial set of necessary and sufficient conditions.

Some characterizations of skeptical hypotheses suggest that an additional condition of the following sort is also required:

> SH7.   In order for a skeptical hypothesis SK to raise a significant skeptical challenge to S's putative knowledge that p, SK must depict a situation that is subjectively indistinguishable from S's actual situation.

Duncan Pritchard (2002a), for example, writes: "Roughly, a skeptical hypothesis is an error-possibility that is incompatible with the knowledge that we ascribe to ourselves but which is also subjectively indistinguishable from normal circumstances (or, at least, what we take normal circumstances to be), such as that we might be currently experiencing a very vivid dream." The requirement of subjective indistinguishability, however, is both too strong and redundant in certain respects. Local skeptical hypotheses such as (3.2), (3.4), (3.6), and (3.8) do not need to depict entire situations (i.e., worlds) that are subjectively indistinguishable from the actual world in order to raise skeptical challenges to particular beliefs. And while we might very well want global skeptical hypotheses to conform to SH7, not all famous skeptical hypotheses have done so. Contrary to Descartes' claim that "there are no certain indications by which we may clearly distinguish wakefulness from sleep," Thomas Hobbes (1982 [1651], pt. 1, ch. 2) and Norman Malcolm (1959, ch. 17) have argued that dreaming may be distinguished from being awake on the basis of the coherence and lack of absurdities in one's waking life, and John Locke (1998 [1689], bk. 4, ch. 2, §2) thought that there were important differences between real pains and merely dreamed ones. Furthermore, the massive deception famously depicted in The Matrix is not subjectively indistinguishable for all characters from a situation where no such deception is occurring. Neo, the movie's hero, is described as always having the feeling that "there is something wrong with the world"—a feeling he sometimes describes as "a splinter in his mind"—which leads him to be suspicious about the "reality" he was supposed to accept. According to the narrative of The Matrix, this feeling would not be present if Neo inhabited a normal world. Thus, it seems sufficient for it to be difficult (rather than impossible) to tell that one is not in a skeptical scenario for a skeptical hypothesis to raise a compelling skeptical challenge. SH7 is, therefore, too strong.

Furthermore, in cases where it might seem plausible to require skeptical hypotheses to satisfy SH7, the condition is redundant. The function of subjective indistinguishability is to render subjects unable to tell that they are in one kind of situation rather than another, in light of the epistemic resources available to them. But by satisfying SH4 through SH6, a skeptical hypothesis will already have explained how we can possess a certain body of evidence for a given proposition yet fail to know that proposition on the basis of that evidence. In other words, the epistemic inability that is of concern in situations of subjective indistinguishability is already incorporated into conditions SH4 through SH6, and so no additional condition is needed.

Call a hypothesis that satisfies SH4, SH5, and SH6 a "defeating explanation." Just as one cannot know that p if one possesses an undefeated defeater for one's belief that p, one cannot know that p if one possesses an undefeated defeating explanation for one's belief that p. And just as ordinary defeaters come in rebutting and undermining varieties (Pollock 1986, pp. 38–9), so too do defeating explanations. A rebutting defeating explanation is one that incorporates the negation of what one believes into its explanatory account, while an undercutting defeating explanation targets the epistemological basis for one's belief, regardless of the truth value of that belief. Knowledge may be closed under known entailment. But according to the explanationist perspective defended here, it is also closed under defeated explanatory defeat. In other words:

DED.   If S knows that p and S knows that SK is a defeating explanation of S's belief that p, then S knows (or is in a position to know) that not-SK.[10]

Call the conjunction of SH4, SH5, SH6, and DED "the explanationist approach to skeptical challenges."

According to the explanationist approach, premise (1.2) of the canonical skeptical argument—the claim that I do not know that not-SK—should be seen as being supported by my alleged inability to rule out a skeptical explanation of my evidence and belief that competes with a commonsense one. Premise (1.1)—the claim that if I know that p, then I know that not-SK—is best understood as an abbreviated substitution instance of DED rather than ECP1. The explanationist approach does not involve the rejection of closure or underdetermination principles. It merely maintains (i) that they are too narrow to account for the full range of skeptical challenges and (ii) that by focusing on logical rather than explanatory considerations they fail to locate the key factors that drive effective skeptical challenges. Notably, the explanationist approach to skeptical challenges and its preferred epistemic principle, DED, can also survive common attacks on closure, since the falsity of ECP1 does not obviously entail the falsity of DED.

The explanationist approach intuitively explains the force behind familiar skeptical challenges and effortlessly handles all of the problem cases that were raised against the quasi-logical approach. Consider how it accounts for skeptical challenges to putatively necessarily true beliefs. Freudian or evolutionary cum cognitive scientific explanations of religious belief all seek to explain (à la SH4) the evidence religious subjects have for their beliefs and (à la SH5) how they could have those beliefs and yet fail to have knowledge. And they provide explanations of those beliefs that compete with commonsense explanations of them (à la SH6). As I will discuss in some detail in Section 4, the explanationist approach can also easily accommodate and explain the challenge posed by dreaming skepticism.

---

[10] Someone might want to argue that DED is a restricted version of the following "defeater elimination principle" (due to Cohen 2002, p. 314) that should be endorsed: If S knows that p on the basis of R, and D is a potential defeater of R as a reason to believe p, then S knows that D is false. This principle treats defeating explanations as simply one species of defeater and maintains that all potential defeaters that are known to be such must be defeated in order to have knowledge. This principle appears to be both broader and stronger than either ECP1 or DED, but I do not have the space to examine its merits here.

4. The Explanationist Approach to Dreaming Skepticism

In what I hope will be a striking demonstration of the strength of the explanationist approach to skeptical challenges, I will show that it can explain the skeptical force of dreaming skeptical arguments without entailing the dreaded KK-principle—something the quasi-logical approach has never succeeded in doing. As we noted above, since dreaming that I am sitting at my desk is compatible with me actually sitting at my desk, there will be no entailment between either of these claims and the negation of the other. This means that epistemic closure and underdetermination principles cannot be used to underwrite a dreaming skeptical argument, as they only concern relations of known mutual entailment between putatively known propositions and the negations of skeptical hypotheses. On the occasions when the inadequacy of closure and under-determination principles is acknowledged, philosophers generally do one of two things. One is to mention the difficulty but fail to offer a better alternative epistemic principle. The other is to suggest stronger replacement principles that seem to entail the KK-principle.

Consider the following variant of ECP1:

> ECP2.   If S knows that p, and S knows that p and q are incompatible, then S knows (or is in a position to know) that not-q.

If dreaming that I am sitting at my desk is not incompatible with sitting at my desk, but it is incompatible with knowing that I am sitting at my desk, a seemingly obvious replacement for ECP2 that might account for the challenge of dreaming skepticism is the following:

> ECP3.   If S knows that p, and S knows that q is incompatible with S's knowing that p, then S knows (or is in a position to know) that not-q.

Mark Steiner (1979), Barry Stroud (1984, chs. 1 and 3), Ernest Sosa (1997, p. 411, 1999, p. 145), and Vogel (2004, sec. 3), *inter alia*, maintain that a principle like ECP3 is needed to articulate the challenge of dreaming skepticism. However, those who suggest that ECP3 is needed to explain dreaming skepticism often express dissatisfaction with it as an epistemic principle. For example, immediately after articulating his version of the principle, Stroud (1984, p. 30) writes, "I will not speculate further on the qualifications or emendations needed to make the principle less implausible."

The most commonly noted reason for dissatisfaction with ECP3 stems from the fact that one of the things that is incompatible with S's knowing that p is S's not knowing that p. Substituting "S does not know that p" for q (and cancelling the double-negation) yields the following:

> ECP4.   If S knows that p, and S knows that S's not knowing that p is incompatible with S's knowing that p, then S knows (or is in a position to know) that S knows that p.

It is often assumed that principles like ECP4 directly entail the KK-principle (first articulated by Hintikka 1962):

KK1.   If *S* knows that *p*, then *S* knows (or is in a position to know) that *S* knows that *p*.

However, KK1 follows from ECP4 only if the following is also true:

KK2.   *S* knows that *S*'s not knowing that *p* is incompatible with *S*'s knowing that *p*.

But KK2 is false.

The reason KK2 is false is that epistemic principles like ECP1 through ECP4, UP, DED, KK1, and KK2 should all be understood as universally quantified generalizations. They are intended to cover every subject and every proposition. Like many authors, I have simply left out the quantifiers for ease of exposition. But it is important to return to them in the present instance. Because most people have never considered the incompatibility represented in KK2, they have no beliefs about the matter and thus the universal generalization expressed by KK2 is false. This means that a fully general KK-principle cannot be deduced from ECP3 or ECP4.

Now, of course, many philosophers have considered the matter in KK2 and have formed beliefs concerning it. This implies that if ECP4 is true, then KK1 will be true for them as well. This will still be a fairly devastating result, as the widely maligned KK-principle seems to impose too heavy an epistemic burden even on the knowledge of sophisticated philosophers.[11] Nevertheless, it is important to understand the ways in which KK1 fails to follow directly from ECP3 or ECP4.

Consider now the explanationist approach to dreaming skeptical challenges. Dreaming hypotheses of the sort described by Descartes and Moore are undercutting defeating explanations rather than rebutting defeating explanations. The explanationist constraints on skeptical hypotheses require only that the hypotheses (i) explain *S*'s evidence for believing *p*, (ii) explain how *S* could believe that *p* without knowing that *p*, and (iii) compete with commonsense explanations of *S*'s belief on which *S* knows that *p*. Dreaming skeptical hypotheses satisfy each of the conditions, and these conditions intuitively and elegantly explain the force of dreaming skeptical challenges.

Furthermore, there is nothing in the explanationist constraints on skeptical hypotheses that imply KK1—even when its scope is restricted to philosophers who satisfy KK2. ECP3 requires that in order to know that *p* I must know (or be in a position to know) the falsity of any proposition incompatible with my not knowing that *p*. DED merely requires that I know (or be in a position to know) the falsity of any defeating explanation of my belief that *p*. Because not every proposition that is incompatible with my knowing that *p* will be a defeating explanation, the set of propositions that I must know to be false is thus significantly smaller on DED than on ECP3. In circumstances where one is faced with an undermining defeating explanation of one's belief,

---

[11] For criticisms of KK-principle, cf. Alston (1980), Sorensen (1988), Williamson (2000, ch. 4), and Hawthorne (2004).

one may need to be in a position to know that one knows that *p* in order to know that *p*. But DED does not demand that one be in this kind of strong epistemic position with respect to any proposition that one hopes to know.

The explanationist approach to skeptical challenges thus explains the full range of compelling skeptical hypotheses without relying upon closure and underdetermination principles that are too weak to explain dreaming skeptical challenges and without bringing in alternative epistemic principles that are too strong. I conclude that the explanatory merits of the explanationist approach to skeptical challenges are significant.

5. Skepticism about Explanationism

Above I noted that many philosophers harbor doubts about whether the explanatory virtues or explanatory features of hypotheses have any "specifically epistemic, truth-conducing value." I would now like to consider this kind of skepticism about explanationism in light of the arguments I presented above about the role that the explanatory features of skeptical hypotheses play in enabling them to present significant skeptical challenges to our knowledge. I contend that skepticism about explanationism is inconsistent with appreciating the force of compelling skeptical hypotheses.

Explanationism comes in many varieties and grades of strength. Some versions of explanationism concern only ampliative inference, while others concern all of epistemic justification. Lycan (2002, p. 417) famously distinguished the following versions of explanationism about inference:

Weak Explanationism: the view that explanatory inference can epistemically justify a conclusion.
Sturdy Explanationism: the view that explanatory inference can epistemically justify a conclusion and can do so without being derived from some other more basic form of ampliative inference.
Ferocious Explanationism: the view that explanatory inference can epistemically justify a conclusion, that it can do so without being derived from some other more basic form of ampliative inference, and that no other form of ampliative inference is basic.
Holocaust Explanationism: the view that all inference and reasoning, including deductive as well as ampliative, is derived from explanatory inference.[12]

Some philosophers have argued that all of epistemic justification is ultimately a matter of explanatory considerations (e.g., Harman 1986; Lycan 1988; Moser 1989; Conee and Feldman 2008; McCain 2014; Poston 2014). Lycan (1988, p. 133), for example, writes, "Whatever ultimately justifies a belief is a matter of the explanatory contribution of that belief." In contrast to these strong forms of explanationism about epistemic jus-

[12] Yes, he named a philosophical perspective on inference after the Holocaust.

tification, one might consider the following, weaker version (analogous to Lycan's weak explanationism about inference above):

WE1.   The explanatory features of a belief can epistemically justify that belief.

I want to focus my discussion on the following, generalized version of WE1 that covers the full range of cognitive attitudes that one can take to a proposition in light of one's evidence:

WE2.   The explanatory features of a proposition can determine whether one should believe the proposition, disbelieve the proposition, or suspend judgment about the proposition.

I will argue that one cannot accept the explanationist approach to skeptical challenges articulated in the sections above and reject WE2.

Consider the justification-neutralizing power that skeptical hypotheses are widely thought to enjoy. In Dretske's (1970, p. 1016) classic discussion of the zebra case, he writes:

> Do you know that these animals are not mules cleverly disguised by the zoo authorities to look like zebras? If you are tempted to say "Yes" to this question, think a moment about what reasons you have, what evidence you can produce in favor of this claim. The evidence you *had* for thinking them zebras has been effectively neutralized, since it does not count toward their *not* being mules cleverly disguised to look like zebras.

If explanatory considerations are "merely practical bonbons of no specifically epistemic, truth-conducing value"—i.e., if WE2 is false—how is it that my epistemic justification for believing that the animal in the pen is a zebra (or that I am sitting at my desk) is somehow called into question by the explanatory relations this proposition bears to the hypothesis that the animal in the pen is a mule cleverly disguised to look like a zebra (or to the hypothesis that I am dreaming I am sitting at my desk)? The failure of the quasi-logical approach to skeptical challenges means that the justification-neutralizing power of skeptical hypotheses cannot be explained in terms of the logical relations they bear to ordinary propositions. But if logical relations are set aside, there does not seem to be any good alternative to believing that it is explanatory relations that are doing the work. Consider the following situation:

(4.1)   S's evidence, E, justifies S in believing that p.
(4.2)   SK explains how S could possess E, believe that p, but fail to know that p.
(4.3)   SK does not entail not-p.

I do not see how (4.1) through (4.3) could be true and WE2 false at the same time. Thus, I conclude that contemporary skeptical challenges that rely upon skeptical hypotheses presuppose central tenets of explanationism.

To come full circle, recall the explanationist response to skepticism, according to which explanatory considerations provide one with sufficient epistemic justification for rejecting skeptical hypotheses and accepting commonsense propositions, on the grounds that the latter are simpler, have greater explanatory power, lead to greater explanatory coherence in one's overall belief set, or are more conservative than the former. As we noted above, this response has never enjoyed widespread acceptance in the epistemological community. However, I have argued that this same epistemological community has failed to properly articulate the structure of skeptical challenges that are based upon dreaming hypotheses or that challenge necessarily true beliefs. Furthermore, I have argued that one cannot reject explanationism and at the same time provide an account of how skeptical hypotheses pose significant epistemological challenges to our putative knowledge of the external world. In light of the arguments presented here, I do not see how any kind of general skepticism about the explanationist response to skepticism can be warranted. One might have doubts about this or that version of the explanationist response, but the suggestion that the guiding idea behind this class of responses to skepticism is somehow fatally flawed appears untenable.

It is hoped that my articulation and defense of the explanationist approach to skeptical challenges will not only deepen philosophers' understanding of the skeptical challenges themselves but will also lead many of them to rethink their prior assessment of the merits of the explanationist response to skepticism.

References

Alston, W. P. (1980) "Level Confusions in Epistemology." *Midwest Studies in Philosophy* 5: 135–50.

Beebe, J. R. (2009) "The Abductivist Reply to Skepticism." *Philosophy and Phenomenological Research* 79: 605–36.

Beebe, J. R. (2010) "Constraints on Sceptical Hypotheses." *Philosophical Quarterly* 60: 449–70.

Beebe, J. R. (2011) "*A priori* Skepticism." *Philosophy and Phenomenological Research* 83: 583–602.

BonJour, L. (1998) *In Defense of Pure Reason.* Cambridge: Cambridge University Press.

BonJour, L. (1999) "Foundationalism and the External World." *Philosophical Perspectives* 13: 229–49.

BonJour, L. (2003) "A Version of Internalist Foundationalism." In L. BonJour and E. Sosa (eds), *Epistemic Justification: Internalism vs. Externalism, Foundations vs. Virtues.* Malden, MA: Blackwell, pp. 3–96.

Brueckner, A. (1985) "Skepticism and Epistemic Closure." *Philosophical Topics* 13: 89–117.

Brueckner, A. (1994) "The Structure of the Skeptical Argument." *Philosophy and Phenomenological Research* 54: 827–35.

Cohen, S. (1998) "Two Kinds of Skeptical Argument." *Philosophy and Phenomenological Research* 58: 143–59.

Cohen, S. (2002) "Basic Knowledge and the Problem of Easy Knowledge." *Philosophy and Phenomenological Research* 65: 309–29.

Conee, E. and Feldman, R. (2008) "Evidence." In Q. Smith (ed.), *Epistemology: New Essays.* Oxford: Oxford University Press, pp. 83–104.

DeRose, K. (1999) "Introduction: Responding to Skepticism." In K. DeRose and T. A. Warfield (eds), *Skepticism: A Contemporary Reader*. Oxford: Oxford University Press, pp. 1–24.

Dretske, F. (1970) "Epistemic Operators." *Journal of Philosophy* 67: 1007–23.

Dretske, F. (1981) "The Pragmatic Dimension of Knowledge." *Philosophical Studies* 40: 363–78.

Greco, J. (2008) "Skepticism about the External World." In J. Greco (ed.), *The Oxford Handbook of Skepticism*. Oxford: Oxford University Press, pp. 108–28.

Harman, G. (1986) *Change in View*. Cambridge, MA: MIT Press.

Hawthorne, J. (2004) *Knowledge and Lotteries*. Oxford: Oxford University Press.

Hintikka, J. (1962) *Knowledge and Belief*. Ithaca, NY: Cornell University Press.

Hobbes, T. (1651/1982) *Leviathan*. New York: Penguin.

Locke, J. (1689/1998) *An Essay Concerning Human Understanding*. New York: Penguin Classics.

Lycan, W. G. (1988) *Judgement and Justification*. Cambridge: Cambridge University Press.

Lycan, W. G. (2002) "Explanation and Epistemology." In P. K. Moser (ed.), *The Oxford Handbook of Epistemology*. Oxford: Oxford University Press, pp. 408–33.

Malcolm, N. (1959) *Dreaming*. London: Routledge and Kegan Paul.

McCain, K. (2014) *Evidentialism and Epistemic Justification*. New York: Routledge.

McCain, K. (Forthcoming) "In Defense of the Explanationist Response to Skepticism." *International Journal for the Study of Skepticism*.

Moore, G. E. (1959) "Proof of an External World." *Philosophical Papers*. London: Allen and Unwin.

Moser, P. K. (1989) *Knowledge and Evidence*. Cambridge: Cambridge University Press.

Pollock, J. (1986) *Contemporary Theories of Knowledge*. Totowa, NJ: Rowman and Littlefield.

Poston, T. (2014) *Reason and Explanation: A Defense of Explanatory Coherentism*. New York: Palgrave Macmillan.

Pritchard, D. (2002a) "Contemporary Skepticism." In J. Fieser and B. Dowden (eds), *Internet Encyclopedia of Philosophy*, <www.iep.utm.edu/skepcont/>.

Pritchard, D. (2002b) "Recent Work on Radical Skepticism." *American Philosophical Quarterly* 39: 215–57.

Pritchard, D. (2005) "The Structure of Sceptical Arguments." *Philosophical Quarterly* 55: 37–52.

Russell, B. (1912) *The Problems of Philosophy*. Oxford: Oxford University Press.

Sorensen, R. A. (1988) *Blindspots*. Oxford: Oxford University Press.

Sosa, E. (1997) "Reflective Knowledge in the Best Circles." *Journal of Philosophy* 94: 410–30.

Sosa, E. (1999) "Skepticism and the Internal/External Divide." In J. Greco and E. Sosa (eds), *The Blackwell Guide to Epistemology*. Malden, MA: Blackwell, pp. 145–57.

Steiner, M. (1979) "Cartesian Scepticism and Epistemic Logic." *Analysis* 39: 38–41.

Stroud, B. (1984) *The Significance of Philosophical Scepticism*. Oxford: Clarendon Press.

van Fraassen, B. (1980) *The Scientific Image*. Oxford: Oxford University Press.

Vogel, J. (1990a) "Are There Counterexamples to the Closure Principle?" In M. D. Roth and G. Ross (eds), *Doubting: Contemporary Perspectives on Skepticism*. Dordrecht: Reidel, pp. 13–27.

Vogel, J. (1990b) "Cartesian Skepticism and Inference to the Best Explanation." *Journal of Philosophy* 87: 658–66.

Vogel, J. (2004) "Skeptical Arguments." *Philosophical Issues* 14: 426–55.

Williamson, T. (2000) *Knowledge and Its Limits*. Oxford: Oxford University Press.

12

Scepticism about Inference to the Best Explanation

Ruth Weintraub

1. Introduction

Scepticism about inference to the best explanation (henceforth, IBE) is far less often discussed than scepticism about another ampliative form of inference, (enumerative) induction.[1] This chapter aims to redress the imbalance by giving scepticism about IBE the attention it, too, deserves.

The chapter is structured as follows. In the initial sections, I explain how I use the term 'IBE' (Section 2), the way I view the exchange with the sceptic (Section 3), and how we might respond to a sceptical challenge as I construe it (Sections 4–5). In Section 6, I present five sceptical arguments aimed to undermine the legitimacy of inferring the best explanation of some data, and consider some responses to them, drawing on insights gleaned from the vast literature on induction. My conclusion is that IBE, even to the observable, may be in a worse epistemic position than induction, because there are arguments that target it which do not have inductive analogues, and the converse isn't true.

2. Inference to the Best Explanation

It might seem that in order to understand the nature of IBE and its sceptical challenge, we need first to know what explanation is. And this requires (at least) a paper on its own, because there is substantive disagreement about the nature of explanation. Do we increase our understanding of a phenomenon by showing it is likely to happen (Hempel, 1965, p. 337),[2] by making it more familiar (Bridgman, 1962, p. 37), by relating it to something that is already understood (Scriven, 1970, p. 202), by unifying what

---

[1] The qualification 'enumerative' is required for those (Swinburne, 1974, p. 1; Howson, 2000; Lipton, 2004) who use the term 'induction' more liberally, to denote all non-deductive inferences. Enumerative induction denotes only inferences from a sample to the entire population or to the next case. I shall use the term 'induction' narrowly, so that 'enumerative induction' is a pleonasm.

[2] The D-N model (Hempel, 1965, pp. 335–76) is an extreme special case. By deducing the *explanandum* from initial conditions and a set of laws, we show that it was *certain* to happen.

we have to accept (Kneale, 1949, p. 91), or by providing causal information—either about causal history (if the *explanandum* is a particular event) or (if the *explanandum* is a causal regularity) about the mechanism linking cause and effect (Lipton, 2004)?

Perhaps, by way of yet another possibility, explanatoriness is simply a measure of plausibility given the evidence? Harman (1965, p. 91) thinks a hypothesis is explanatory insofar as it is 'a better, simpler, more plausible (and so forth) hypothesis... [i.e., if] it is a better hypothesis in the light of all the evidence'. Lycan (1988, p. 130), too, takes simplicity, coherence, fruitfulness, and initial plausibility all to contribute to the explanatoriness of a hypothesis.[3]

For the purpose of this chapter we may bypass the question about the nature of explanation. Each analysis picks out a notion of explanation that can be plugged into an inference rule, IBE, whose legitimacy may be denied. All the arguments I will consider are insensitive to the precise nature of explanation.

There's a distinction that cuts across the distinction between different construals of IBE engendered by the different conceptions of explanation. It pertains to IBE's *conclusions*. For every reading of 'explanation', there is a strong form of IBE, which allows us to infer to the (probable) truth of the best explanation, and a weak form of IBE, call it *modest IBE* (or constructive-empiricist IBE), which allows us to infer to the (probable) truth of the best explanation's *observable content* (van Fraassen, 1980). Where the difference between the weak and the strong versions matters, I shall point it out.

3. Scepticism

The sceptic impugns beliefs or inference rules. He claims, for instance, that a belief is unjustified or fails to constitute knowledge. Such defects he typically takes to be *necessary*: our beliefs about the external world, Descartes' sceptic claims, *cannot* constitute knowledge. Hume claims that induction (1777, pp. 35–6) and deduction (1739, pp. 180–3) cannot be justified. Of course, the two flaws (pertaining to beliefs and to inference rules) are related: a belief sustained by a faulty inference rule is—for this reason—itself epistemologically defective.

The only sceptic that we need to take seriously is one who adduces arguments in support of his position. This means that we may ignore sceptics who merely *challenge* us to vindicate a belief or an inference rule, whereas it is (epistemically) incumbent on us to respond (in one of the ways I will adumbrate in Section 4) to the sceptic's arguments.

4. Reasoned Responses to Scepticism: Typology

The reasoning sceptic adduces arguments in support of his position. Our response, too, must be reasoned. But there are *several* reasoned ways of responding to a sceptical

---

[3] Confusingly, Lycan's list of the features which enhance explanatory goodness (1988, p. 130) includes explanatoriness (in addition to simplicity and fitting what else one already believes). Here, Lycan seems to be using the term 'explanatoriness' more narrowly.

argument. We can distinguish, first, between *combative* and *acquiescent* responses. A combative response attempts to *rebut* the sceptical arguments adduced in support of the sceptical claim. I will consider combative responses to five sceptical arguments against IBE (Sections 6–10).

In contrast to a combative response to scepticism, an *acquiescent* response *accepts* the sceptical claim. This might engender distress, but it needn't. Perhaps—initial appearances notwithstanding—the sceptical claim doesn't deny us something valuable. To take the sting out of the sceptical claim instead of denying it is to *dissolve* the problem. And this can be done in two ways. The first, which I mention only to put aside, is illustrated by Sextus Empiricus, the ancient Greek sceptic. He is unperturbed by the sceptical conclusion, not because he thinks it doesn't impugn our cognitive endeavours, but rather, because he is not *interested* in them. He aims to attain peace of mind (*ataraxia*), an aim that belief in the sceptical conclusion may actually promote (OP, book I, ch. XII). But even if (contentiously) Sextus' psychological claim is sound, it is irrelevant if our perspective is *epistemological*.

Even if our interest is cognitive, we might shrug aside, or at least belittle the significance of, a sceptical claim: we might think that it pertains to an epistemological concept ('knowledge', 'justification') that is *needlessly* stringent. Our beliefs, which never count as 'known' when assessed in the light of our concept of knowledge, can, nonetheless, measure up to the requirements of the concept that is well-worth falling under (cognitively speaking)—justification.

The sceptical arguments against IBE that I will consider in this chapter target its *justification*. I will assume that our concept of justification reflects something significant. So it might seem that from an epistemological perspective, the only way to avoid epistemological defeat is to adopt a combative response. But in fact, there is yet another acquiescent response, a way of taking the (cognitive) sting out of a sceptical argument that brands as irrational a practice (use of an inference rule, in our case). The sceptical claim, the thought is, doesn't constitute a threat to our cognitive endeavours, because the practice it brands as irrational isn't ours. I will label this way of dissolving the sceptical challenge *eliminativist*, and will consider its application to IBE in Section 5.[4] This order of discussion makes sense: there is a point (other than by way of an intellectual exercise) in attempting to rebut a sceptical argument only if it targets something that is dear to us.

5. Eliminativism about IBE

Plausibly, some putative cases of IBE involve only deduction and (enumerative) induction. When we infer upon seeing human footprints in the sand that a human being walked here, Fumerton (1980, p. 597) suggests, we are not really inferring to the best explanation of the phenomenon, but rather, (implicitly) invoking a generalization to

---

[4] There are also eliminativist responses to scepticism about *induction* (Norton, 2003; Okasha, 2001).

the effect that human footprints are typically caused by humans walking, a generalization which is derived by induction from observable premises.

But even if we grant that such examples involve only induction and deduction, this, I will argue, can't *always* be the case. Scientists infer from *observed* phenomena to theories about the *unobservable* (electrons, quarks, black holes, etc.). And *this* kind of inference cannot be reduced to a set of inductive and deductive inferences. Induction and deduction cannot engender a substantive (non-tautological) conclusion about entities or properties not mentioned in the premises. So, for instance, when Newton inferred his laws of motion, which included claims about (unobservable) masses and forces from his observations about motions of planets and the tides, he must have been invoking *another* rule of inference.

Fumerton (1980, p. 599) argues that it is possible to justify at least some theoretical statements inductively. Even if this is true, it falls short of the eliminativist claim, because it only purports to do away with *some* invocations of IBE. And, anyway, I don't think it *is* true. Fumerton argues that '[r]eaching a conclusion inductively based on one's *entire* body of evidence usually involves countless inductive arguments the premises of which confirm countless hypotheses to varying degrees of probability' (1980, p. 599, original italics). This is clearly true, but irrelevant. If the scientist only infers *inductively*, I don't see how he can reach from statements (no matter how many) pertaining to observable entities a conclusion relating to entities *of a different kind*. Fumerton suggests that 'though theoretical entities are themselves unobservable, their defining properties may be such that we have observation of other things having those properties. This opens the door to ... indirect inductive reasoning' (1980, p. 599). Again, I don't see how induction and deduction are supposed to warrant conclusions about the unobservable. Even if 'having speed' is a defining property of electrons, and some observable things have speed too, how does this license an inductive inference to the existence of electrons?[5]

Inference to the unobservable isn't restricted to scientific contexts. When we ascribe beliefs and desires (and perhaps pains) to others on the basis of their behaviour, we are inferring to the unobservable, and, therefore, invoking IBE.

Can *modest* IBE, the inference to the (probable) truth of the observable content of theories, be eliminated in favour of induction? The answer is 'No'. True, from Craig's (1953) theorem it follows that if a theory is recursively axiomatizable, so is its observable content. And this might suggest that we need not traffic with the unobservable, and can make do with induction. But both claims are false. First, it is not our practice to believe precisely the restricted theory: it is horribly complex, even if effectively specifiable. Second, even if we do believe only the restricted theory, the question arises as to the

---

[5] Fumerton argues that statements about the unobservable that cannot be grounded inductively should be construed instrumentally, so that they make claims only about observable entities. But this is an invitation to *reinterpret* theoretical terms. And even if (very contentiously) such an interpretation is required, it cannot warrant an eliminativist claim to defang the sceptical argument against IBE, targeting, as it does, our *current* linguistic practice.

justification of this (restricted) belief. We know one way of getting to the restricted theory: derive the unrestricted theory by invoking IBE, and then restrict it by invoking Craig's theorem. This derivation *isn't* inductive. And the rule ('modest IBE') that it invokes is as vulnerable, it is easy to check, to the arguments I will consider against IBE as is the strong version of IBE. The onus is on the eliminativist to show that the axioms of the restricted theory can be confirmed *inductively*. Pending such a demonstration, we are justified in thinking that explanatory considerations are invoked in our inferential practice; that some form of IBE is indispensable.

My case against the possibility of eliminating IBE rests on judiciously chosen examples: theoretical statements of science, and belief and desire ascriptions. Some philosophers (Jackson, 1977; Vogel, 1990) think that our beliefs about ordinary material objects are inferred from beliefs about our experience ('appearances'). My belief that there is a tree in front of me best explains it appearing to me that there is a tree in front of me. If this were true, we would have another ineliminable invocation of IBE (one the giving up of which incurs a cost). But the suggestion is very contentious. It is perhaps more plausible to suppose that our beliefs about ordinary-sized external objects (trees, houses, etc.) are *basic*: based on our experience, but not *inferred* from beliefs about it. This response is clearly unavailable in the case of scientific theories. No one thinks that beliefs about electrons are basic! So I can conclude that we haven't (yet) managed to eliminate IBE.

Here is a second eliminativist response to scepticism about IBE, adapted from Popper (1972). Only deduction, Popper claims, is employed in scientific reasoning. Science progresses by formulating bold hypotheses and subjecting them to rigorous tests. The only mode of reasoning it invokes, Popper argues, is deduction. Predictions ('observation statements') are deductively derived from hypotheses, and when observationally refuted, the theory is *deductively* falsified: if $T \vdash E$, then $\sim E \vdash \sim T$. We may now cheerfully concede the sceptical conclusion—that non-deductive reasoning (including IBE) is irrational. Since science only invokes deduction, its rational credentials are not thereby impugned.

Popper's eliminativist attempt to take the sting out of the sceptical claim is also unsuccessful. Even on Popper's account, the scientific method essentially invokes non-deductive reasoning.[6] Indeed, Popperian scientific practice involves ineliminable invocations of IBE. A theory's Achilles' heel must be located so as to subject it to *severe* tests. But a theory's prediction is more vulnerable the *less likely* it is. And judgements about likelihood are based on non-deductive reasoning. Suppose I predict that the sun will rise only on Tuesdays from now on. Watching out for the sun on Wednesday morning (but not on Tuesday morning) is a severe test, because my prediction about that day contravenes my current supposition that the sun rises daily, which is *inductively* based—inferred from past sunrises.

---

[6] This is persuasively argued by Newton-Smith (1981, ch. III).

Because we are interested in scepticism about IBE, such invocations of induction are irrelevant in our context (although they impugn Popper's deductivism). But Popperian scientific reasoning also involves IBE. When Eddington tested General Relativity, he (sensibly) chose a prediction that was incompatible with Newtonian Mechanics, which made it *unlikely*. He was relying on the (probable) truth (or empirical adequacy) of Newtonian Mechanics, which was warranted via IBE.

There are invocations of non-deductive reasoning at a later stage in the scientific practice as well. First, we do not rest content with having 'on our books' a theory that has withstood rigorous attempts of refutation. If we rely on theories in action, we must believe their empirical consequences to be (very probably) true. When an engineer boards a plane, he isn't (typically) merely testing the theory that guided its construction. He is *assuming* the plane will land safely. And unlike the inference from the existence of a falsifying instance to the falsity of the theory, the inference from a theory's having withstood rigorous tests to it continuing to do so is not deductively valid. But while this constitutes another counter-example to Popper's deductivism, it doesn't necessarily impugn eliminativism about IBE. Even when a theory has unobservable content, we may reason inductively to the success of the next (observable) prediction on the basis of previous ones.[7]

But there *is* an ineliminable invocation of IBE in addition to that involved in theory testing. We can only partake of our theories' non-practical virtues (verisimilitude, explanatory strength) if we take them to be (probably) true. And this belief is sometimes inferred via IBE. Van Fraassen (1980) thinks we can use a theory to explain without believing it is true. But this seems to me false. 'There are tides because of gravity but I am agnostic about the existence of gravity' is pragmatically self-defeating. Of course, *really* to explain, a theory must be true. But belief in its truth is necessary for it to explain to *us*.

The eliminativist response to scepticism about IBE fails. So, given that justification matters, we must attempt to rebut the sceptical arguments. But, I shall argue, some of the work has already been done. Having grappled for so long with inductive scepticism, we can bring to bear the insights thereby gleaned when contending with arguments for scepticism about IBE. We don't have to consider the argument against IBE *tabula rasa*. Although there is no agreed upon solution to the problem of induction, the main sceptical arguments, the major division lines between contending solutions and the basic assumptions underlying them have become clear, and these lessons can all be usefully applied to the study of the less familiar sceptical challenge to IBE. (Who says there's no progress in philosophy?)

---

[7] I argued above that IBE is required to arrive at a theory's observable content from observations. But in accounting for our reliance on a theory in action, we are concerned with the inference to the truth of a *particular* claim a theory makes about the observable, a claim that we *know* to belong to the theory's observable content (because we have deduced it from the theory, for instance). And for *this*, induction may suffice.

6. Five Sceptical Argument against IBE

6.1 First argument against IBE

How can we be justified in inferring statements about material bodies from statements about our ideas, Berkeley (1710, §18) asks? Our belief cannot be justified by the senses, because they only yield knowledge of what is perceived by them. And it cannot be justified inferentially (by 'reason'). What 'reason can induce us to believe the existence of bodies without the mind, from what we perceive, since [there is no] necessary connection betwixt them and our ideas'?

This argument relies only on the fact that there is no necessary connection between the premises of the argument and its conclusion; that the former may be true while the latter is false. And since induction, too, isn't deductively valid, the argument works equally well against it. But we already know how to rebut this argument. Justification is *fallible*. So the sceptic is mistaken in (implicitly) assuming that non-deductive grounds do not provide good reasons. Taking a risk, we will remind the sceptic, may be eminently reasonable. It is rational, for instance, to take a medicine one thinks is very likely, although uncertain, to cure one of a horrible disease, even if it sometimes (but sufficiently rarely) has pretty unpleasant side-effects and fails to cure.

6.2 Second argument against IBE

It is natural to suppose that Hume's argument against induction (1777, pp. 35–6) may be straightforwardly adapted so as to target IBE. The natural thought fails, I will argue, but there is a kindred argument that targets both inference rules. Consider the simple modification of Hume's argument:

(1) The conclusion of an IBE isn't logically entailed by its premises.
(2) Every IBE assumes that nature is explicable.
(3) The principle of explicability must be warranted if IBE is to be justified.
(4) The principle of explicability cannot be justified *a priori*. (from 1)
(5) The principle of explicability cannot be justified *a posteriori*, since such a justification would be circular.
(6) *Conclusion*: IBE is unjustified.

As it stands, the argument isn't clear: how are we to construe the principle of explicability? An answer will be forthcoming by considering Hume's principle of nature's uniformity, which he invokes against induction. Ayer (1972, pp. 20–1) thinks that the principle, as Hume construes it, is too strong: when conjoined with a statement of some observed regularity, it logically entails predictions about unobserved cases. But, Ayer points out, we do *not* take a counter-example to an inductive prediction to refute the principle of uniformity: we continue to extrapolate from observed cases to unobserved ones even after encountering a white raven. Ayer concludes that there is no principle of uniformity to be invoked.

Ayer's reaction is precipitate. We should, instead, substitute a weakened and vague, but still substantive, principle, according to which observed cases are *sufficiently often* similar to unobserved ones. From this principle, no *particular* inductive prediction follows even when conjoined with observations of positive cases, so no single counter-example to a generalization falsifies nature's uniformity. But the principle is nonetheless substantive, and fulfils an important function: justifying the inference from an observed regularity to an unobserved case. Even if the inference isn't deductively valid, it is reasonable, since sufficiently many like it have true conclusions. Unlike deduction, induction sometimes leads us astray. But—thanks to the (weakened) uniformity principle—this is atypical.

If the claim that nature is uniform is equivalent to the claim that induction is (sufficiently) reliable, the sceptic can, by analogy, construe nature's explicability as the claim that IBE is (sufficiently) reliable. But he isn't yet home and dry. Hume argues, plausibly, that induction's reliability is established in a circular way, because it is inferred inductively from its reliability hitherto. But is establishing IBE's reliability circular in this way? It would be if we justified it by inferring it as the best explanation of something. But what is IBE's reliability supposed to explain? Perhaps it explains its success hitherto. Perhaps, alternatively, the reliability of IBE is established *inductively*, on the basis of its past success.

The sceptic needn't commit himself to the precise way IBE's reliability is to be established. He can allow for all the possibilities consistent with the (plausible) assumption that it must invoke *some* inference rule by complicating the argument somewhat:

(1) If an inference rule is justified, it must be justifiably believed to be reliable. (Call this the 'internalist' premise.[8])
(2) IBE is ampliative.
(3) IBE's reliability is empirical. (from 2)
(4) The belief that IBE is reliable is (very) general, so it cannot be *basic*. (from 3)
(5) IBE's reliability must be justified inferentially. (from 4)
(6) The inference rule I_1, *via* which IBE's reliability is derived must be justifiably believed to be reliable. (from 1)

...

(7) The justification of IBE involves a chain of inference rules, each giving rise to the belief that the previous one is reliable.

---

[8] I add yet another way of using the term 'internalism'. It is hitherto used in (at least) four different ways. There is, first, the doctrine which requires for the justification of a belief (or inference rule) a second-order justified belief about the factors that make first-order justification possible (van Cleve, 2003). This usage is closest to mine. The difference is that unlike mine, it doesn't require the second-order belief to be justified *independently* of the first-order belief. According to the second variety of internalism, commonly known as *mentalism*, epistemic status supervenes on facts about the subject's mental states. According to the third, commonly known as *accessibilism*, the subject must be able to tell by reflection that his belief (inference rule) is justified. According to the fourth, the subject constitutes the conditions of justification: a belief (inference rule) is justified just in case the subject thinks it is.

(8) The chain of inference rules required for the justification of IBE is either circular, infinite, or it terminates.

(9) None of the possibilities canvassed in (8) makes for genuine justification.

(10) *Conclusion*: IBE is unjustified.

Note that by substituting 'induction' for 'IBE', we get the much more familiar argument against induction. (Here, the justification of reliability involves a (very small) circle: the reliability of induction is established by induction.) And we now have a perspicuous argument against IBE, with two contentious premises, (1) and (9), to whose discussion I now turn. My aim is not to rebut the argument, but, more modestly, to note some relevant insights gleaned from the study of inductive scepticism and add some new ones, and to show that the argument against IBE is as cogent as its inductive counterpart.

The first way of rebutting the argument is to reject the internalist premise. I will only mention, without dwelling on, the (externalist) ways of so doing that are familiar, and pay closer attention to those that aren't. First, reliabilists think that a belief (inference rule) is justified if given true premises, it yields true conclusions sufficiently often (van Cleve, 1984),[9] or if it is grounded in virtuous (reliable) cognitive dispositions (Greco, 2000). The agent doesn't—contrary to the internalist premise—need to believe this so to be justified in inferring via IBE.

Second, Goodman (1955, p. 64) thinks that the rules of inference we employ and particular inferences we make, whether deductive or non-deductive, are adequately justified by 'being brought into agreement with each other', the 'agreement achieved... [being] the only justification needed for either'. No belief about reliability is required.

Finally (and less familiarly), Boyd (1984) claims (in passing) that without IBE, science wouldn't be possible. This defence of IBE seems to me ineffectual. The indispensability to science of IBE confers on it justification only if science has value. (Compare: to murder someone, I must obtain the wherewithal.) And the justification is *epistemic* only if scientific inquiry is epistemically worthwhile. It is irrelevant, for instance, that it provides an intellectual challenge. But it is precisely the *epistemic* worthiness of (IBE-based) science that the sceptic denies!

This concludes my discussion of the first strategy: rejecting the internalist premise. If we think that the reliability of IBE does have to be justifiably believed for its invocation to be justified, we must uphold one of the three possibilities rejected in premise (9). The infinitist possibility seems to be ruled out, even if (Klein, 2006) it is legitimate *in principle*: induction and IBE are the only ampliative inference rules that seem relevant. Neither does a terminating chain of justification make for genuine justification here: the reliability of the first ampliative inference rule in the chain, I_1, is established, if at all, deductively, because it is the *first* ampliative rule in the chain. But the reliability

---

[9] The formulation is intentionally vague. Does 'sufficiently often' mean 'more often than not', 'most of the time', 'more than any other method'? The choice doesn't matter, because what is at stake is the *very idea* that the reliability of an inference rule could constitute justification.

of an ampliative rule *cannot* be established deductively. So I_j isn't justifiably believed to be reliable. And since we have conceded the internalist requirement, we must admit that a terminating chain doesn't make for genuine justification.

We are left with the possibility of circular justification: IBE's reliability established, on the basis of its past successes, either by IBE or by induction, whose reliability is established inductively, on the basis of *its* past successes.

I will merely point out, without attempting to assess, two objections that might be levelled against circular justifications of reliability. The first is familiar. It concedes that such justifications are not trivial, as are circular justifications that invoke the proposition that is to be justified. *Any* proposition can be 'justified' if allowed as a premise in a justificatory argument. By contrast, even if we are allowed to invoke an inference rule in its own justification, success is not guaranteed. If induction had failed abysmally in the past, we wouldn't have the requisite premise for reasoning inductively to its reliability. But rule-circular justification is still *insufficiently discriminating*. Crazy rules can be invoked in their own defence (Salmon, 1957, p. 46). Consider the anti-inductive rule, AI, which allows us to infer from 'All observed As have been Bs' that the next A will *not* be a B. We can construct an argument utilizing AI to show that AI will be successful:

$$\frac{\text{Most applications of AI have not been successful}}{\text{(probably) the next use of AI will be successful}} \quad \text{AI}$$

Here's a less familiar objection to rule circularity, which is most clearly seen in the case of IBE. It pertains to the requisite premises from which IBE's reliability is to be inferred (either inductively or by IBE). The success of past invocations of IBE cannot be established (because the conclusions are unobservable). This is in marked contrast to our (inductively-based) inferences that the sun will rise the following day, which we have tested innumerable times—by waiting a day and seeing the sun rising. This is a serious concern. But initial appearances to the contrary, it also arises with respect to the inductive justification of induction whose conclusions are *general* ('The sun *always* rises'). To be sure, there are (relatively) few instances of such inductions being known to fail (by particular instances of the generalizations being found to be false). But this only allows us to infer (inductively) that generalizations inferred inductively will not be falsified. And that's different from their being *true*, which we need in order to show the reliability of induction to generalizations. This means that with respect to circular justification, too, induction and IBE are on a par epistemologically: in neither case can the reliability of the inference rule be established, even if the inference rule is allowed to be invoked.

6.3 Third argument against IBE

IBE, van Fraassen (1989, pp. 142–3) argues, takes us to the best explanation among those we have considered. And even if explanatoriness is a guide to truth, we have no reason to suppose that the true theory is among the theories we have considered. So 'our selection may well be the best of a bad lot'. IBE, he concludes, is unjustified.

Here is Lipton's (1993) rendition of the argument:

1)   We can reliably rank the competing theories we have generated with respect to likelihood of truth. [*ranking* assumption (which the sceptic grants)]

2)   We have no reason for thinking a true theory is included among the theories we have considered. [*no-privilege* assumption]

Conclusion: We have no warrant for making absolute probability assessments (for instance, that a theory is more likely to be true than false). In particular, we have no reason for thinking our best explanation is likely to be true.

This argument is easily rebutted. As Lipton (1993) notes, the first premise is inconsistent with the conclusion. If, as the first premise has it, we can reliably compare the likelihood of any two theories we consider (on the basis of their explanatoriness), we *can* reliably assess absolute probabilities. Given the relative ranking of T and ~T, we can either judge that the higher-ranking one is likely to be true (its probability is greater than 1/2), or—if the two are tied—that they both have probability 1/2.

The argument that Lipton (ingeniously) rebuts is based on one way of interpreting the sceptic's concession that explanatoriness is a guide to truth. But by invoking a *different* construal, the sceptic can adduce an argument that is invulnerable to the objection. The sceptic means that the true theory is more explanatory than its rivals. Were the true theory among those we consider, we would rank it highest (and infer to its truth).[10] But this does not mean, as Lipton's sceptic thinks, that we can compare probabilities of theories *that we consider* on the basis of explanatory considerations, and justifiably suppose that *our* topmost theory is likely to be true.

Here is a better rendition of the Bad Lot argument:

1)   The best (and good enough) explanation *out of all possible explanations* of a phenomenon is true. [assumption conceded by the sceptic]

2)   We have no reason for thinking a true theory is included among the theories we have considered. [*no-privilege* assumption]

Conclusion: We have no warrant for making absolute probability assessments (for instance, that a theory is more likely to be true than false). In particular, we have no reason for thinking our best explanation is likely to be true.

In this argument, the first premise *is* consistent with the conclusion. If the true theory were included among those we considered, it would be chosen. But the sceptic thinks we have no reason for thinking it *is* included.

To contend with the argument, we have to rebut the *no-privilege* premise or deny that the sceptical conclusion follows from the argument's premises. All the commentators

---

[10]   The Bad Lot argument concedes an assumption that even friends of IBE don't accept: that explanatoriness is a *perfect* guide to truth. Of course, this strengthens the argument: even if we are allowed to make this overly optimistic assumption, IBE will be unjustified.

accept that *no-privilege* renders IBE unjustified. The sceptics (van Fraassen, 1989; Wray, 2008; Khalifa, 2010) conclude that IBE is unjustified. Friends of IBE (Lipton, 1993; Psillos, 1999) attempt to rebut *no-privilege*.

I think the prospects for establishing *privilege* are not good. I will focus, instead, on the possibility of faulting the inference from *no-privilege* to the sceptical conclusion. The inference presupposes that the IBE-derived conclusion is justified only if we are justified in believing that a true theory is included among the theories we have considered. But why does the sceptic think we must be justified in believing this? The answer must be that, given the first premise (that explanatoriness is perfectly correlated with truth), it justifies our belief that when we invoke IBE, we will choose a true theory, i.e. that IBE is reliable. But to require the (justified) belief in IBE's reliability for its justified invocation is to impose the internalist requirement. This means that (without knowing it) we have already considered this way of contending with the Bad Lot argument (when we considered the Humean argument against IBE (Section 6.2)).

The Bad Lot argument has no inductive analogue. The sceptic here concedes that explanatoriness is a (perfect) guide to truth, and invokes the fact that we only consider a small subset of the explanations. We have no reason, he argues, for thinking that there aren't better explanations that we haven't considered. And the analogous argument for induction is hopeless. We *know* that a generalization is the most *uniform* state description. For instance, a world in which all ravens are black is more uniform, everything else being equal, than one in which two thirds of the ravens are black. There's no need even to consider the alternatives.

It might seem as if the Bad Lot argument doesn't work against IBE to the *observable*. The problem it discerns seems to be engendered by the fact that there are many explanatory theories that we haven't even formulated because they are couched in unfamiliar terminology. Newton *coined* the term 'mass', so scientists before him *couldn't* consider his theory. That is why, in applying IBE, they chose the best from a 'bad lot'. But when we infer to the best *observable* explanation, we are considering alternative theories that are all couched in observable terminology: concerned with the relationship between heat, flames, and distances, for instance. And here, we might suppose, there is no under-consideration.

Alas, the (anti-realist) insouciance is unwarranted. True, there are explanations that we haven't considered because they cannot be formulated in our current language. But we never consider all the explanations in our *own* language. When the detective hits upon the idea that the butler is the illegitimate and unacknowledged son of the host, and murdered him having born him a strong grudge, the explanation that occurred to no one else is couched in everyday language. Even more prosaically, when it occurs to me that my spectacles are in the fridge (where I absent-mindedly put them), the explanation does not involve any new theoretical terms. In both cases, the explanatory knack takes a different form from Newton's. And our limitation renders even IBE to the observable vulnerable to the under-consideration argument.

6.4 Fourth argument against IBE

The Bad Lot argument has another version, whose conclusion is that IBE is *unreliable*, rather than (as most sceptical arguments have it) *unjustified*. Thus, van Fraassen argues (1989, p. 146) that it is very likely that among the explanations we haven't considered there are many that are as good as, or even better than, the best explanation among those we have considered. Since (on the basis of explanatory considerations) our explanation isn't more likely than some of the unconsidered ones, its likelihood of being true is inversely proportionate to the number of its serious rivals. IBE, van Fraassen concludes, isn't reliable.

Like the previous version of the Bad Lot argument, and for much the same reason, this argument has no inductive analogue. The sceptic concedes that explanatoriness is a (perfect) guide to truth, but argues that it is very likely that there are many equally good, even better, explanations that we haven't considered. But we know (for sure) that the generalization ('All ravens are black', for instance) is the most *uniform* state description. So if the inductive sceptic concedes, analogously, that uniformity is a (perfect) guide to truth, induction is (perfectly) reliable! So, this argument constitutes a challenge only to IBE (even to the observable).

6.5 Fifth argument against IBE

Stanford (2006) adduces historical evidence in support of the claim that scientists choose from theories to which there are unconsidered rivals that explain equally well or even better. For instance, the evidence available to proponents of Aristotelian science was equally well (or even better) explained by Newtonian science (that they didn't consider). We are *always*, he claims, in a state of 'transient underdetermination': there are alternatives to our best scientific theories that are *presently unconceived* by us and as explanatory of our available evidence (Stanford 2006, p. 17). This means that IBE engenders choices that aren't justified: if a theory has a rival that is equally likely, its probability is below the threshold for justified belief.

Stanford's 'new induction', too, has no plausible inductive analogue, for the (by now) familiar reason: we *always* conceive of the generalization when we extrapolate, and we know it is more uniform than any rival, even unconsidered.

Unlike the previous arguments against IBE, Stanford's targets only IBE *to the unobservable*. But it has an analogue that targets IBE to the observable: we have often failed to think of all the plausible observable explanations (Section 6.3).

7. Conclusion

Since the eliminativist response to scepticism about IBE isn't viable (Section 5), our only way of avoiding capitulation to the sceptic is to rebut his arguments. In this chapter I attempted to compare the strength of the sceptical case against IBE with that of its inductive counterpart. I considered one argument that targets both inference rules

(Section 6.1), one argument (Hume's) that can be transformed into an equally cogent argument against IBE (Section 6.2), and three arguments which target only IBE and have no inductive analogues (Sections 6.3–6.5). These three may show (if they cannot be rebutted) that IBE, even to the observable, is more vulnerable to scepticism than induction.[11]

References

Ayer, A.J. 1972. *Probability and Evidence*. London: Macmillan.

Berkeley, G. 1710. *Principles of Human Knowledge*, in *The Works of Georg Berkeley*, 9 vols, Luce, A.A. and Jessop, T.E. (eds). London: Nelson, 1948–57.

Boyd, R. 1984. 'Current Status of Scientific Realism', in *Scientific Realism*, Leplin, J. (ed.). Berkeley: University of California Press.

Bridgman, P.W. 1962. *A Sophisticate's Primer of Relativity*. Middletown, CT: Wesleyan University Press.

Craig, W. 1953. 'On Axiomatizability within a System'. *Journal of Symbolic Logic* 18, pp. 30–2.

Fumerton, R. 1980. 'Induction and Reasoning to the Best Explanation'. *Philosophy of Science* 47, pp. 589–600.

Goodman, N. 1955. *Fact, Fiction and Forecast*. New York: Bobbs-Merrill.

Greco, J. 2000. *Putting Skeptics in Their Place*. Cambridge: Cambridge University Press.

Harman, G. 1965. 'The Inference to the Best Explanation'. *Philosophical Review* 74, pp. 88–95.

Hempel, C.G. 1965. 'Aspects of Scientific Explanation', in *Aspects of Scientific Explanation*. New York: Free Press.

Howson, C. 2000. *Hume's Problem*. Oxford: Clarendon Press.

Hume, D. 1739. *A Treatise of Human Nature*, edited by L.A. Selby-Bigge, 2nd edition. Oxford: Clarendon Press, 1978.

Hume, D. 1777. *Enquiries Concerning Human Understanding*, edited by L.A. Selby-Bigge, 3rd edition. Oxford: Clarendon Press, 1975.

Jackson, F. 1977. *Perception*. Cambridge: Cambridge University Press.

Khalifa, K. 2010. 'Default Privilege and Bad Lots: Underconsideration and Explanatory Inference'. *International Studies in the Philosophy of Science* 24, pp. 91–105.

Klein, P. 2006. 'Contemporary Responses to Agrippa's Trilemma', in *The Blackwell Guide to Epistemology*, Greco, J. and Sosa, E. (eds). Oxford: Blackwell.

Kneale, W. 1949. *Probability and Induction*. Oxford: Clarendon Press.

Lipton, P. 1993. 'Is the Best Good Enough?' *Proceedings of the Aristotelian Society* 93, pp. 89–104.

Lipton, P. 2004. *Inference to the Best Explanation*. London: Routledge.

Lycan, W.G. 1988. *Judgement, and Justification*. Cambridge: Cambridge University Press.

Newton-Smith, W. 1981. *The Rationality of Science*. London: Routledge and Kegan Paul.

Norton, J. 2003. 'A Material Theory of Induction'. *Philosophy of Science* 70, pp. 647–70.

Okasha, S. 2001. 'What Did Hume Really Show about Induction?' *Philosophical Quarterly* 51, pp. 307–27.

Popper, K.R. 1972. *The Logic of Scientific Discovery*, 4th edition. London: Hutchinson.

[11] I am very grateful to the editors for their insightful, constructive, and painstaking comments. This research was supported by the Israeli Science Foundation (grant number 366/14).

Psillos, S. 1999. *Scientific Realism: How Science Tracks Truth*. London: Routledge.

Salmon, W.C. 1957. 'Should We Attempt to Justify Induction?' *Philosophical Studies* 8, pp. 33–48.

Sextus Empiricus (OP). 1933. *Outlines of Pyrrhonism*, translated by R.G. Bury. London: W. Heinemann.

Stanford, P.K. 2006. *Exceeding Our Grasp: Science, History and the Problem of Unconceived Alternatives*. Oxford: Oxford University Press.

Swinburne, R. 1974. *The Justification of Induction*. Oxford: Oxford University Press.

van Cleve, J. 1984. 'Reliability, Justification and Induction'. *Midwest Studies in Philosophy* 9, pp. 555–67.

van Cleve, J. 2003. 'Is Knowledge Easy – or Impossible? Externalism as the Only Alternative to Skepticism', in *The Skeptics*, Luper, S. (ed.). Farnborough: Ashgate, 2003.

van Fraassen, Bas C. 1980. *The Scientific Image*. Oxford: Clarendon Press.

van Fraassen, Bas C. 1989. *Laws and Symmetry*. Oxford: Oxford University Press.

Vogel, J. 1990. 'Cartesian Skepticism and Inference to the Best Explanation'. *Journal of Philosophy* 87, pp. 658–66.

Wray, K.B. 2008. 'The Argument from Underconsideration as Grounds for Anti-Realism: A Defence'. *International Studies in the Philosophy of Science* 22, pp. 317–26.

13

External World Skepticism and Inference to the Best Explanation

Susanna Rinard

1. Introduction

Many philosophers regard the conclusions of skeptical arguments as obviously false. The interest of such arguments, it is often said, lies in the challenge of figuring out exactly which premise is false, and why.

I have argued elsewhere (Rinard 2013) that such an attitude is wrong-headed. We cannot determine in advance of serious inquiry that skeptical arguments will ultimately fail. If, on reflection, the premises are sufficiently plausible, and the argument valid, the rational response may be to accept the conclusion. This view pulls the rug out from under responses to skepticism that rely essentially on the Moorean attitude just described—responses that simply presuppose that the skeptic's premises aren't all true. Such responses involve identifying which premise they think is false, and arguing against views that identify a different premise as false. Such responses fail to rule out the possibility that all the premises, and the conclusion, are true, and so they fall short of a complete and satisfying response to skepticism.

One sort of response to skepticism that doesn't obviously fall into this category is that involving Inference to the Best Explanation (IBE). My aim in this chapter is to explore a variety of IBE responses to external world skepticism. I identify three potential problems for such approaches. The first two pose genuine challenges, but I'm not certain that they undermine all possible IBE responses.

However, the third line of concern, in my view, gets to the heart of the matter. I make a case for the claim that explanatory goodness is not a guide to the truth. If so, this is a fatal blow to any IBE response to skepticism (and, indeed, any use of IBE in general). My argument rests on a principle, in the spirit of the principle of indifference, which some may reject. So I cannot hope to convince everyone. However, the line of reasoning I give identifies what I take to be the central flaw in IBE responses to skepticism. At the very end of the chapter I briefly mention my own preferred approach to skepticism: the skeptic is right that our evidence does not support our external world beliefs, but pragmatic and moral considerations justify us in holding them anyway.

2. An Argument for External World Skepticism

There are many different ways of formulating the argument for external world skepticism. What I take to be one of the strongest versions begins with the following premise:

(1)   One's basic evidence about the external world is restricted to propositions about the way the external world appears.[1]

Before stating the second premise, we'll need to define some terms. Let "Normal" refer to the Normal World hypothesis, according to which the external world is largely as you believe it to be. Let "BIV" refer to the hypothesis that you are a radically deceived brain in a vat, with sensory experiences exactly like those you actually have. The second premise, as follows, is motivated by the observation that Normal and BIV both entail that you have the sensory experiences that you do:

(2)   Propositions about the way the external world appears are evidentially neutral between Normal and BIV.

Premise (3) denies that there is any asymmetry in extra-evidential considerations:

(3)   Neither Normal nor BIV is intrinsically more worthy of belief, independently of one's evidence.

From (1)–(3) we can derive the subconclusion that one neither knows, nor is justified in believing, that BIV is false. In order to derive full-on external world skepticism, we just need one more premise—the so-called closure principle:

(4)   If one neither knows nor is justified in believing Q, and one knows that P entails Q, then one neither knows nor is justified in believing P.[2]

These four premises entail the skeptic's conclusion:

(5)   For many external world propositions P, one neither knows nor is justified in believing P.

I'll be exploring IBE responses to this argument that involve denying premise (2). According to responses of this kind, facts about how the external world appears to us are not evidentially neutral between Normal and BIV, but rather favor Normal over BIV, because Normal provides a *better explanation* of these facts than BIV does. Indeed, according to the IBE response, Normal is the *best explanation* of our actual

---

[1]   Some, such as McCain (2014, chapter 2), take evidence to be mental states, rather than propositions.
[2]   Although most accept it, Dretske (1970) and Nozick (1981) are two prominent deniers of the closure principle for knowledge. Note, however, that it is far less common, and far more implausible, to deny closure for justification. The argument for skepticism about justification remains intact even if closure for knowledge is rejected.

sensory experiences, and this means we should believe it—or, at the very least, we should be more confident of it than its negation.

There are many different ways of developing a response of this kind, which differ in their accounts of why Normal is the best explanation. The first view I'll consider is as follows:

> The Continuity View: Normal is the best explanation of our sensory experiences in virtue of the *continuity* and *regularities* that they exhibit over time.[3]

In the next few sections I'll explore several different potential lines of concern I have about an IBE response to external world skepticism based on the Continuity View.

3. First Line of Concern about the Continuity-Based IBE Response

I'll begin by highlighting two assumptions that I will be granting, for the sake of argument, in this section. I'll question both assumptions in subsequent sections. The first assumption I'll grant is that explanatory goodness is, in the relevant sense, a guide to truth. That is, if some hypothesis really is the best explanation of some data, then those data do, in fact, support that hypothesis over all alternative hypotheses. The second assumption I'll grant is that Normal is in fact the best explanation of the continuity and regularities that our experiences exhibit over time.

I will argue that *even if* both of these claims are true, there may still be a serious problem for the continuity-based IBE response. The source of the potential problem is an argument for skepticism about the past that is perfectly parallel to the argument given in Section 2 for skepticism about the external world.

To construct this argument, first let "BIV(NoPast)" refer to the hypothesis that you are a brain in a vat who was brought into existence just a second or so ago, complete with apparent memories that are indistinguishable from all your real memories in Normal.[4] (Normal, again, is the hypothesis that the external world, and the past, are largely as you believe them to be.) Now we can construct the argument for skepticism about the past by replacing "BIV" with "BIV(NoPast)" and "the external world" with "the past." The result is as follows:

> (1*)   One's basic evidence about the past is restricted to propositions about the way the past appears (i.e. the way one seems to remember things having been).
>
> (2*)   Propositions about the way the past appears are evidentially neutral between Normal and BIV(NoPast).

---

[3] Strategies along these lines are pursued by Bertrand Russell (1921) and Laurence BonJour (1999) and (2003).

[4] Perhaps the most famous skeptical hypothesis concerning the past is Russell's (1921, 159), in which the world sprang into existence five minutes ago, complete with a group of people who seem to remember what we actually remember.

(3*) Neither Normal nor BIV(NoPast) is intrinsically more worthy of belief, independently of one's evidence.

(4*) If one neither knows nor is justified in believing Q, and one knows that P entails Q, then one neither knows nor is justified in believing P.

(5*) Therefore, for many propositions P about the past, one neither knows nor is justified in believing P.

This argument is perfectly parallel to the argument for skepticism about the external world, and the premises of each seem equally plausible. So, I claim, one who accepts the argument for external world skepticism given in Section 2 should also accept this argument for skepticism about the past.

Now, the external world skeptic, who, we are supposing, is also a skeptic about the past, suspends judgment on all propositions about the past. In particular, they suspend judgment on whether or not they've had any experiences at all in the past, and so, they suspend judgment on whether or not those experiences (if any) have exhibited continuity and regularities over time.

What this means is that an IBE response to external world skepticism that is based on the Continuity View will not be dialectically effective against the external world skeptic who also accepts skepticism about the past (which, I've just argued, they should). I take this to be a serious problem for the continuity-based IBE response.

Although this problem does undermine the IBE response based on the Continuity View as formulated above, there might be an alternative strategy in the vicinity that would avoid this difficulty. I'll briefly consider two such possibilities, Proposal 1 and Proposal 2.

Proposal 1 claims that one's sensory experiences *at a single time* exhibit continuity and regularities, and that Normal is the best explanation of these for the same reason that Normal is the best explanation of continuity and regularities in experience *over time*. Even the skeptic about the past has beliefs about their current experiences; so, if the proponent of the IBE response to skepticism can successfully make the case that *those* are best explained by Normal, for precisely the same reason that Normal best explains regularities in experience over time, then they will have avoided this particular problem.[5]

Proposal 2 claims that at a single time we *seem* to remember having had experiences that were continuous and regular over time, and that this is best explained by the hypothesis that we actually had such experiences, for the same reason that Normal is the best explanation of the existence of continuity and regularities in experience over time.[6] If this is right, then even someone who starts out skeptical about the past should infer from their current apparent memories that they did in fact have experiences that

---

[5] A similar proposal can be found in McCain (2014, chapter 6).

[6] Thanks to Andrew Graham for suggesting this to me. Further discussion of similar views can be found in Ted Poston (2016).

were continuous and regular over time; and, if the best explanation of those is Normal, then, they should then infer that Normal is (probably) true.

At this point, I think we cannot evaluate either Proposal 1 or Proposal 2 without more detail about why we should think that Normal is the best explanation of the existence of continuity and regularities in our experiences over time.

This is because we need to know what the reasons are for thinking that Normal is the best explanation of continuity and regularities in experience over time before we can ascertain whether reasons *of the same kind* support thinking that (as Proposal 1 claims) Normal is the best explanation of continuity and regularities in our experiences *at a single time*, and (as Proposal 2 claims) that the best explanation of our apparent memories as of having had continuous and regular experiences over time is that we actually had such experiences.

So, it is to this question that I turn in Section 4. There I will cast doubt on the claim that Normal is the best explanation of the existence of continuity and regularities in our experiences over time. This, in turn, casts doubt on the idea that either Proposal 1 or Proposal 2 will be capable of grounding a satisfactory IBE response to skepticism.

4. Second Line of Concern about the Continuity-Based IBE Response

In this section I will continue to grant, for the sake of argument, that explanatory goodness is a guide to truth, and so, a hypothesis that better explains the data is better supported by them. However, I'll question whether Normal really is the best explanation of the continuity and regularities that our experiences exhibit over time.

It would be natural to begin by comparing Normal with BIV. Many reasons have been given for the claim that Normal is a better explanation than BIV.[7] For example, it's been claimed that BIV is more complicated than Normal; that it's more ad hoc than Normal; that BIV is in some sense parasitic on Normal; and so forth. While I'm not persuaded by these claims, I won't focus on that here. Rather, I'll suggest that, *even if* we grant that Normal is a better explanation than BIV, there is still another hypothesis that may be an even better explanation than Normal.

First, I'll make some remarks about the explanatory strategy motivating Normal. The phenomenon to be explained is the presence of continuities and regularities *in experience*. In order to explain this phenomenon, Normal postulates a new realm of objects of an entirely different sort—external, mind-independent physical objects— that exhibit continuities and regularities that correspond to the continuities and regularities in experience. For example, to explain the fact that my visual experience, over a period of time, has a stable round black component, we postulate the existence of a ball that remains round and black over that period of time. The postulated roundness

---

[7] For example, Jonathan Vogel (1990, 2005, and elsewhere) has given detailed arguments for this claim.

and blackness of a ball over time is a continuity in the realm of external physical objects that corresponds to the continuity in my visual experience of a component that remains round and black over time.

But now consider this: if the regularities in our *experience* call out for explanation, don't the regularities in the postulated material objects also call out for explanation in precisely the same way? Why is it that the ball remains round and black over time? So far, Normal gives us no explanation. We seem not to have made much explanatory progress.

One strategy would be to supplement Normal by reiterating the same strategy. Perhaps we should postulate a *third* realm of objects of a still different sort, with continuities and regularities that correspond to, and hence explain, the continuities and regularities in the postulated physical objects. Of course, *those* regularities will call out for explanation as well—and so we will need to implement the strategy again, postulating yet another realm of an entirely different sort of objects, and so on ad infinitum.

This is absurd. We do not think that the existence of infinitely many realms of different sorts of objects, with continuities mirroring each other, is the best explanation of the continuities in our experience. When defenders of Normal are faced with the question of how best to explain the continuities they postulate in the physical objects they postulate, they do not in fact re-apply their original strategy. Instead, they postulate *laws* governing the physical objects in order to explain the regularities they exhibit.

But if this is the strategy one should adopt at this stage, we should question why we bothered to take the step of postulating a new realm of objects (external physical objects) in the first place. If, at some point, we're going to have to postulate laws operating in a certain realm to explain the continuities in that realm, why not simplify matters and just postulate laws that operate in the realm of experience to explain the continuities exhibited in experience? This approach is both quantitatively and qualitatively ontologically simpler than Normal. We can, it seems, do exactly the same explanatory job while doing without an entire realm of objects, the external material objects. Call this the Idealist Alternative. My suggestion is that Normal is *not* the best explanation of the continuities and regularities in our experiences, because the Idealist Alternative is a better one.

I'll consider two objections. According to the first, we can postulate *more* continuity in the external objects than is present in our experiences. For example, suppose our visual field contains an immobile brown square shape in the middle, and a round black shape on the left that is slowly moving to the right. The round black shape approaches the brown shape, and then vanishes bit by bit from the right to the left, and then reappears bit by bit, from the right to the left, on the right-hand side of the brown square. The black round component of our experience is discontinuous in a certain way over time. But if we explain it by postulating external material objects, we can suppose that the black round ball corresponding to it is not discontinuous in a parallel way over time. It remains exactly as black, and as round, the entire time. We explain the discontinuity in our experience by supposing that it moved *behind* the brown shape.

Since we can postulate more continuity in the physical objects than is present in our experience, the thought goes, the laws we could postulate governing the physical objects would be *simpler* than laws we might postulate governing our experience. This, the objector claims, makes Normal simpler than the Idealist Alternative.

I am happy to agree that Normal may be a bit simpler *in this respect*. But it is more complicated in a number of other respects. As we have already seen, the Idealist Alternative is much more ontologically simple, in two respects: it postulates a much smaller total number of objects, and also fewer *kinds* of objects. Moreover, it is nomologically simpler in at least the following respect: Normal needs to postulate not only laws governing the physical objects themselves, but also laws governing the interface between the physical objects and the experiences. The mere existence of a round black ball does not, by itself, explain why I have a round black component in my experience. We require an account of how the one gives rise to the other. But no such complication is involved in the Idealist Alternative. We need only postulate laws governing the experiences themselves, and not also laws connecting regularities in another realm with regularities in experience. In short, although I agree that Normal is simpler in at least one respect, I claim that the Idealist Alternative is simpler in many others.

The second objection points out that, at this stage, we do not have much of an idea about what laws of experience would look like and how they would operate. In particular, whereas we can postulate mechanistic laws that govern physical objects, it's unclear how we might do so in the realm of experience.

I agree that we have, to date, a much more fully fleshed out system of laws that could govern physical objects than laws that could govern experiences. However, it seems to me the best explanation of this is simply that an enormous amount of intellectual effort has been expended on the first task, but not the second. There have not been any serious, sustained attempts to develop a system of laws governing experience. I see no principled reason to think that such a system would be harder to develop than a system of laws governing physical objects. Of course, some such reason may be discovered, were inquiry in this direction to be pursued; but at this stage, I see no reason to suspect as much.

It's also worth pointing out that, presumably, those defending an IBE response to skepticism would want to vindicate the ordinary external world beliefs of pre-scientific humans as well as scientifically sophisticated modern humans. There was a time when very little progress had been made in developing a system of laws that might govern physical objects. Presumably a defender of the IBE response would want to claim that even for people living at that time, Normal was a better explanation than the Idealist Alternative. But I see no reason to think so.

Let's take stock. Thus far in this section I have given reasons for doubting that Normal is the unique best explanation of the continuities and regularities in our experience over time. The Idealist Alternative, I have suggested, compares favorably with Normal; at the very least, it is not obvious that Normal is a superior explanation.

At this point, it is worth considering the prospects for an IBE response to skepticism that focuses on some feature of our experiences *other than* the continuities and regularities that they exhibit. In particular, we can imagine an IBE response grounded in the *representational nature* of our experiences. On this proposal—which I'll call the "Representational View" (in contrast to the Continuity View), Normal is the best explanation of the fact that our experiences represent the world as being one way rather than another. Normal is the best explanation because a set of beliefs according to which the world is the way it is represented as being is, other things equal, more coherent than a set of beliefs according to which the world is *not* the way it is represented as being. Coherence is an explanatory virtue; so, given that our experiences represent the world as containing external objects with certain features, the hypothesis that there are such objects will have greater explanatory virtue than alternatives.[8]

In Section 5 I will present what I take to be the deepest problem for IBE responses to external world skepticism. This problem will apply to the Representational View as well as the Continuity View. Indeed, it applies to any IBE response to skepticism whatsoever.

5. Third Line of Concern: Explanatory Goodness Is Not a Guide to Truth

In this section I will evaluate a claim that, in Sections 3 and 4, I have taken for granted: the claim that explanatory goodness is a guide to truth, and so, we should believe the hypothesis that best explains our evidence—or, at least, be more confident of it than its negation. I will make a case for the claim that explanatory goodness is not a guide to truth in this sense. If I am right, then even if Normal is the best explanation of our experiences—either in virtue of their continuity and regularities over time, or their representational content, or some other feature—it does not follow that we should be more confident that Normal is true than false.

First, I'll make a terminological clarification. On one reading, "best explanation" just means something like "explanation most likely to be true." On this reading, it could not fail to be the case that we have the most reason to believe the hypothesis that is, in fact, the best explanation. This is not what I mean by "best explanation." In the sense I intend, a hypothesis is the best explanation just in case it has the greatest overall balance of explanatory virtues, which may include simplicity, parsimony, elegance, coherence, and so forth. My claim will be that these features are not guides to truth.

I'll begin by considering the following version of the Principle of Indifference:

The Principle of Indifference (POI):   If P is a finite natural partition over the space of epistemic possibilities (i.e. possibilities compatible with what you know with certainty), then you should assign equal credence to each cell of P.

---

[8] Paul Moser (1989, 161) seems to have in mind some idea along these lines.

The requirement that P be a *natural* partition is crucially important. If we allowed any finite partition whatsoever, then POI would give radically inconsistent recommendations across the board. There are many finite partitions of a set of possibilities, and it is impossible to assign equal credence to the cells of each with a single credence function that respects the laws of probability.

Some may worry that the requirement that P be a natural partition is not enough. It is well known that in some cases there are multiple partitions, each of which seems natural, and none more natural than any of the others, but which are such that no probability function assigns the same value to the cells of each (see, for example, van Fraassen 1989). Such cases are indeed puzzling. One possibility is that there is, in fact, a unique genuinely natural partition in such cases; our inability to identify which one it is constitutes a limitation on our part, perhaps one that can be overcome with further reflection on the cases. Another possibility, though, is that there is no unique natural partition in such cases. If so, then our principle needs to be modified to take this into account. The simplest way to do so is to weaken the principle so that it remains silent on such cases:

The Principle of Indifference (POI): Let P be a finite natural partition over the space of epistemic possibilities (i.e. possibilities compatible with what you know with certainty). If there is no other finite natural partition P', such that no coherent credence function assigns equal credence to each cell of P, and equal credence to each cell of P', then you should assign equal credence to each cell of P.

Henceforth "POI" will refer to this weaker formulation of the principle, which I endorse. Although the principle just stated remains silent in the case of multiple natural partitions, I'll note here one possible proposal, which has a subjectivist flavor. One might hold that, in such cases, rationality requires you to simply make a judgment call; to just pick one of the partitions and assign equal credence to each cell of that partition. It is permissible to pick any one of the natural partitions, but it is required that you do pick one.

I'll now illustrate POI by applying it to a few cases. First, suppose that all you know with certainty is that ball B is either red, green, or blue. Partition P groups the possibilities this way: (1) B is red; (2) B is green; (3) B is blue. Partition P', on the other hand, groups the possibilities in this way: (1) B is red; (2) B is not red. Intuitively, P is natural, but P' is not.[9] So, POI recommends that we assign equal credence (namely, one third) to each of the three cells of partition P. This means that we will assign unequal credence to the cells of P': one third to "B is red," and two thirds to "B is not red."

What I want to emphasize here—this will be important later—is that this corresponds to the fact that in some intuitive sense, there are *more ways* for ball B to be non-red than red. I say "in *some* intuitive sense" because we have to be careful here. It is not the case that there are more *maximally fine-grained possibilities* in which B is non-red than in which B is red, because there are, arguably, uncountably infinitely many of both. So when I say that there are more ways for B to be non-red than red, by "ways" I do not mean "maximally fine-grained possibilities." Nonetheless, I think we can identify an intuitive sense in which there are more ways for B to be non-red than red, and that this corresponds to the fact that we should be more confident that B is non-red than red.

Here's another example. Suppose that all you know is that X's height, h, is between 5 feet and 6 feet. Partition P has two cells: (1) h is in [5, 5.5]; (2) h is in [5.5, 6]. Partition P' also has two cells: (1) h is in [5, 5.01]; (2) h is in [5.01, 6]. Again, P is natural but P' is not. So, according to POI, we should assign equal credence (namely, one half) to each of the two cells in P, but unequal credence to each of the two cells in P'. We should be more confident that h is in [5.01, 6] than [5, 5.01]. Again, this corresponds to the fact that there are more ways for h to be in the former than the latter. Again, by "ways" here I do not mean "maximally fine-grained possibilities," since, again, there are uncountably many of these in both partition cells of both partitions.

One thing I have done so far is use the Principle of Indifference to help identify an intuitive sense in which, for some proposition P, there can be more ways in which P is true than ways in which P is false. However, having identified this sense, we can use it directly in the statement of epistemic principles, without explicit mention of POI. For example, consider the following principle: (*) If, given what you know with certainty, there are more ways for P to be true than for P to be false, then you should be more confident of P than its negation.

If we wish, we can apply (*) to cases directly, without first thinking explicitly about natural partitions and POI. Sometimes doing so will be more straightforward. For example, suppose that all I know with certainty is that my birthday present is in a box covered with red wrapping paper. I should be more confident that my birthday present is not a beetle than that it is a beetle, because there are more ways for it to be a non-beetle than a beetle. Similarly, I should be more confident that my birthday present is non-red than that it is red, because there are more ways for it to be non-red than red. And so forth. Had we insisted on explicitly using POI, reaching these verdicts would have been more involved. What exactly is the natural partition, one of whose cells is "My birthday present is a beetle," such that I should have equal credence in every cell of that partition? There may well be one; and reflection may enable us to identify it; but it is not obvious what exactly it is. What is clear, though, is that there are more ways for my birthday present to be a non-beetle than a beetle, and so, (*) allows us to conclude that I should be more confident that it is a non-beetle than a beetle.

It is worth emphasizing the fact that this is all relative to what the agent knows with certainty. For example, consider some arbitrary unknown object X, and let p be the percentage of the surface area of X that is red. Let H be the hypothesis that p is in [0.5, 0.8]. If the agent knows nothing about X, then she should be more confident that H is false than true, because there are more ways for p to be outside [0.5, 0.8] than inside it.

But suppose the agent learns, with certainty, E, namely, the proposition that p is either in [0.5, 0.8] or in [0.00001, 0.00002]. Now she should be more confident that H is true than false, because, given E, there are more ways for p to be inside [0.5, 0.8] than outside.

Having set up the basic approach, let us now apply it to the crucial question at hand. Given the sensory experiences that I have—given, for example, that I have an appearance as of a brown desk, a chalkboard, bookshelves, etc.—how confident should I be that my appearances are veridical, i.e., that Normal is true? According to the approach I've been developing, it will suffice to answer that question to answer this one: given that I have the sensory experiences that I do, are there more ways for my experiences to be veridical, or non-veridical?

The hypothesis that my experiences are veridical imposes strong constraints on the world. There must be a brown desk just there; a blackboard of a certain size on a wall opposite the desk; bookshelves, with particular books in a particular order, on the other wall; and so forth and so on. The hypothesis that my experiences are *not* veridical, however, leaves open all sorts of possibilities. There could be a black desk, or a green one, or no desk at all. There could be a chair instead, or no furniture; there could be a circus, or an orange tree, or a three-headed dog, or ... practically anything at all, *except* a brown desk, bookshelves, etc. of a very particular sort arranged in a very particular way. Since there are more ways for the world to be such that my experiences are non-veridical than veridical, I should, according to the approach developed here, be more confident that things are *not* the way they seem than that they are.

Once again, by "ways" here I do not mean "maximally fine-grained possibilities." There are uncountably infinitely many maximally fine-grained possibilities in which my experiences are veridical, and uncountably infinitely many maximally fine-grained possibilities in which my experiences are non-veridical. But, as we have seen in several examples already, that does not settle whether, in the intuitive sense identified above, there are more ways for my experiences to be non-veridical or veridical.

It may be helpful to compare the birthday present example with the case of the veridicality (or not) of sense experience. Supposing, as we did above, that I know the wrapping paper is red, I should be more confident that the birthday present it contains will be non-red than red, because there are many more ways for it to be non-red than red. In other words, I should be more confident that the color of the wrapping paper and the color of my birthday present will not match than that they will match. Similarly, I should be more confident that the nature of the external world will not match my sensory experiences than that it will match, since there are many more ways for it to not match than to match.

On this approach, then, I should not be more confident that Normal is true than false, *even granting* that Normal is the best explanation of my experiences, in the sense that it scores highest overall with respect to explanatory virtues like simplicity, coherence, parsimony, etc. Given the basic evidence of my senses, there are more ways for Normal to be false than true. So, even if Normal has the most explanatory virtues,

I should not be more confident that it's true than false—and so, these explanatory virtues are not guides to truth.

I have just outlined what I take to be the central problem for any IBE response to external world skepticism. It is worth noting that this problem is not unique to IBE responses—it undermines a number of other anti-skeptical views as well. For example, consider Pryor's (2000) dogmatism view, according to which an appearance as of P gives one prima facie justification for believing P. The considerations described above also undermine this view, and many others.

My own view, which I present in Rinard (2017), is that the skeptic is right in one important way, but wrong in another. The skeptic is right that the evidence of our senses does not support ordinary external world beliefs. But the skeptic is wrong to infer from this that we *should not* have ordinary beliefs. In my view, there are moral and pragmatic considerations that justify our having ordinary beliefs. We should believe the Normal World Hypothesis not because it is the best explanation of our sensory experiences, but rather because our doing so is good for ourselves and others.

Acknowledgments

Many thanks, for helpful questions, comments, and suggestions to Andrew Graham, Kevin McCain, Ted Poston, Miriam Schoenfield, participants in a TalkShop at the Harvard Philosophy Department, and audience members at the 2015 Orange Beach Epistemology Workshop.

References

BonJour, Laurence (1999) "Foundationalism and the External World," *Philosophical Perspectives* 13: 229–49.

BonJour, Laurence (2003) "A Version of Internalist Foundationalism," in *Epistemic Justification: Internalism vs Externalism, Foundations vs Virtues*. Oxford: Blackwell.

Dretske, Fred (1970) "Epistemic Operators," *Journal of Philosophy* 67: 1007–23.

McCain, Kevin (2014) *Evidentialism and Epistemic Justification*. London: Routledge.

Moser, Paul (1989) *Knowledge and Evidence*. Cambridge: Cambridge University Press.

Nozick, Robert (1981) *Philosophical Explanations*. Cambridge: Cambridge University Press.

Poston, Ted (2016) "Acquaintance and Skepticism about the Past," in B. Coppenger and M. Bergmann (eds), *Intellectual Assurance: Essays on Traditional Epistemic Internalism*. New York: Oxford University Press.

Pryor, James (2000) "The Skeptic and the Dogmatist," *Nous* 34(4): 517–49.

Rinard, Susanna (2013) "Why Philosophy Can Overturn Common Sense," in T. Gendler and J. Hawthorne (eds), *Oxford Studies in Epistemology*, Volume 4. Oxford: Oxford University Press.

Rinard, Susanna (2017) "No Exception for Belief," *Philosophy and Phenomenological Research* 94(1): 121–43.

Russell, Bertrand (1921) *Problems of Philosophy*. London: Home University Library.

van Fraassen, Bas (1989) *Laws and Symmetry*. Oxford: Clarendon Press.

Vogel, Jonathan (1990) "Cartesian Skepticism and Inference to the Best Explanation," *Journal of Philosophy* 87(11): 658–66.

Vogel, Jonathan (2005) "The Refutation of Skepticism," in M. Steup and E. Sosa (eds), *Contemporary Debates in Epistemology*. Oxford: Blackwell.

PART V

Applications of Inference to the Best Explanation

PART A

Applications of Inference
to the Best Explanation

14

Explanation, Confirmation, and Hempel's Paradox

William Roche

1. Introduction

Each of the following conditions has some prima facie plausibility:

Converse Consequence Condition (CCC): For any propositions E, H^*, and H, if (i) E confirms H^* and (ii) H^* is entailed by H, then E confirms H.

Entailment Condition (EC): For any propositions E and H, if E entails H, then E confirms H.

Special Consequence Condition (SCC): For any propositions E, H^*, and H, if (i) E confirms H^* and (ii) H^* entails H, then E confirms H.

The same is true of the condition:

Non-Triviality Condition (NTC): For some propositions E and H, E does not confirm H.

Hempel (1965a), though, shows (in effect) that CCC, EC, and SCC together entail that NTC is false:

Hempel's Argument

(1) E entails E for any proposition E.
Thus
(2) E confirms E for any proposition E. [by (1) and EC]
(3) E is entailed by E & H for any propositions E and H.
Thus
(4) E confirms E & H for any propositions E and H. [by (2), (3), and CCC]
(5) E & H entails H for any propositions E and H.

Thus

(6)   *E* confirms *H* for any propositions *E* and *H*. [by (4), (5), and SCC]

So, given that there is no questioning NTC (on any legitimate sense of confirmation), one or more of CCC, EC, and SCC should be rejected.[1] This is "Hempel's paradox."[2]

An intriguing possibility is that one or more of CCC, EC, and SCC should be modified in terms of explanation. Suppose, for example, CCC is modified so that "entailed" is replaced by "entailed and explained." Then (3) in Hempel's Argument would need to be modified accordingly. (3), though, would be false if it were so modified. For, as is uncontroversial, it is false that *E* is explained by *E* & *H* for any propositions *E* and *H*. Perhaps, then, Hempel's paradox can be solved by modifying one or more of CCC, EC, and SCC in terms of explanation.

It would not be enough—to solve Hempel's paradox—to simply modify one or more of CCC, EC, and SCC in terms of explanation so that Hempel's Argument fails. The new condition or conditions would need to be correct.

Does Hempel's paradox admit of a satisfactory solution? Are conditions such as CCC, EC, and SCC correct when modified in terms of explanation? These are the main questions of the chapter.

My interest in the second question is due in part to my interest in the first question. If the answer to the second question is affirmative, then perhaps so too is the answer to the first question.

There is more. Inference to the Best Explanation (IBE) is standardly construed in terms of categorical beliefs. The core idea can be put as follows:

Idea 1 (I_1):   Inferences to categorical beliefs (at least some of them) should be guided by explanatory considerations.

This idea has a counterpart concerning degrees of belief:

Idea 2 (I_2):   Changes in degrees of belief (at least some of them) should be guided by explanatory considerations.

I_2 has received much attention in recent years.[3] One issue discussed is whether I_2 runs counter to Bayesian conditionalization. Some researchers, for example, flesh out I_2 by

---

[1] The names "Converse Consequence Condition," "Entailment Condition," and "Special Consequence Condition" are Hempel's (1965a, pp. 31–2). Hempel formulates CCC, EC, and SCC in terms of sentences (as opposed to propositions) where *E* is an observation report. This difference in formulation is inconsequential for my purposes. The name "Non-Triviality Condition" is mine. Hempel (1965a, p. 32) discusses NTC but does not give it a name.

[2] This paradox is of course distinct from the ravens paradox (which sometimes goes by the name "Hempel's paradox"). See Hempel (1965a).

[3] I have in mind here the growing literature on Bayesianism and IBE. See, e.g., van Fraassen (1989), Douven (1999, 2011a), Okasha (2000), Lipton (2001, 2004), Salmon (2001a, 2001b), Psillos (2004), Tregear

developing an explanation-based alternative to Bayesian conditionalization. But there are alternative, and less radical, ways in which I_2 might be true. Suppose the answer to the second of the two main questions of the chapter is affirmative and so conditions such as CCC, EC, and SCC are correct when modified in terms of explanation. Suppose, further, conditions such as CCC, EC, and SCC should be understood in terms of changes in degrees of belief. Then it follows that I_2 is true.

The remainder of the chapter is organized as follows. In Section 2, I take a step towards solving Hempel's paradox by, in part, disambiguating CCC, EC, and SCC and showing that EC is true whereas CCC and SCC are false. In Section 3, I modify CCC and SCC in terms of explanation. The new conditions are CCC_E and SCC_E. I also note that CCC_E and SCC_E are correct only if explanatory relations place constraints on probability distributions. In Section 4, I consider three obvious candidate constraints (placed by explanatory relations on probability distributions). I argue that CCC_E and SCC_E are incorrect when understood in terms of those candidate constraints. In Section 5, I turn to Schupbach's (Chapter 4) articulation and defense of IBE in terms of power (explanatory power). I use it to construct a fourth candidate constraint. I argue that it too is no help for CCC_E and SCC_E. In Section 6, I set out the basics of Bayesian causal networks and construct a fifth candidate constraint. This constraint improves on the four constraints considered in Sections 4 and 5 in a certain key respect. I argue, though, that in the end the fifth candidate constraint too does not help with CCC_E and SCC_E. In Section 7, I return to Hempel's paradox and set out my preferred solution. In Section 8, I conclude.

2. A Step towards Solving Hempel's Paradox

It is standard in Bayesian confirmation theory to distinguish between *absolute* confirmation and *incremental* confirmation.[4] The distinction is this:

Absolute Confirmation: For any propositions E and H, E absolutely confirms H if and only if $Pr(H \mid E) > t$ where t is the threshold for high probability and $1 > t \geq 0.5$.

Incremental Confirmation: For any propositions E and H, E incrementally confirms H if and only if $Pr(H \mid E) > Pr(H)$.

Note that I have in mind probabilities understood as "degrees of belief" (or "rational degrees of belief").

(2004), Iranzo (2008), Huemer (2009), Weisberg (2009), Roche and Sober (2013, 2014), McCain and Poston (2014), Poston (2014, ch. 7) and Douven and Wenmackers (2017).

[4] See Carnap (1962, preface to 2nd ed.) on "concepts of firmness" and "concepts of increase in firmness."

Suppose the sense of confirmation at issue is absolute confirmation. Then CCC should be rejected while EC and SCC should be accepted. First, suppose a card is randomly drawn from a standard (and well-shuffled) deck of cards. Let E be the proposition that the card drawn is a Heart, H^* be the proposition that the card drawn is a Red, and H be the proposition that the card drawn is a Diamond. E confirms H^*, since $\Pr(H^* \mid E) = 1 > \mathbf{t}$. H^* is entailed by H. But E does not confirm H, since $\Pr(H \mid E) = 0 < 0.5 \leq \mathbf{t}$. So CCC is false. Second, for any propositions E and H, if E entails H, then $\Pr(H \mid E) = 1 > \mathbf{t}$. So EC is true. Third, for any propositions E, H^*, and H, if H^* entails H, then $\Pr(H \mid E) \geq \Pr(H^* \mid E)$. It follows that for any propositions E, H^*, and H, if $\Pr(H^* \mid E) > \mathbf{t}$ and H^* entails H, then $\Pr(H \mid E) > \mathbf{t}$. So SCC is true.[5]

Suppose, instead, the sense of confirmation at issue is incremental confirmation. Then each of CCC, EC, and SCC should be rejected. Return to the card case above. First, note that E confirms H^*, since $\Pr(H^* \mid E) = 1 > 1/2 = \Pr(H^*)$, and H^* is entailed by H, but E does not confirm H, since $\Pr(H \mid E) = 0 < 1/4 = \Pr(H)$. So CCC is false. Second, let E be some contingent proposition and H be a logically true proposition. Then E entails H but $\Pr(H \mid E) = 1 = \Pr(H)$. So EC is false. Third, let E be the proposition that the card drawn is not a Heart, H^* be the proposition that the card drawn is a Diamond, and H be the proposition that the card drawn is a Red. Then E confirms H^*, since $\Pr(H^* \mid E) = 1/3 > 1/4 = \Pr(H^*)$, and H^* entails H, but E does not confirm H, since $\Pr(H \mid E) = 1/3 < 1/2 = \Pr(H)$. So SCC is false.

The situation is a bit different if the sense of confirmation at issue is incremental confirmation and CCC, EC, and SCC are understood as restricted to propositions with non-extreme unconditional probabilities (i.e., unconditional probabilities greater than 0 and less than 1). CCC and SCC are still false, but EC is true. This is because for any propositions E and H with non-extreme unconditional probabilities, if E entails H, then $\Pr(H \mid E) = 1 > \Pr(H)$. Hence CCC and SCC should be rejected while EC should be accepted.

There are senses of confirmation (or evidential support) in addition to absolute confirmation and incremental confirmation.[6] But none of them is such that EC should be rejected while CCC and SCC should be accepted.[7] And none

---

[5] Hempel (1965a) holds that CCC should be rejected while EC and SCC should be accepted. It does not follow, however, that he has in mind absolute confirmation. There is reason to believe, in fact, that he does *not* have in mind absolute confirmation. His "satisfaction criterion of confirmation" (1965a, sec. 9), which is motivated in part by appeal to EC and SCC, is obviously inadequate when absolute confirmation is at issue. Let E be the proposition that Tweety is a raven and Tweety is black. Let H be the proposition that all ravens are black. Suppose $\Pr(H)$ is low. Hempel's satisfaction criterion of confirmation implies that E confirms H. Clearly, though, $\Pr(H \mid E)$ is low (not high) and thus is less than $\mathbf{t}$. See Carnap (1962, sec. 87) and Huber (2008) for discussion of how to understand Hempel on confirmation.

[6] See Douven (2011b), Roche (2012b, 2015), and Roche and Shogenji (2014a) for discussion.

[7] Suppose CCC and SCC are true. Take some propositions E and H^* such that E confirms H^*. Then:

(1) E confirms H^*.
(2) H^* is entailed by $H^* \mathrel{\&} H$ for any propositions H^* and H.

of them is such that SCC should be rejected while CCC and EC should be accepted.[8, 9]

I leave it for future investigation whether there are (interesting) senses of confirmation on which CCC and EC should be rejected while SCC should be accepted, or on which EC and SCC should be rejected while CCC should be accepted. I want to focus on confirmation in the sense of incremental confirmation. This is in part because of my interest in whether I_2 is true. So hereafter, unless otherwise noted, all talk of confirmation should be understood in terms of incremental confirmation.

No condition concerning incremental confirmation has any plausibility unless it is understood as restricted to propositions with non-extreme unconditional probabilities. So hereafter, unless otherwise noted, all talk of confirmation should be understood as restricted to propositions with non-extreme unconditional probabilities.

Thus

(3) E confirms H^* & H for any propositions E, H^*, and H such that E confirms H^*. [by (1), (2), and CCC]

(4) H^* & H entails H for any propositions H^* and H.

Thus

(5) E confirms H for any propositions E, H^*, and H such that E confirms H^*. [by (3), (4), and SCC]

Thus one or both of CCC and SCC should be rejected. This argument is noted in Skyrms (1966, p. 238).

[8] Suppose CCC and EC are true. Then:

(1) E entails $E \vee H$ for any propositions E and H.

Thus

(2) E confirms $E \vee H$ for any propositions E and H. [by (1) and EC]

(3) $E \vee H$ is entailed by H for any propositions E and H.

Thus

(4) E confirms H for any propositions E and H. [by (2), (3), and CCC]

Thus one or both of CCC and EC should be rejected. This argument would fail if EC were understood as restricted to propositions with non-extreme unconditional probabilities. Suppose $H = {\sim}E$. Then E entails $E \vee H$, but, since $E \vee H$ has an extreme unconditional probability (of 1), it does not follow by EC that E confirms $E \vee H$. The argument, though, can be readily modified to show that CCC and EC together entail an absurdity along the lines of the negation of NTC. See Skyrms (1966, p. 238) and Moretti (2003). Le Morvan (1999) gives an alternative argument for the thesis that CCC and EC together entail that NTC is false. Le Morvan's argument, like the argument above, would fail if EC were understood as restricted to propositions with non-extreme unconditional probabilities. But, it seems, Le Morvan's argument, unlike the argument above, cannot be readily modified to show that CCC and EC together entail an absurdity along the lines of the negation of NTC. See Moretti (2003) for discussion.

[9] The same is true with respect to SCC and the condition:

Converse Entailment Condition (CEC): For any propositions E and H, if H entails E, then E confirms H.

Suppose CEC and SCC are true. Take some propositions E and H. Then:

(1) E & H entails E for any propositions E and H.

Thus

(2) E confirms E & H for any propositions E and H. [by (1) and CEC]

(3) E & H entails H for any propositions E and H.

Thus

(4) E confirms H for any propositions E and H. [by (2), (3), and SCC]

Thus one or both of CEC and SCC should be rejected. This argument is noted in Tuomela (1976).

Given the focus on confirmation in the sense of incremental confirmation, and given the restriction to propositions with non-extreme unconditional probabilities, it follows that EC is true whereas CCC and SCC are false. Thus CCC and SCC should be rejected while EC should be accepted.

This is a step towards solving Hempel's paradox. But it is not enough. Each of CCC and SCC, it seems, contains a kernel of truth in that many cases where (i) E confirms H^* and (ii) H^* is entailed by or entails H are cases where, it seems, E confirms H. Some such cases are cases where E is a piece of observational evidence and H^* and H are scientific hypotheses one of which is more general than the other. Other such cases are more ordinary. Suppose, for example, a card is randomly drawn from a standard and well-shuffled deck of cards. Let E be the proposition that Smith testified that the card drawn is a Heart, H^* be the proposition that the card drawn is a Heart, and H be the proposition that the card drawn is the Jack of Hearts. Then, given certain rather natural ways of filling in the details, E confirms H^*, H^* is entailed by H, and, as per CCC, E confirms H. Now let E be the proposition that Smith testified that the card drawn is the Jack of Hearts, H^* be the proposition that the card drawn is the Jack of Hearts, and H be the proposition that the card drawn is a Jack. Then the case is such that E confirms H^*, H^* entails H, and, as per SCC, E confirms H. So I want to find some adequate replacement conditions for CCC and SCC (conditions similar to them in content but not open to counterexample).

I turn now to the possibility that CCC and SCC should be modified in terms of explanation.

3. CCC$_E$ and SCC$_E$

Consider:

CCC$_E$:   For any propositions E, H^*, and H, if (i) E confirms H^*, (ii) H^* is entailed by H, and (iii) H^* is explained by H, then E confirms H.

SCC$_E$:   For any propositions E, H^*, and H, if (i) E confirms H^*, (ii) H^* entails H, and (iii) H^* explains H, then E confirms H.

CCC$_E$ is CCC but with the added condition that H^* is explained by H. SCC$_E$, in turn, is SCC but with the added condition that H^* explains H.[10]

The expressions "explained by" and "explains" in CCC$_E$ and SCC$_E$ should be understood as short for "potentially explained by" and "potentially explains." The latter

[10] Brody (1968, 1974) considers a condition similar to CCC$_E$ but (a) without the constraint that H^* is entailed by H and (b) restricted to confirmation by positive instances. See Koslow (1970), Martin (1972, 1975), Gower (1973), and Tuomela (1976) for relevant discussion.

expressions, in turn, should be understood so that H^* can be potentially explained by or potentially explain H even if neither H^* nor H is true.[11]

If all explanation is deductive, then CCC_E and SCC_E are equivalent to the following:

CCC_{E^*}:   For any propositions E, H^*, and H, if (i) E confirms H^* and (ii) H^* is explained by H, then E confirms H.

SCC_{E^*}:   For any propositions E, H^*, and H, if (i) E confirms H^* and (ii) H^* explains H, then E confirms H.

It is a matter of controversy, though, whether all explanation is deductive.[12] So I shall assume just that some explanation is deductive and focus on CCC_E and SCC_E.[13]

There is a clear respect in which CCC_E and SCC_E improve on CCC and SCC. If Hempel's Argument were modified in terms of CCC_E and SCC_E, then it would have a false premise. It would read:

Hempel's Argument*
(1) E entails E for any proposition E.
Thus
(2) E confirms E for any proposition E. [by (1) and EC]
(3) E is entailed and explained by E & H for any propositions E and H.
Thus
(4) E confirms E & H for any propositions E and H. [by (2), (3), and CCC_E]
(5) E & H entails and explains H for any propositions E and H.
Thus
(6) E confirms H for any propositions E and H. [by (4), (5), and SCC_E]

Each of (3) and (5) is false.

Is it the case, though, that CCC_E and SCC_E are correct? The answer is negative if explanatory relations place no constraints on probability distributions. This can be seen as follows. Take some H^* and H such that H^* is explained by H or vice versa. Suppose explanatory relations place no constraints on probability distributions. Then the fact that H^* is explained by H or vice versa does not rule out any probability distributions and thus does not rule out any probability distributions on which CCC or SCC fails. It follows that neither CCC_E and SCC_E is correct.

The key question, then, is this: What constraints, if any, do explanatory relations place on probability distributions?

---

[11] See Lipton (2004, ch. 4) for helpful discussion of the distinction between actual and potential explanations.
[12] For helpful discussion, and for additional references, see Salmon (1998, ch. 9).
[13] The assumption is not that there are cases of explanation where the explanans explains the explanandum *by virtue of* entailing the explanandum. The assumption is simply that there are cases of explanation where *in fact* the explanans entails the explanandum.

4. Entailment, High Probability, and Increase in Probability

There are some obvious candidate constraints (placed by explanatory relations on probability distributions). Consider (where, as above, t is the threshold for high probability and $1 > t \geq 0.5$):

> (c1)   For any propositions E and H, H explains E only if H entails E and thus $\Pr(E \mid H) = 1$.
>
> (c2)   For any propositions E and H, H explains E only if $\Pr(E \mid H) > t$.
>
> (c3)   For any propositions E and H, H explains E only if $\Pr(E \mid H) > \Pr(E)$.

Each of (c1), (c2), and (c3) has some intuitive appeal. And each of them can be found in the literature.[14]

Are CCC_E or SCC_E correct when understood in terms of (c1), (c2), and (c3)? Suppose the only (non-trivial) constraints placed by explanatory relations on probability distributions are (c1), (c2), and (c3) (and their implications). Then whether CCC_E and SCC_E are correct hinges on whether the following are correct:

> CCC_{Ec1-c3}:   For any propositions E, H^*, and H, if (i) E confirms H^*, (ii) H^* is entailed by H, and (iii) $\Pr(H^* \mid H) = 1$, $\Pr(H^* \mid H) > t$, and $\Pr(H^* \mid H) > \Pr(H^*)$, then E confirms H.
>
> SCC_{Ec1-c3}:   For any propositions E, H^*, and H, if (i) E confirms H^*, (ii) H^* entails H, and (iii) $\Pr(H \mid H^*) = 1$, $\Pr(H \mid H^*) > t$, and $\Pr(H \mid H^*) > \Pr(H)$, then E confirms H.

But CCC_{Ec1-c3} is equivalent to CCC, and SCC_{Ec1-c3} is equivalent to SCC. This follows from the fact that in each case condition (iii) holds if condition (ii) holds. So, since CCC and SCC are incorrect, it follows that so too are CCC_{Ec1-c3} and SCC_{Ec1-c3}.

The same is true, of course, if some but not all of (c1), (c2), and (c3) are set aside. And for the same reason: the entailment condition holds only if the posterior probability in question equals 1, is greater than t, and is greater than the prior probability in question.

I turn now to a fourth candidate constraint.

5. Explanatory Power

Schupbach (Chapter 4) sets out and defends a version of IBE on which explanatoriness is fleshed out in terms of power (explanatory power), where the degree to which H has power over E is measured by:

$$e(E,H) = \frac{\Pr(H \mid E) - \Pr(H \mid \sim E)}{\Pr(H \mid E) + \Pr(H \mid \sim E)}$$

---

[14] (c1) and (c2) can be found in Hempel (1965b). (c3) can be found in Salmon (1965). For discussion of the rather extensive extant literature on explanation, and for further references, see Salmon (2006) and Woodward (2014).

$e(E, H)$'s range is from -1 to 1 (inclusive). H's power over E is positive if $e(E, H) > 0$. H has no power over E if $e(E, H) = 0$. H's power over E is negative if $e(E, H) < 0$.

It is not important for my purposes whether Schupbach's defense of his version of IBE, which includes his defense of e, succeeds. But a certain part of that defense is important for my purposes.

Schupbach notes that positive power places constraints (some of which are rather significant) on probability distributions. If $e(E, H) > 0$, then:

(a) $\dfrac{\Pr(H|E) - \Pr(H|\sim E)}{\Pr(H|E) + \Pr(H|\sim E)} > 0$

(b) $\Pr(H|E) - \Pr(H|\sim E) > 0$

(c) $\dfrac{\Pr(E|H)}{\Pr(E)} > \dfrac{\Pr(\sim E|H)}{\Pr(\sim E)}$

(d) $\Pr(E|H) - \Pr(E|H)\Pr(E) > \Pr(E) - \Pr(E|H)\Pr(E)$

(e) $\Pr(E|H) > \Pr(E)$

(f) $\Pr(E|H) > \Pr(E|\sim H)$

(g) $\Pr(H|E) > \Pr(H)$

It is worth noting that (a)–(g) follow from positive power as measured by all of the main measures of power in the literature.[15]

Now consider:

(c4)   For any propositions E and H, H explains E only if $e(E, H) > 0$.

Suppose the only (non-trivial) constraint placed by explanatory relations on probability distributions is (c4) (and its implications). Then whether CCC_E and SCC_E are correct hinges on whether the following are correct:

CCC_{Ec4}:   For any propositions E, H^*, and H, if (i) E confirms H^*, (ii) H^* is entailed by H, and (iii) $e(H^*, H) > 0$, then E confirms H.

SCC_{Ec4}:   For any propositions E, H^*, and H, if (i) E confirms H^*, (ii) H^* entails H, and (iii) $e(H, H^*) > 0$, then E confirms H.

Are CCC_{Ec4} and SCC_{Ec4} correct?

It is true that positive power places constraints on probability distributions. It is true in particular that if $e(E, H) > 0$, then (a)–(g) above all hold. It is also true, though, that any constraints placed on probability distributions by $e(H^*, H)$'s being positive are also placed on probability distributions by H^*'s being entailed by H, and that, similarly, any constraints placed on probability distributions by $e(H, H^*)$'s being positive are also

---

[15] Here I have in mind the measures noted in Schupbach (2011) and Crupi and Tentori (2012).

placed on probability distributions by H^*'s entailing H. This is because any case where H^* is entailed by H is a case where $e(H^*, H)$ is positive (in fact equal to 1), and because any case where H^* entails H is a case where $e(H, H^*)$ is positive (in fact equal to 1). It follows that CCC_{Ec4} is equivalent to CCC and that SCC_{Ec4}, in turn, is equivalent to SCC. So CCC_{Ec4} and SCC_{Ec4}, as with CCC and SCC, are incorrect.[16]

The situation is this. It is not enough, for CCC_E and SCC_E to be correct, that explanatory relations place some constraints on probability distributions. It needs to be the case that they place constraints on probability distributions over and above the constraints already placed on probability distributions by entailment relations. This is why CCC_{Ec1-c3}, SCC_{Ec1-c3}, CCC_{Ec4}, and SCC_{Ec4} all fail.

The candidate constraint constructed in Section 6 improves on (c1)–(c4) in that it provides a way of understanding CCC_E and SCC_E on which they are not equivalent to CCC and SCC. I turn now to that candidate constraint.

6. Causation and Screening-Off

It is not implausible prima facie that some explanations are non-causal in that the explanans phenomenon is not a cause of the explanandum phenomenon.[17] Clearly, though, many explanations are causal. I want to focus on CCC_E and SCC_E understood so that the kind of explanation at issue is causal. I want to do this because, arguably, causes screen-off in a sense to be explained below and because screening-off places constraints on probability distributions over and above the constraints already placed on probability distributions by entailment. In Section 6.1, I explain the basics of Bayesian causal networks. In Section 6.2, I construct a fifth candidate constraint placed by explanatory relations on probability distributions. This constraint, (c5), involves screening-off. In Section 6.3, I return to CCC and SCC and examine why they fail. In Section 6.4, I evaluate CCC_E and SCC_E when understood in terms of (c5).

6.1 Bayesian causal networks

A Bayesian causal network consists of a set of variables $\mathbf{V} = \{V_1, V_2, \ldots, V_n\}$, a directed acyclic graph $\mathbf{G}$ over $\mathbf{V}$, and a probability distribution Pr over $\mathbf{V}$. $\mathbf{G}$ consists of nodes and directed edges (or arrows). Each node is a variable in $\mathbf{V}$ (and each variable in $\mathbf{V}$ is a node). Each directed edge connects exactly two nodes. A directed edge from one node to another indicates that the former is a direct cause of the latter. One node is a parent of another (and the latter is a child of the former) if and only if there is a directed edge from the former to the latter. One node is an ancestor of another (and the latter is a

---

[16] It does not help, of course, to modify CCC and SCC in terms of all four of the candidate constraints on the table at this point. The resulting conditions—CCC_{Ec1-c4} and SCC_{Ec1-c4}—are equivalent to CCC and SCC and thus are incorrect.

[17] See Lipton (2009) for discussion.

descendant of the former) if and only if there is a directed path (or series of directed edges), aligned tip-to-tail linking intermediate nodes, from the former to the latter. G is acyclic in that no node is an ancestor of itself.

An example of **V** (taken from Neapolitan 2004) is the set $\{B, C, F, H, L\}$, where:

Variable	Values
B	b_1 = bronchitis is present
	b_2 = bronchitis is not present
C	c_1 = chest X-ray is positive
	c_2 = chest X-ray is not positive
F	f_1 = fatigue is present
	f_2 = fatigue is not present
H	h_1 = there is a history of smoking
	h_2 = there is not a history of smoking
L	l_1 = lung cancer is present
	l_2 = lung cancer is not present

Here each variable is binary. But this is not required in general.

An example of **G** (also taken from Neapolitan 2004) is the directed acyclic graph in Figure 14.1. Here H is a parent of each of B and L, B is a parent of F, L is a parent of each of F and C, and though H is not a parent of F or of C, H is an ancestor of F and of C.

It is crucial to note that not just any probability distribution is admissible in a Bayesian causal network. Pr should be such that for any variable V_i in **V** with parents

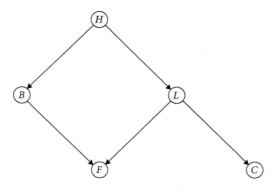

Figure 14.1 Directed acyclic graph.

(direct causes) and non-descendants (non-effects) in $\mathbf{G}$, V_i's parents screen-off V_i's non-descendants from V_i, that is, render V_i's non-descendants probabilistically irrelevant to V_i. This assumption is (a version of a condition) called "the causal Markov condition." Take the example above. Each of L and C is a non-descendant of B. So, by the causal Markov condition, Pr should be such that each of L and C is screened-off from B by H in that $\Pr(b_i \mid h_j \& l_k) = \Pr(b_i \mid h_j)$ for any i, j, k and $\Pr(b_i \mid h_j \& c_k) = \Pr(b_i \mid h_j)$ for any i, j, k. If, say, it is given that h_1 (there is a history of smoking), then Pr should be such that none of l_1 (lung cancer is present), l_2 (lung cancer is not present), c_1 (chest X-ray is positive), and c_2 (chest X-ray is not positive) has any impact on the probability of b_1 (bronchitis is present) or the probability of b_2 (bronchitis is not present).

It might seem strange to talk of variables as causes. Bear in mind, though, that there are ways of understanding Bayesian causal networks on which variables are not causes. Consider, say, the directed edge in the example above from H to B. This edge need not be understood as indicating that H is a direct cause of B. It could instead be understood as indicating that a given subject's having (or not having) a history of smoking is a direct cause of her having (or not having) bronchitis. This in no way runs counter to the idea that causes (in singular causation) are events.[18]

Much more can be said about all this.[19] The important point for my purposes, though, is the idea that causal relations and screening-off walk hand in hand. This idea is important in that screening-off places rather significant constraints on probability distributions.

6.2 (c5)

Consider:

(c5)   For any propositions E and H, H explains E only if there are partitions of propositions Γ, Γ^*, and Γ^{**} such that (a) H is a member of Γ, (b) E is a member of Γ^*, (c) Γ^{**} is a set of non-descendants of E, and (d) Γ screens-off Γ^{**} from Γ^*.

This condition is like the causal Markov condition except that it is framed in terms of explanation (as opposed to causation) and partitions (as opposed to variables).[20]

Return to the example above where $\mathbf{V} = \{B, C, F, H, L\}$. It is straightforward to reinterpret that example in terms of (c5). Let P be the patient at issue. Take H, B, L, and the fact that H screens-off L from B. H is the partition $\{P$ has a history of smoking, P does not have a history of smoking$\}$, B is the partition $\{P$ has bronchitis, P does not have bronchitis$\}$, and L is the partition $\{P$ has lung cancer, P does not have lung cancer$\}$.

---

[18] See Ehring (2009) for discussion of the relata of causation.

[19] For helpful introductory discussions of Bayesian causal networks, and for additional references, see Hitchcock (2009, 2012) and Williamson (2009).

[20] A similar condition can be motivated by appeal to Salmon's Statistical-Relevance (S-R) model of explanation (1971). See Woodward (2014) for discussion of that model.

H screens-off L from B in that given either member of the partition {P has a history of smoking, P does not have a history of smoking}, neither member of the partition {P has lung cancer, P does not have lung cancer} has any impact on the probability of either member of the partition {P has bronchitis, P does not have bronchitis}.

Now suppose the only (non-trivial) constraints placed by explanatory relations on probability distributions are (c5) and its implications. Then whether CCC_E and SCC_E are correct hinges on whether the following are correct:

CCC_{Ec5}: For any propositions E, H^*, and H, if (i) E confirms H^*, (ii) H^* is entailed by H, and (iii) there are partitions Γ, Γ^*, and Γ^{**} such that (a) H is a member of Γ, (b) H^* is a member of Γ^*, (c) Γ^{**} is a set of non-descendants of H^*, and (d) Γ screens-off Γ^{**} from Γ^*, then E confirms H.

SCC_{Ec5}: For any propositions E, H^*, and H, if (i) E confirms H^*, (ii) H^* entails H, and (iii) there are partitions Γ, Γ^*, and Γ^{**} such that (a) H^* is a member of Γ, (b) H is a member of Γ^*, (c) Γ^{**} is a set of non-descendants of H, and (d) Γ screens-off Γ^{**} from Γ^*, then E confirms H.

CCC_{Ec5} improves on CCC_{Ec1-c3} and CCC_{Ec4} in that CCC_{Ec5} is not equivalent to CCC. SCC_{Ec5}, in turn, improves on SCC_{Ec1-c3} and SCC_{Ec4} in that SCC_{Ec5} is not equivalent to SCC. But are CCC_{Ec5} and SCC_{Ec5} correct?

I turn now to the issue of why CCC and SCC fail. Getting clear on that issue will help in evaluating CCC_{Ec5} and SCC_{Ec5}.

6.3 Why CCC and SCC fail

Take any three propositions E, H^*, and H. Then (by a proof given in Shogenji Forthcoming) it follows that:

$$(1) \quad \Pr(H|E) - \Pr(H) = \begin{pmatrix} \left[\Pr(H|H^*) - \Pr(H)\right]\left[\Pr(H^*|E) - \Pr(H^*)\right] + \\ \left[\Pr(H|\sim H^*) - \Pr(H)\right]\left[\Pr(\sim H^*|E) - \Pr(\sim H^*)\right] + \\ \Pr(H^*|E)\left[\Pr(H|H^* \& E) - \Pr(H|H^*)\right] + \\ \Pr(\sim H^*|E)\left[\Pr(H|\sim H^* \& E) - \Pr(H|\sim H^*)\right] \end{pmatrix}$$

Let the first addend on the right side of (1) be "A", the second addend on the right side of (1) be "B", the third addend on the right side of (1) be "C", and the fourth addend on the right side of (1) be "D." Any case where the antecedent of CCC or SCC holds is a case where A and B are positive. It does not follow, though, that any case where the antecedent of CCC or SCC holds is a case where the left side of (1) is positive, for it might be that in some such cases the sum of C and D is negative and greater than or equal to in absolute value the sum of A and B.

Suppose, for example, a card is randomly drawn from a standard and well-shuffled deck of cards. Let E be the proposition that the card drawn is a Club, the Ace of Spades, or the Ace of Hearts, H^* be the proposition that the card drawn is not a Spade, and H be the proposition that the card drawn is a Red. Then the antecedent of CCC holds and thus the sum of A and B in (1) is positive:

$$(2) \quad \left[\frac{2}{3}-\frac{1}{2}\right]\left[\frac{14}{15}-\frac{3}{4}\right]+\left[0-\frac{1}{2}\right]\left[\frac{1}{15}-\frac{1}{4}\right]=\frac{11}{90}$$

The sum of C and D in (1), though, is negative:

$$(3) \quad \left[\frac{14}{15}\right]\left[\frac{1}{14}-\frac{2}{3}\right]+\left[\frac{1}{15}\right]\left[0-0\right]=-\frac{5}{9}$$

The right side of (3) is greater than in absolute value the right side of (2) and thus the left side of (1) is negative. So this case is a counterexample to CCC.

Now let E be the proposition that the card drawn is not a Heart, H^* be the proposition that the card drawn is a Diamond, and H be the proposition that the card drawn is a Red. Then the antecedent of SCC holds and thus the sum of A and B in (1) is positive:

$$(4) \quad \left[1-\frac{1}{2}\right]\left[\frac{1}{3}-\frac{1}{4}\right]+\left[\frac{1}{3}-\frac{1}{2}\right]\left[\frac{2}{3}-\frac{3}{4}\right]=\frac{1}{18}$$

The sum of C and D in (1), however, is negative:

$$(5) \quad \left[\frac{1}{3}\right]\left[1-1\right]+\left[\frac{2}{3}\right]\left[0-\frac{1}{3}\right]=-\frac{2}{9}$$

Since the right side of (5) is greater than in absolute value the right side of (4), it follows that the left side of (1) is negative. So this case is a counterexample to SCC.

It is clear, then, why CCC and SCC fail. Their antecedents leave it open that the sum of C and D in (1) is negative and greater than or equal to in absolute value the sum of A and B.

The crucial question vis-à-vis CCC_{Ec5} and SCC_{Ec5} is whether the same is true of their antecedents. If yes, then CCC_{Ec5} and SCC_{Ec5}, as with CCC and SCC, are open to counterexample. If no, then CCC_{Ec5} and SCC_{Ec5}, unlike CCC and SCC, hold without exception.

6.4 Why CCC_{Ec5} and SCC_{Ec5} fail

First, consider the following variant of (1):

$$(1^*) \quad \Pr(H^*|E)-\Pr(H^*)=\left(\begin{array}{l}\left[\Pr(H^*|H)-\Pr(H^*)\right]\left[\Pr(H|E)-\Pr(H)\right]+\\ \left[\Pr(H^*|\sim H)-\Pr(H^*)\right]\left[\Pr(\sim H|E)-\Pr(\sim H)\right]+\\ \Pr(H|E)\left[\Pr(H^*|H\&E)-\Pr(H^*|H)\right]+\\ \Pr(\sim H|E)\left[\Pr(H^*|\sim H\&E)-\Pr(H^*|\sim H)\right]\end{array}\right)$$

It follows from (1*) that:

$$(6) \quad \Pr(H^*|E) - \Pr(H^*) = \begin{pmatrix} \left[\Pr(H^*|H) - \Pr(H^*|\sim H)\right]\left[\Pr(H|E) - \Pr(H)\right] + \\ \Pr(H|E)\left[\Pr(H^*|H \& E) - \Pr(H^*|H)\right] + \\ \Pr(\sim H|E)\left[\Pr(H^*|\sim H \& E) - \Pr(H^*|\sim H)\right] \end{pmatrix}$$

Let the first addend on the right side of (6) be "A*," the second addend on the right side of (6) be "B*," and the third addend on the right side of (6) be "C*." Suppose the antecedent of CCC_{Ec5} holds and thus the left side of (6) is positive and the first multiplicand in A* is positive. Suppose $\Gamma = \{H, \sim H\}$ and E is included in Γ^{**} (a set of non-descendants of H^*). It follows that:

$$(7) \quad \Pr(H^*|H \& E) = \Pr(H^*|H)$$

$$(8) \quad \Pr(H^*|\sim H \& E) = \Pr(H^*|\sim H)$$

By (7) and (8) it follows that the sum of B* and C* in (6) equals zero. Hence, as the left side of (6) is positive and the first multiplicand in A* in (6) is positive, it follows that the second multiplicand in A* in (6) is positive and so E confirms H. Hence it is not the case that the sum of C and D in (1) is negative and greater than or equal to in absolute value the sum of A and B in (1).

Next, suppose the antecedent of SCC_{Ec5} holds. Suppose $\Gamma = \{H^*, \sim H^*\}$ and E is included in Γ^{**} (a set of non-descendants of H). Then:

$$(9) \quad \Pr(H|H^* \& E) = \Pr(H|H^*)$$

$$(10) \quad \Pr(H|\sim H^* \& E) = \Pr(H|\sim H^*)$$

By (9) and (10) it follows that each of C and D in (1) equals zero and thus the left side of (1) equals the sum of A and B in (1). Given that the first two conditions in SCC_{Ec5} hold and thus the sum of A and B in (1) is positive, it follows that the left side of (1) is positive and thus E confirms H.

There are conditions, then, under which CCC_{Ec5} and SCC_{Ec5} hold without exception. CCC_{Ec5} holds without exception under the condition that $\Gamma = \{H, \sim H\}$ and E is included in Γ^{**} (a set of non-descendants of H^*). SCC_{Ec5}, in turn, holds without exception under the condition that $\Gamma = \{H^*, \sim H^*\}$ and E is included in Γ^{**} (a set of non-descendants of H).

There is a potential problem however. The antecedents of CCC_{Ec5} and SCC_{Ec5} leave it open that E is not included in Γ^{**}. Do CCC_{Ec5} and SCC_{Ec5} hold without exception even when E is not included in Γ^{**}?[21]

---

[21] The antecedents of CCC_{Ec5} and SCC_{Ec5} leave it open not just that E is not included in Γ^{**} but also that the screening-off partition is not binary. So another question is whether CCC_{Ec5} and SCC_{Ec5} hold without exception even when the screening-off partition is not binary. I want to set aside this question and focus on the question of whether CCC_{Ec5} and SCC_{Ec5} hold without exception even when E is not included in Γ^{**}. If, as I argue below, the answer to this question is negative, then it does not matter for my purposes whether CCC_{Ec5} and SCC_{Ec5} hold without exception even when the screening-off partition is not binary.

It turns out that the answer is negative and that because of this the antecedents of CCC_{Ec5} and SCC_{Ec5} leave it open that the sum of C and D in (1) is negative and greater than or equal to in absolute value the sum of A and B in (1) (see Appendices for proof). Hence CCC_{Ec5} and SCC_{Ec5}, as with CCC and SCC, are incorrect.

CCC_{Ec5} and SCC_{Ec5} could be modified by adding a condition to their antecedents to the effect that E is included in Γ^{**}. This would shield them from counterexample. But then they would be open to a different objection. Their ranges of application would be too restrictive. Neither CCC_{Ec5} nor SCC_{Ec5} would have application in cases where E is not included in Γ^{**}.

It does not follow that there is no plausible way of understanding explanation on which an explanatory relation between H^* and H suffices to close off the possibility that the sum of C and D in (1) is negative and greater than in absolute value the sum of A and B in (1). I suspect, though, that this is the case.

I want to set aside CCC_E and SCC_E and turn to a different approach.

7. A Solution to Hempel's Paradox

It follows from (1) in Section 6.3 that:

CCC_{SO}: For any propositions E, H^*, and H, if (i) E confirms H^*, (ii) H^* is entailed by H, and (iii) $\{H^*, \sim H^*\}$ screens-off E from H in that $\Pr(H \mid H^* \& E) \geq \Pr(H \mid H^*)$ and $\Pr(H \mid \sim H^* \& E) \geq \Pr(H \mid \sim H^*)$, then E confirms H.

SCC_{SO}: For any propositions E, H^*, and H, if (i) E confirms H^*, (ii) H^* entails H, and (iii) $\{H^*, \sim H^*\}$ screens-off E from H in that $\Pr(H \mid H^* \& E) \geq \Pr(H \mid H^*)$ and $\Pr(H \mid \sim H^* \& E) \geq \Pr(H \mid \sim H^*)$, then E confirms H.

Any case where (i) and (ii) in CCC_{SO} or SCC_{SO} hold is a case where A and B in (1) are positive. So, given that any case where (iii) in CCC_{SO} or SCC_{SO} holds is a case where C and D in (1) are non-negative, any case where the antecedent of CCC_{SO} or SCC_{SO} holds is a case where the left side of (1) is positive. Hence CCC_{SO} and SCC_{SO} hold without exception.

Note that with both CCC_{SO} and SCC_{SO} the screening-off partition involves the *middle* proposition H^* in the chain: E, H^*, H. Note also that with both CCC_{SO} and SCC_{SO} the screening-off at issue is *negative*-impact screening-off opposed to *no*-impact screening-off. Thus the use of "$\geq$" instead of "$=$."[22]

---

[22] CCC_{SO} and SCC_{SO} are special cases of a more general condition:

> For any propositions E, H^*, and H, if (i) E confirms H^*, (ii) H^* confirms H, and (iii) $\{H^*, \sim H^*\}$ screens-off E from H in that $\Pr(H \mid H^* \& E) \geq \Pr(H \mid H^*)$ and $\Pr(H \mid \sim H^* \& E) \geq \Pr(H \mid \sim H^*)$, then E confirms H.

This condition is stronger than the condition:

> For any propositions E, H^*, and H, if (i) E confirms H^*, (ii) H^* confirms H, and (iii) $\{H^*, \sim H^*\}$ screens-off E from H in that $\Pr(H \mid H^* \& E) = \Pr(H \mid H^*)$ and $\Pr(H \mid \sim H^* \& E) = \Pr(H \mid \sim H^*)$, then E confirms H.

I gave a case in Section 2 where a card is randomly drawn from a standard and well-shuffled deck of cards, E is the proposition that Smith testified that the card drawn is a Heart, H^* is the proposition that the card drawn is a Heart, and H is the proposition that the card drawn is the Jack of Hearts. I also gave a case where instead E is the proposition that Smith testified that the card drawn is the Jack of Hearts, H^* is the proposition that the card drawn is the Jack of Hearts, and H is the proposition that the card drawn is a Jack. In each case, given certain rather natural ways of filling in the details, E confirms H^*, H^* is entailed by or entails H, and $\{H^*, \sim H^*\}$ screens-off E from H in that $\Pr(H \mid H^* \& E) \geq \Pr(H \mid H^*)$ and $\Pr(H \mid \sim H^* \& E) \geq \Pr(H \mid \sim H^*)$. So, given CCC_{SO} and SCC_{SO}, in each case E confirms H.

Is it the case that $\{H^*, \sim H^*\}$ screens-off E from H if and only if $\{H, \sim H\}$ screens-off E from H^*? And, regardless, is it the case that CCC_{SO} and SCC_{SO} would hold without exception if they were modified so that the screening-off partition involved the last, as opposed to the middle, proposition H?

The answer to each question is no. But the following conditions hold without exception:

$\text{CCC}_{\text{SO*}}$:   For any propositions E, H^*, and H, if (i) E confirms H^*, (ii) H^* is entailed by H, and (iii) $\{H, \sim H\}$ screens-off E from H^* in that $\Pr(H^* \mid H \& E) \leq \Pr(H^* \mid H)$ and $\Pr(H^* \mid \sim H \& E) \leq \Pr(H^* \mid \sim H)$, then E confirms H.

$\text{SCC}_{\text{SO*}}$:   For any propositions E, H^*, and H, if (i) E confirms H^*, (ii) H^* entails H, and (iii) $\{H, \sim H\}$ screens-off E from H^* in that $\Pr(H^* \mid H \& E) \leq \Pr(H^* \mid H)$ and $\Pr(H^* \mid \sim H \& E) \leq \Pr(H^* \mid \sim H)$, then E confirms H.

Recall (6) from Section 6.4. Any case where the antecedent of $\text{CCC}_{\text{SO*}}$ or the antecedent of $\text{SCC}_{\text{SO*}}$ holds is a case where the left side of (6) is positive, the first multiplicand in A^* is positive, and the sum of B^* and C^* is non-positive. It follows that any case where the antecedent of $\text{CCC}_{\text{SO*}}$ or the antecedent of $\text{SCC}_{\text{SO*}}$ holds is a case where the second multiplicand in A^* is positive and thus E confirms H.[23]

Note that the screening-off at issue in $\text{CCC}_{\text{SO*}}$ and $\text{SCC}_{\text{SO*}}$ is *positive*-impact screening-off as opposed to *negative*-impact screening-off. If, instead, the screening-off at issue in $\text{CCC}_{\text{SO*}}$ and $\text{SCC}_{\text{SO*}}$ were negative-impact screening-off, then $\text{CCC}_{\text{SO*}}$ and $\text{SCC}_{\text{SO*}}$ would be open to counterexample.[24]

The first of these conditions is established in Roche (2012a). The second is established in Shogenji (2003). See Roche (2014) for discussion of the first condition in the context of peer disagreement. See Roche and Shogenji (2014a) for discussion of the first condition in the context of Moore's proof of the existence of a material world. See Sober (2015, ch. 5) for discussion of the first condition in the context of the problem of evil. See Roche and Shogenji (2014b) for discussion of the second condition in the context of degree of confirmation.

[23] $\text{CCC}_{\text{SO*}}$ and $\text{SCC}_{\text{SO*}}$ are due essentially to Tomoji Shogenji (personal correspondence).

[24] Similarly, if the screening-off at issue in CCC_{SO} and SCC_{SO} were positive-impact screening-off, then CCC_{SO} and SCC_{SO} would be open to counterexample.

CCC_{SO}, SCC_{SO}, $CCC_{SO'}$, and SCC_{SO*} together have a rather large range of application. There are cases where H^* is entailed by H, but there are also cases where H^* entails H. There are cases where the screening-off partition involves the middle proposition H^* and the screening-off is negative-impact screening-off, but there are also cases where the screening-off partition involves the last proposition H and the screening-off is positive-impact screening-off.

The situation is this. CCC and SCC are false. But each, it seems, contains a kernel of truth. This is borne out by CCC_{SO}, SCC_{SO}, $CCC_{SO'}$, and SCC_{SO*}.

My solution to Hempel's paradox is now complete. EC should be accepted while CCC and SCC should be rejected in favor of CCC_{SO}, SCC_{SO}, $CCC_{SO'}$, and SCC_{SO*}.

My solution is similar to the proposed explanation-based solution considered in Section 6 in that it involves the notion of screening-off. But with my solution on hand there is simply no need to appeal to explanation.

8. Conclusion

I noted in Section 1 that the core idea behind IBE can be put as follows:

I_1:   Inferences to categorical beliefs (at least some of them) should be guided by explanatory considerations.

I also noted that this idea, which concerns categorical beliefs, has a counterpart concerning degrees of belief:

I_2:   Changes in degrees of belief (at least some of them) should be guided by explanatory considerations.

One way to flesh out I_2 is to modify conditions such as CCC, EC, and SCC in terms of explanation. There is no need, though, to modify EC in terms of explanation, for EC holds without exception (given the assumption that E and H have non-extreme unconditional probabilities). And, it seems, neither CCC nor SCC is correct when understood in terms of explanation.

It does not follow, of course, that I_2 is incorrect. There are other ways to flesh out I_2. McCain and Poston (2014, Chapter 8), for instance, flesh out I_2 in terms of the notion of resiliency.[25]

The overall picture, however, is clearer. If changes in degrees of belief should be guided by explanatory considerations, then, it seems, this is not because conditions such as CCC, EC, and SCC should be modified in terms of explanation.

---

[25] See Roche and Sober (2014) for discussion.

Appendix A: Counterexample to CCC_{Ec5}

Consider the following probability distribution:

E	H^*	H	P	Pr	E	H^*	H	P	Pr
T	T	T	T	$\frac{1}{29}$	F	T	T	T	$\frac{2}{39}$
T	T	T	F	$\frac{407473}{2841072}$	F	T	T	F	$\frac{8247278569811}{16455562695936}$
T	T	F	T	$\frac{1}{156}$	F	T	F	T	$\frac{1}{531}$
T	T	F	F	$\frac{1}{314}$	F	T	F	F	$\frac{1}{128}$
T	F	T	T	0	F	F	T	T	0
T	F	T	F	0	F	F	T	F	0
T	F	F	T	$\frac{2}{37}$	F	F	F	T	$\frac{3}{56}$
T	F	F	F	$\frac{5}{592}$	F	F	F	F	$\frac{320066869}{2383827712}$

It can be readily verified that:

(a) $Pr(H^* \mid E) = 0.75 > 0.749662 \approx Pr(H^*)$
(b) $Pr(H^* \mid H) = 1$
(c) $Pr(H^* \mid H \& P) = Pr(H^* \mid H) = 1$
(d) $Pr(H^* \mid H \& \sim P) = Pr(H^* \mid H) = 1$
(e) $Pr(H^* \mid \sim H \& P) = Pr(H^* \mid \sim H) \approx 0.072$
(f) $Pr(H^* \mid \sim H \& \sim P) = Pr(H^* \mid \sim H) \approx 0.072$
(g) $[Pr(H \mid H^*) - Pr(H)][Pr(H^* \mid E) - Pr(H^*)] \approx 0.0000823$
(h) $[Pr(H \mid \sim H^*) - Pr(H)][Pr(\sim H^* \mid E) - Pr(\sim H^*)] \approx 0.000246$
(i) $Pr(H^* \mid E) [Pr(H \mid H^* \& E) - Pr(H \mid H^*)] \approx -0.0191$
(j) $Pr(\sim H^* \mid E) [Pr(H \mid \sim H^* \& E) - Pr(H \mid \sim H^*)] = 0$

Given (a), E confirms H^*. Suppose, consistent with (b), H entails H^*. Let $\Gamma = \{H, \sim H\}$, $\Gamma^* = \{H^*, \sim H^*\}$, and $\Gamma^{**} = \{P, \sim P\}$, where P and $\sim P$ are non-descendants of H^*, and where E is a descendant of H^* and thus is not a member of Γ^{**}. Then, given (c)–(f), there are partitions Γ, Γ^*, and Γ^{**} such that H is a member of Γ, H^* is a member of Γ^*, Γ^{**} is a set of non-descendants of H^*, and Γ screens-off Γ^{**} from Γ^*. Thus the antecedent of CCC_{Ec5} holds. But, given (g)–(j), it is not the case that E confirms H. Thus the consequent of CCC_{Ec5} does not hold. Thus CCC_{Ec5} is false. QED.

Appendix B: Counterexample to SCC_{Ec5}

Consider the following probability distribution:

E	H^*	H	P	Pr	E	H^*	H	P	Pr
T	T	T	T	$\frac{1}{16}$	F	T	T	T	$\frac{5}{43}$
T	T	T	F	$\frac{1}{8}$	F	T	T	F	$\frac{1888104014513}{6010675728640}$
T	T	F	T	0	F	T	F	T	0
T	T	F	F	0	F	T	F	F	0
T	F	T	T	$\frac{1}{310}$	F	F	T	T	$\frac{1}{11}$
T	F	T	F	$\frac{2}{41}$	F	F	T	F	$\frac{1}{5}$
T	F	F	T	$\frac{1}{146}$	F	F	F	T	$\frac{1}{256}$
T	F	F	F	$\frac{27051}{7422640}$	F	F	F	F	$\frac{157449879}{6353779840}$

It can be readily verified that:

(a) $\Pr(H^* \mid E) = 0.75 > 0.618 \approx \Pr(H^*)$
(b) $\Pr(H \mid H^*) = 1$
(c) $\Pr(H \mid H^* \& P) = \Pr(H \mid H^*) = 1$
(d) $\Pr(H \mid H^* \& \sim P) = \Pr(H \mid H^*) = 1$
(e) $\Pr(H \mid \sim H^* \& P) = \Pr(H \mid \sim H^*) \approx 0.897$
(f) $\Pr(H \mid \sim H^* \& \sim P) = \Pr(H \mid \sim H^*) \approx 0.897$
(g) $[\Pr(H \mid H^*) - \Pr(H)][\Pr(H^* \mid E) - \Pr(H^*)] \approx 0.00518$
(h) $[\Pr(H \mid \sim H^*) - \Pr(H)][\Pr(\sim H^* \mid E) - \Pr(\sim H^*)] \approx 0.00837$
(i) $\Pr(H^* \mid E) [\Pr(H \mid H^* \& E) - \Pr(H \mid H^*)] = 0$
(j) $\Pr(\sim H^* \mid E) [\Pr(H \mid \sim H^* \& E) - \Pr(H \mid \sim H^*)] = -0.0163$

Given (a), E confirms H^*. Suppose, consistent with (b), H^* entails H. Let $\Gamma = \{H^*, \sim H^*\}$, $\Gamma^* = \{H, \sim H\}$, and $\Gamma^{**} = \{P, \sim P\}$, where P and $\sim P$ are non-descendants of H, and where E is a descendant of H and thus is not a member of Γ^{**}. Then, given (c)–(f), there are partitions Γ, Γ^*, and Γ^{**} such that H^* is a member of Γ, H is a member of Γ^*, Γ^{**} is a set of non-descendants of H, and Γ screens-off Γ^{**} from Γ^*. Thus the antecedent of SCC_{Ec5} holds. But, given (g)–(j), it is not the case that E confirms H. Thus the consequent of SCC_{Ec5} does not hold. Thus SCC_{Ec5} is false. QED.

Acknowledgments

Thanks to Kevin McCain, Ted Poston, and Michael Roche for helpful comments on prior versions of the chapter. Special thanks to Tomoji Shogenji for extensive and very helpful input on the project.

References

Brody, B. (1968). Confirmation and explanation. *Journal of Philosophy*, 65, 282–99.

Brody, B. (1974). More confirmation and explanation. *Philosophical Studies*, 26, 73–5.

Carnap, R. (1962). *Logical foundations of probability* (2nd ed.). Chicago: University of Chicago Press.

Crupi, V. and Tentori, K. (2012). A second look at the logic of explanatory power (with two novel representation theorems). *Philosophy of Science*, 79, 365–85.

Douven, I. (1999). Inference to the best explanation made coherent. *Philosophy of Science*, 66 (Proceedings), S424–S435.

Douven, I. (2011a). Abduction. In E. Zalta (ed), *The Stanford Encyclopedia of Philosophy* (spring ed.), <http://plato.stanford.edu/archives/spr2011/entries/abduction/>.

Douven, I. (2011b). Further results on the intransitivity of evidential support. *Review of Symbolic Logic*, 4, 487–97.

Douven, I. and Wenmackers, S. (2017). Inference to the best explanation versus Bayes's rule in a social setting. *British Journal for the Philosophy of Science*, 68, 535–70.

Ehring, D. (2009). Causal relata. In H. Beebee, C. Hitchcock, and P. Menzies (eds), *The Oxford handbook of causation* (pp. 387–413). Oxford: Oxford University Press.

Gower, B. (1973). Martin on explanation and confirmation. *Analysis*, 33, 107–9.

Hempel, C. (1965a). Studies in the logic of confirmation. In C. Hempel, *Aspects of scientific explanation and other essays in the philosophy of science* (pp. 3–46). New York: Free Press.

Hempel, C. (1965b). Aspects of scientific explanation. In C. Hempel, *Aspects of scientific explanation and other essays in the philosophy of science* (pp. 331–496). New York: Free Press.

Hitchcock, C. (2009). Causal modelling. In H. Beebee, C. Hitchcock, and P. Menzies (eds), *The Oxford handbook of causation* (pp. 299–314). Oxford: Oxford University Press.

Hitchcock, C. (2012). Probabilistic causation. In E. Zalta (ed.), *The Stanford Encyclopedia of Philosophy* (winter ed.), <http://plato.stanford.edu/entries/causation-probabilistic/>.

Huber, F. (2008). Hempel's logic of confirmation. *Philosophical Studies*, 139, 181–9.

Huemer, M. (2009). Explanationist aid for the theory of inductive logic. *British Journal for the Philosophy of Science*, 60, 345–75.

Iranzo, V. (2008). Bayesianism and inference to the best explanation. *Theoria*, 23, 89–106.

Koslow, A. (1970). Comment. *Minnesota Studies in the Philosophy of Science*, 5, 104–7.

Le Morvan, P. (1999). The Converse Consequence Condition and Hempelian qualitative confirmation. *Philosophy of Science*, 66, 448–54.

Lipton, P. (2001). Is explanation a guide to inference? A reply to Wesley C. Salmon. In G. Hon and S. Rakover (eds), *Explanation: Theoretical approaches and applications* (pp. 93–120). Dordrecht: Kluwer.

Lipton, P. (2004). *Inference to the best explanation* (2nd ed.). London: Routledge.

Lipton, P. (2009). Causation and explanation. In H. Beebee, C. Hitchcock, and P. Menzies (eds), *The Oxford handbook of causation* (pp. 619–31). Oxford: Oxford University Press.

McCain, K. and Poston, T. (2014). Why explanatoriness is evidentially relevant. *Thought*, 3, 145–53.

Martin, M. (1972). Confirmation and explanation. *Analysis*, 32, 167–9.

Martin, M. (1975). Explanation and confirmation again. *Analysis*, 36, 41–2.

Moretti, L. (2003). Why the Converse Consequence Condition cannot be accepted. *Analysis*, 63, 297–300.

Neapolitan, R. (2004). *Learning Bayesian networks*. Upper Saddle River, NJ: Pearson Prentice Hall.

Okasha, S. (2000). Van Fraassen's critique of inference to the best explanation. *Studies in the History and Philosophy of Science*, 31, 691–710.

Poston, T. (2014). *Reason and explanation: A defense of explanatory coherentism*. New York: Palgrave Macmillan.

Psillos, S. (2004). Inference to the best explanation and Bayesianism: Comments on Ilkka Niiniluoto's "Truth-seeking by abduction." In F. Stadler (ed.), *Induction and deduction in the sciences* (pp. 83–91). Dordrecht: Kluwer.

Roche, W. (2012a). A weaker condition for transitivity in probabilistic support. *European Journal for Philosophy of Science*, 2, 111–18.

Roche, W. (2012b). Transitivity and intransitivity in evidential support: Some further results. *Review of Symbolic Logic*, 5, 259–68.

Roche, W. (2014). Evidence of evidence is evidence under screening-off. *Episteme*, 11, 119–24.

Roche, W. (2015). Evidential support, transitivity, and screening-off. *Review of Symbolic Logic*, 8, 785–806.

Roche, W. and Shogenji, T. (2014a). Confirmation, transitivity, and Moore: The screening-off approach. *Philosophical Studies*, 168, 797–817.

Roche, W. and Shogenji, T. (2014b). Dwindling confirmation. *Philosophy of Science*, 81, 114–37.

Roche, W. and Sober, E. (2013). Explanatoriness is evidentially irrelevant, or inference to the best explanation meets Bayesian confirmation theory. *Analysis*, 73, 659–68.

Roche, W. and Sober, E. (2014). Explanatoriness and evidence: A reply to McCain and Poston. *Thought*, 3, 193–9.

Salmon, W. (1965). The status of prior probabilities in statistical explanation. *Philosophy of Science*, 32, 137–46.

Salmon, W. (1971). Statistical explanation. In W. Salmon, R. Jeffrey, and J. Greeno (eds), *Statistical explanation and statistical relevance* (pp. 29–87). Pittsburgh, PA: University of Pittsburgh Press.

Salmon, W. (1998). *Causality and explanation*. Oxford: Oxford University Press.

Salmon, W. (2001a). Explanation and confirmation: A Bayesian critique of inference to the best explanation. In G. Hon and S. Rakover (eds), *Explanation: Theoretical approaches and applications* (pp. 61–91). Dordrecht: Kluwer.

Salmon, W. (2001b). Reflections of a bashful Bayesian: A reply to Lipton. In G. Hon and S. Rakover (eds), *Explanation: Theoretical approaches and applications* (pp. 121–36). Dordrecht: Kluwer.

Salmon, W. (2006). *Four decades of scientific explanation*. Pittsburgh, PA: University of Pittsburgh Press.

Schupbach, J. (2011). Comparing probabilistic measures of explanatory power. *Philosophy of Science*, 78, 813–29.

Shogenji, T. (2003). A condition for transitivity in probabilistic support. *British Journal for the Philosophy of Science*, 54, 613–16.

Shogenji, T. (Forthcoming). Mediated confirmation. *British Journal for the Philosophy of Science*.

Skyrms, B. (1966). Nomological necessity and the paradoxes of confirmation. *Philosophy of Science*, 33, 230–49.

Sober, E. (2015). *Ockham's razors: A user's manual*. Cambridge: Cambridge University Press.

Tregear, M. (2004). Utilising explanatory factors in induction? *British Journal for the Philosophy of Science*, 55, 505–19.

Tuomela, R. (1976). Confirmation, explanation, and the paradoxes of transitivity. In R. Bogdan (ed.), *Local induction* (pp. 319–28). Dordrecht: D. Reidel.

van Fraassen, B. (1989). *Laws and symmetry*. Oxford: Oxford University Press.

Weisberg, J. (2009). Locating IBE in the Bayesian framework. *Synthese*, 167, 125–43.

Williamson, J. (2009). Probabilistic theories. In H. Beebee, C. Hitchcock, and P. Menzies (eds), *The Oxford handbook of causation* (pp. 185–212). Oxford: Oxford University Press.

Woodward, J. (2014). Scientific explanation. In E. Zalta (ed.), *The Stanford Encyclopedia of Philosophy* (winter ed.), <http://plato.stanford.edu/entries/scientific-explanation/>.

15

The Spirit of Cromwell's Rule

Timothy McGrew

I beseech you in the bowels of Christ,
think it possible that you may be mistaken.

Oliver Cromwell, Letter to the General Assembly
of the Church of Scotland, 1650

In general, there is a degree of doubt, and caution,
and modesty, which, in all kinds of scrutiny and decision,
ought for ever to accompany a just reasoner.

David Hume, *Philosophical Essays*

Bayesians, broadly speaking, like their formulas; explanationists are fond of their intuitions. That is not to say, of course, that explanationists cannot be rigorous or that Bayesians cannot have useful insights. But one of the challenges for those who think that the two projects can be brought together is to explain what is going on when they seem to come apart. This chapter is a short essay in that direction, inspired by Peter Lipton's pious wish that "the farmer and the cowmen should be friends."[1] But before we can effect a reconciliation, we need to find a problem.

1. Problem

In his book *Making Decisions*, Dennis Lindley proposes what he humorously dubs "Cromwell's rule": For any hypothesis H with knowledge base K, your $P(H|K)$ is 1 if and only if K logically implies the truth of H.[2]

Cromwell's rule is not a consequence of the axioms of probability. But it seems to be a reasonable constraint on prior probability assignments given the fact that conditionalization—the only method available to the orthodox Bayesian for changing

---

[1] See Lipton's contribution, "What Good Is an Explanation?" to G. Hon and S. Rakover, eds, *Explanation: Theoretical Approaches and Applications* (Dordrecht: Springer, 2001), pp. 43–59.

[2] Dennis Lindley, *Making Decisions*, 2nd ed. (London: John Wiley and Sons, 1985), p. 104.

THE SPIRIT OF CROMWELL'S RULE 243

probability assignments—is unable to shift extreme probabilities of 0 or 1, no matter how much evidence is brought to bear on H. The dogmatism involved in assigning a probability of 0 or 1 to something that is not logically forced upon one seems epistemically undesirable. It ought to be possible, on any matter where one's belief is not logically guaranteed to be true, to change one's mind.

That change of mind, in turn, is associated with a process known as the Swamping of the Priors. This can be seen most easily in the odds form of Bayes's Theorem:

$$P(H|E_1 \& \ldots \& E_n)/P(\sim H|E_1 \& \ldots \& E_n) = P(H)/P(\sim H) * P(E_1 \& \ldots \& E_n|H)/ \\ P(E_1 \& \ldots \& E_n|\sim H)$$

$$= P(H)/P(\sim H) * P(E_1|H)/ P(E_1|\sim H) * \ldots * P(E_n|H \& E_1 \& \ldots \& E_{n-1})/ \\ P(E_n|\sim H \& E_1 \& \ldots \& E_{n-1}).$$

Each successive term on the right side of the equation contributes its component of the evidence to the cumulative case for H. In order to make our examples even more tidy, we can stipulate that the likelihoods are independent of one another, which vastly simplifies the form of the right side:

$$= P(H)/P(\sim H) * P(E_1|H)/P(E_1|\sim H) * \ldots * P(E_n|H)/P(E_n|\sim H).$$

Finally, we can stipulate that each piece of evidence has at least some minimal positive relevance to H: for all m, $P(E_m|H)/P(E_m|\sim H) \geq k > 1$, where k is a real number. It now follows that for any finite presumption against H, a sufficient number of pieces of evidence of this sort can overcome the unfavorable prior and push the probability of H higher than any desired level short of 1.

This is all very well if the assumptions are satisfied. But are they? Consider a simple urn experiment. You are told that you have a choice of one of two outwardly indistinguishable urns; you are further told that one contains 75 red balls and 25 black ones, while the other contains the reverse: 25 red balls and 75 black ones. You select an urn at random, and someone who reports the results to you is allowed to sample with replacement 1,000 times. As it happens, the report on a thousand-fold sample says that 501 red balls and 499 black ones were drawn. What should you conclude?

The standard answer is that you should use an indifference prior (after all, you could not distinguish the urns by external examination) and then compare the likelihoods. In this case, a quick bit of algebra tells us that the posterior odds are 9 to 1, so the probability that you have been drawing from the mostly red urn is 0.9. So runs the calculation.[3]

But this is a conclusion that no one in his right mind should accept. The reported result is, so our pre-theoretical intuitions tell us, overwhelmingly better explained by

---

[3] Under the assumption of independence, the likelihood ratios are all multiplied, and 499 of the 3/1 ratios (because 75/25 = 3/1) are exactly canceled by the 499 ratios with the value 1/3. That leaves only two left over, so 3/1 * 3/1 = 9/1 for the posterior odds. We turn these new odds back into a probability by the standard transformation of A/B odds to A/(A+B) for a posterior probability.

the assumption that *something was wrong with the boundary conditions*—that one of the urns was really stocked with a 50–50 mix, or that the sampling procedure alternated draws from one urn with draws from the other, or that your reporter is playing a practical joke on you.[4]

2. Two Analyses

How should the Bayesian analyze cases of this sort? One way, suggested by the preceding paragraph, is to think of them in terms of partition failure. In supposing that fair draws from urn 1 and fair draws from urn 2 form a partition, we are not comparing H and ~H; we are comparing H_1 and H_2. The existence of a third alternative—perhaps a "catch-all" (or "none-of-the-above") hypothesis that we could call H_3, or perhaps something that is itself buried in the catch-all—may have seemed antecedently unlikely, but it was possible, and data like those described draw our attention to it. The Cromwellian moral, on this analysis, runs something like this: be slow to assume that the salient alternatives form a partition; do not assign a probability of 1 to their disjunction unless they are truly mutually exclusive and jointly exhaustive.

That antecedent implausibility is an important issue in this analysis. In many situations, regardless of whether we are working with numerical data, we make simplifying assumptions by putting out of consideration alternatives that seem not to be worth our time. Flip a coin: heads or tails? That the coin will stand on edge when it lands is a remote possibility but one that we might concede to be within the realm of believability. That it is heavily biased toward one side or another is also a remote possibility, and one might feel moved to take that possibility seriously by, say, a long run of heads. That it will turn into a pink and purple polka-dotted dragon who offers to write advertising slogans in the sky with his fiery breath for minimum wage is, near as makes no difference, a non-starter. Most people would distrust their eyes (or their sanity) if that seemed to happen, though here again Cromwell's voice beseeches us to think the thing at least logically possible.[5] Fortunately, the dragon hypothesis offers us no purchase on the explanation of a heavy preponderance of tails.

Much hangs, in this sort of analysis, on two points: the initial plausibility or improbability of the overlooked hypothesis, and the explanatory power that hypothesis brings to the data. In probabilistic terms, the lower the prior probability for H_3, the slower rational agents will be to invoke it; and unless the respective likelihoods are higher than those for H_1 and H_2, it will not gain sufficient traction to overtake one of them. So before we employ this analysis, we will need to think hard about how we might grasp—even in rough, comparative terms—the prior and the likelihood for H_3.

[4] Nor are such cases purely hypothetical. A similar line of reasoning casts doubt on the reported data that lie at the foundation of Mendel's work on pea plants. The problem is not that the data fail to support Mendel's hypothesis. Rather, the problem is that the data are *so* good that they raise suspicion of being fudged.
[5] Whether there is such a thing as metaphysical possibility, irreducible to logical possibility plus some very high-level contingent facts, is a question for another day.

It is not enough that one of them be high. One can always gerrymander a hypothesis so as to make it entail everything that its rival would entail; Cartesian deceiver scenarios and conspiracy theories both trade on this possibility. But what they gain in predictive power, they lose in prior plausibility; the very specificity necessary to spell out a hypothesis that gives high likelihood tends to reduce its prior.[6] Neither piggybacking on the predictive power of H_1 nor building one's data directly into H_3 is a reasonable way to come up with a serious competing hypothesis.

The second way to analyze these cases is to think in terms of a buried sub-hypothesis S within ~H—possibly one with a very low prior and therefore not on the epistemic radar at first glance—that can be "exhumed." This analysis is not widely different from the first, but in certain cases it may be more intuitive. Consider the "hypothesis" H that Abraham Lincoln was an actual human being who was president of the United States during the Civil War. This hypothesis explains admirably the various pieces of historical evidence we have regarding Lincoln. There is no clear alternative hypothesis that would make a partition, or even something close to a partition, with this one, so we may lump all of the alternatives into one ungainly heap and call them "~H."

Lurking within ~H there may be a subhypothesis S which, combined with ~H, would yield likelihoods comparable to those for H: $P(E_1 \& \ldots \& E_n|H) \approx P(E_1 \& \ldots \& E_n|\sim H \& S)$. But here again, much depends on the prior probability of (~H & S). In particular, if $P(\sim H \& S) << P(H)$ and $P(E_1 \& \ldots \& E_n|H) \geq P(E_1 \& \ldots \& E_n|\sim H \& S)$, and if there is no other subhypothesis S' such that $P(E_1 \& \ldots \& E_n|\sim H \& S') \geq P(E_1 \& \ldots \& E_n|H)$, then in the long run of pieces of evidence of this type, $P(H|E_1 \& \ldots \& E_n)$ will wind up very close to 1 as the ratio $P(H|E_1 \& \ldots \& E_n)/P(\sim H|E_1 \& \ldots \& E_n)$ approaches the ratio $P(H)/P(\sim H \& S)$.

The sensitivity of both of these analyses to the prior probabilities creates a way for those of a dogmatic turn of mind to do something that certainly violates the spirit, though not the letter, of Cromwell's rule. The process is appealingly simple: pick a low prior for the view you do not wish to come to believe. In a way, this is the converse procedure to the swamping of the priors. For any specified weight of evidence, expressed in a likelihood ratio as $P(E_1 \& \ldots \& E_n|H)/P(E_1 \& \ldots \& E_n|\sim H) = k$ (for some threateningly large value of k), pick a prior probability for H so low that $P(H)/P(\sim H) < 1/k$. If the evidence hasn't come in yet, calculate how much *could* come in before you are dead, or before the sun goes nova, or by some other extravagant criterion, and set your prior accordingly. By back-solving one's prior probabilities, one can always avoid having to change one's mind.[7]

---

[6] Playing a game of chess online against an unrated opponent with a guest login handle, I find myself completely routed in under twenty moves. I suspect him of using a chess computer, which would make the result quite predictable. It may be objected that he could be one of the world's top players incognito and that I, as a mere master, should expect to lose devastatingly to him. True enough—but what are the odds that one of the world's top players would be slumming under an anonymous handle online? Focusing attention on an exceptional subhypothesis under ~H does not actually change the force of the inference for H.

[7] For a case in point, see Jordan Howard Sobel's suggestion (*Logic and Theism* (Cambridge University Press, 2009), chapter VIII) that the prior probability of a miracle be stipulated to be *infinitesimally* small. Hume might have approved—but then, we can hardly expect a Scotsman to give much deference to Cromwell.

There is another kind of case, however, where in a certain limited sense it may be appropriate to put a boundary on how much one might learn from experience. Convergence (however rapid or slow) will occur in the case of certain open-ended evidence sets $E_1, E_2, \ldots$ where some other proposition F screens all of the E_n, singly and jointly, from H, and $P(H|F) < 1$. Then the relevance of the E_n to H is bounded by the support they can give to F, and that support may be insufficient to overcome the antecedent improbability of H: $P(\sim H)/P(H) > P(F|H)/P(F|\sim H)$.

An example may make this phenomenon clearer. While sitting in my favorite chair in my own home, I think that I hear a slight pattering noise—the sort that might be caused by a few raindrops on the window. I put down my book and focus my attention, and a few seconds later I again think that I hear a slight pattering. *Prima facie*, each of my auditory experiences of this type is evidence for the claim that

H: It is raining.

But the value of the evidence is limited by the fact that it is more directly evidence for

F: Something in my vicinity is causing a slight pattering sound.

No matter how many times I pause to listen carefully, auditory evidence of this sort cannot take me any further than F. And the relevance of F to H is limited by the fact that there are alternative hypotheses that would also account for F, e.g.,

M: There is a mouse in the wainscoting.

If I had very strong antecedent reason to doubt that it was raining, the most careful, repeated listening might be insufficient to raise the probability of H above 0.5.

When we are dealing with a case of screening like this one, questions about the rapidity of convergence may become somewhat less urgent. A few thoughtful pauses to listen intently will likely raise the probability of F to something quite close to 1. The exact rate of the diminishing returns after that point is relatively unimportant, since *ex hypothesi* we know the maximum value to which the cumulative likelihood ratios converge in any event.

This is not, strictly speaking, a case where a second-order Cromwell rule is in question. After all, in the case of the pattering noise, I can obtain different kinds of information relevant to H and M. I might get up and look out the window; I might step outside; I might stoop down to see whether the noise seems louder behind the wainscoting; I might knock on the wall to see if that act provokes a flurry of scurrying noises. Such an expansion of the evidence set to incorporate different *kinds* of data might not always be available. Astronomy, where we are limited in our analysis to the radiation we can detect with available instrumentation, is a field that provides illustrations of narrowly limited inputs. In those cases, if the relevant screening holds and there is no other sort of information that we can even imagine receiving, it may be rational to violate a semi-Cromwellian constraint. But when we can readily understand what such an expansion would look like, it is an obvious move to make. And in a wide range of cases, we can.

3. Limiting Cases

But what about really extreme cases? Take the example of the existence of President Lincoln sketched above. Here is a belief so firmly rooted in a wide range of my experiences that my mind boggles at the thought of giving it up. If you were to ask me what it would take to make me abandon it, I would be at a loss to say anything specific or useful. A better explanation of the total evidence I have for it, I suppose. But I would not be able to give you any very good description of what that "better explanation" would look like. Is this a case where even the spirit of Cromwell's rule does not apply?

I think not. It is one thing to confess that one does not know what could rationally induce change in one's belief regarding proposition P but quite another to declare P to be unchangeable. The former stance is one of uncertainty, not about P but about what it would even look like to challenge and overthrow it; the latter is a declaration that nothing could do the job. Such declarations are not necessarily unreasonable. As a neoclassical foundationalist I have a fondness for strong foundational beliefs and would be willing, in defiance of personalism, to go to the mat for the epistemic status of those foundations.[8] But to make the existence of President Lincoln, or my right hand, or the external world, one of those beliefs seems to be a clear category mistake.

One of John Stuart Mill's most suggestive remarks is that our beliefs have no better safeguard to rest on than "a standing invitation to the whole world to prove them unfounded."[9] In the limiting cases where we cannot imagine what that "proof" would look like, the invitation may be accompanied by a wry smile and the recollection of Hume's quip that a man of sense does not go chasing after every story of witches or hobgoblins or fairies to canvass the evidence. But for all that, if we are to apportion our belief to the evidence, we would do well to leave the invitation standing. That much deference, if no more, seems to be due to the spirit of Cromwell's rule.

In order to learn, Charles Peirce once wrote, one must desire to learn; and this desire means that one is not so satisfied with one's present beliefs as to be unwilling to have them challenged. "There follows," he adds, "one corollary, which itself deserves to be inscribed upon every wall in the city of philosophy: Do not block the way of inquiry."[10]

---

[8] I have in mind beliefs of the form "I am experiencing like this," where the demonstrative picks out my occurrent experiential state. For a further discussion, see Timothy McGrew, *The Foundations of Knowledge* (Lanham, MD: Littlefield Adams, 1995).

[9] John Stuart Mill, *On Liberty*, 2nd ed. (Boston, MA: Ticknor and Fields, 1863), p. 43.

[10] Charles S. Pierce, *Collected Papers* (Cambridge, MA: Harvard University Press), 1.135.

16

Bayesianism and IBE

The Case of Individual vs. Group Selection

Leah Henderson

1. Introduction

There is a general question about how to relate two paradigmatic ways of talking about scientific inference. On the one hand, there is the Bayesian approach, according to which scientific inference involves determining which theories are the most probable in the light of the evidence. On the other hand, many episodes of theory evaluation in the history of science appear to be also aptly characterized as cases of Inference to the Best Explanation (IBE). What is the relationship then between IBE and Bayesianism? One possible answer is that the two are incompatible: IBE and Bayesian updating constitute two quite different rules for responding to the evidence (Van Fraassen 1989). However, it has been more common to think that the two approaches must be compatible. How then do the two approaches go together exactly?

To date, a common approach has been to look for places in the Bayesian framework where explanatory considerations could be feasibly brought in to play a role. Various philosophers have then recommended what might be called 'explanationist Bayesianism', according to which (depending on the version) it is either de facto the case, or reasonable, or even essential that explanatory considerations play critical roles in the Bayesian machinery (Lipton 2004; Weisberg 2009; Poston 2014). For example, one common suggestion is that explanatory considerations 'guide' or 'constrain' the assignment of prior probabilities.

An alternative way for IBE and Bayesianism to be compatible was proposed in Henderson (2014). The idea here is that, when the structure of scientific theories is taken properly into account, a Bayesian model of the preference for better explanations can be provided, making only assumptions which are quite natural and independent of any explanatory concerns. The key is to recognize that scientific theory evaluation occurs at multiple levels of generality or abstraction. A general level or framework theory is often regarded as providing a better explanation to the extent that it explains the phenomena without too much reliance of fine-tuning of auxiliary hypotheses. For natural choices of priors, the Bayesian method of calculating likelihoods for general frameworks of theories inherently penalizes such reliance on fine-tuning. Explanatory considerations then need not constrain or guide the Bayesian machinery, particularly the choice of priors. Rather inference to the best explanation emerges automatically from the Bayesian model. I have called this view 'emergent compatibilism', in contrast with the 'constraint-based compatibilism' provided by explanationist Bayesian approaches.

In this chapter, I will re-examine a case of IBE, which has been treated along explanationist Bayesian lines by Elliott Sober (Sober 1990). The case is the influential attack by the biologist George C. Williams (Williams 1966) on group selection hypotheses in evolutionary biology. Williams argues that the hypothesis that natural selection operates only at the level of individual organisms provides a better explanation of altruistic behaviour than a hypothesis also invoking group selection. This is because, on his view, invoking only individual level selection provides a more parsimonious account. Sober argues that the parsimony considerations in Williams' argument make it reasonable to assign a higher prior to the individual selection hypothesis.

I will argue that this case is actually much better modelled according to the emergent compatibilist approach. The explanatory considerations behind Williams' appeal to parsimony are driven by a concern to avoid fine-tuning the theory, and as such are naturally taken into account in the Bayesian likelihoods. There is no need then for explanatory considerations to govern the assignment of priors.

The reassessment of the case of Williams' argument has more general implications. Sober uses this example as part of a wider argument in favour of what we might call a 'local' view of the epistemic significance of parsimony (Sober 1990). In contrast to this view, I suggest that Williams' argument shares basic features with other cases of IBE. The Bayesian treatment of the Williams' case is also similar to the treatment of other cases of IBE, in which explanatory considerations are reflected in the Bayesian likelihood. Thus, consideration of these cases appears to point to a more general characterization of IBE and its evidential relevance, which contrasts with Sober's localism.

2. IBE and Bayesianism

There is a long-standing puzzle over how explanatory and confirmatory aspects of scientific theories are related. On the one hand, it seems that scientific theory evaluation should be driven primarily by how well confirmed various rival theories are in relation to the evidence. Philosophers of science have proposed a number of theories about what such empirical support or confirmation consists in. Prominent among these is the Bayesian approach. In the Bayesian approach, agents' attitudes towards hypotheses in a hypothesis space $\mathcal{H}$ are taken to be degrees of belief which can be represented by a probability distribution (or density) over $\mathcal{H}$. The Bayesian starts from a prior probability distribution $p(T)$ representing initial degrees of belief in theories T in $\mathcal{H}$ and as evidence D is obtained, updates the prior by conditionalization to a 'posterior' distribution $p(T|D)$. The posterior distribution is the conditional probability $p(T|D)$, given by Bayes' rule $p(T \mid D) = \dfrac{p(D \mid T)p(T)}{p(D)}$. In this equation, $p(D|T)$ is the 'likelihood' for the theory T and $p(T)$ is the 'prior'.

On the other hand, scientific arguments for one theory over another often appeal more or less explicitly to explanatory advantages of that theory. For example, Darwin argues very explicitly in *The Origin of the Species* that the theory of natural selection provides a better explanation of 'large classes of facts' than special creationism (Darwin 1962 [1859]). Copernicus argues that his heliocentric theory possesses greater 'harmony' than its Ptolemaic rival, which may be seen as an appeal to a kind of explanatory virtue (Copernicus 1543).

The fact that both Bayesianism and IBE appear to be broadly applicable to cases in the history of science suggests that they are compatible with one another. But then the puzzle is how to characterize their relationship. How does IBE fit in with the evaluation of theories in terms of their empirical support? Is IBE just another way of talking about empirical support? Or is it an additional component which needs to be included in a full account of theory evaluation?

As we saw in the introduction, a dominant approach to reconciling the two paradigms has been to recommend that Bayesianism and IBE should be amalgamated into an explanatorily driven form of Bayesianism. Minimally, this position requires finding roles that explanatory considerations could play in the Bayesian framework. A stronger version requires defending the idea that explanatory considerations are somehow critical or essential for the operation of Bayesianism.

One of the key suggestions about the role for explanatory considerations is that they act as a constraint on the prior probabilities.[1] The task is then to justify giving higher priors to more explanatory hypotheses. One way to tackle this task is by seeking a general reason for the connection. Ted Poston, for example, argues that if more explanatory hypotheses are not given higher priors, inductive scepticism follows (Poston 2014). Others have advocated that simpler hypotheses should be given higher priors (e.g. Jeffreys 1998 [1939]), though there has also been serious opposition to this view (e.g. Kelly 2007). The basic problem, critics urge, is that we have no reason to think the world is more likely to be simple than complex, so no grounds for making that assumption in the prior.

Another approach is to adopt a more local strategy. The idea is that perhaps even without a general connection between parsimony and priors, in certain specific cases, we can justify giving higher priors to more parsimonious hypotheses. In this vein, Sober argues that Williams' appeal to parsimony in his arguments against group-selection hypotheses provides such a local justification for assigning them a lower prior.

---

[1] There have been other suggestions too. For example that explanatory considerations help to determine likelihoods, relevant data, and hypotheses for consideration in the hypothesis space (Lipton 2004), or that explanatory considerations aid in the application of the principle of indifference (Huemer 2009, Poston 2014).

However, there may be no need to defend assigning higher priors to more explanatory hypotheses either on a global or a local level. In Henderson (2014), I have proposed a different kind of Bayesian model of IBE. The key observation here is that scientific theory evaluation can be thought of as occurring at different levels. The overall aim of theory evaluation is to find the best specific theory with all its parameters filled in, which can explain the data. But it is often helpful to think of accomplishing this aim in two stages: on the one hand, finding the right theory at the general level, that is the right general principles or 'framework', and on the other finding the right parameter values or specific auxiliary hypotheses which allow this framework to fit or explain the data. To give a very simple example, suppose we are interested in the relationship between two variables, x and y. One question is what kind of general functional relationship exists between those variables. One 'framework' level theory might be that the relationship between the two is direct proportionality—i.e. $y = ax$, for some constant a. Now suppose a can take integer values from 1 to 10. Then we have ten fitted theories produced by the framework $y = ax$ corresponding to the ten different auxiliary hypotheses about the value for the constant a. Many theories contain open variables which require fixing before they make specific predictions. For instance, Newton's second law of motion says there is a force, F, which is directly proportional to the product of the mass and the acceleration of a body. In a fitted theory for this framework, we fill in a particular force for the case in hand, for example a force obeying Hooke's law if the system is a spring.

My suggestion is that a large part of what makes for better explanation is a feature of the relationship between the specific and framework levels of theory evaluation, in particular how much tuning of specific parameters or auxiliaries is required in order for the higher level framework to produce a good empirical fit to the data (Henderson 2014). One of the key characteristics of a theory that provides a better explanation is that it can account for the phenomena without requiring that the auxiliary hypotheses be highly fine-tuned. The theory will provide a better explanation if it explains from its core principles, rather than relying on such fine-tuning.

When we build a Bayesian model, it is helpful to capture this distinction between frameworks and specific (or fitted) theories. Rather than the standard approach of placing all candidate-specific theories in one big hypothesis space, we can use a hierarchical Bayesian model, in which the hypothesis space is structured into levels (Henderson et al. 2010). Frameworks can compete with one another at a high level of the model, whereas the fitted theories occupy a lower level. Bayesian updating then is carried out at each level of the model.

In order to evaluate the frameworks at the high level, the Bayesian updates a prior probability $p(F_i)$ over the frameworks to a posterior $p(F_i|D)$. The posterior depends on both the prior for the model $p(F_i)$ and on the likelihood $p(D|F_i)$. The likelihood for the framework F_i, $p(D|F_i)$ is an average over the likelihoods for the specific hypotheses h in the framework, weighted by priors for those hypotheses with respect to the framework $p(h|F_i)$. As long as the prior probability is spread roughly evenly over the hypotheses of the model, the average likelihood penalizes less simple, or more fine-tuned explanations. That is because in such cases we have to average over a relatively large number of ill-fitting specific hypotheses which have very low likelihoods. The principles at work here are those that underpin the Bayesian approach to 'model selection', which is known to favour simpler hypotheses (MacKay 2003).[2]

Notice that in this type of Bayesian account, there is no need to postulate that explanatory considerations constrain the priors. There are some assumptions about how the prior probabilities are spread over the specific hypotheses in a framework, but these need not be motivated by anything to do with explanatoriness. Rather they appear to be quite 'natural' assumptions that one would make on independent grounds. Thus, this type of model provides an independently motivated Bayesian account of why better explanations are preferred in scientific inference.

Williams' argument against group selection hypotheses provides an interesting case of an inference to the best explanation. Elliott Sober has treated this case along explanationist Bayesian lines, adopting the local strategy of arguing that there are case-specific reasons to take the explanatory considerations, particularly parsimony, as a guide to the prior probabilities. I will now argue that the case is better treated along the lines of the alternative Bayesian model I have sketched above, where the explanatory considerations naturally emerge from the calculation of the likelihoods in a hierarchical Bayesian model. Thus, the example is one where explanatory considerations emerge from the Bayesian model, rather than being explicitly introduced into it. To see this, let us first review the example.

3. Williams' Argument against Group Selection

The debate over units of selection goes back to the very origin of the theory of natural selection. The theory of natural selection in evolutionary biology can be put in some-what abstract terms as follows. If there are biological entities which reproduce with some variation in traits, such that different variants differ in fitness, and the traits are heritable, then those variants which are more fit are more likely to survive and reproduce. This statement is neutral about which biological entities are the 'units of selection'. In Darwin's original theory of natural selection, the units of selection were individual organisms. The idea then was that organisms generally inherit their traits, but they reproduce with some variation. Organisms which are better adapted to their environment are more 'fit' and out-breed organisms which are less well adapted. Thus over time, and generations, there will be a shift in the population towards organisms with better adapted traits.

---

[2] Sometimes, this is seen by making approximations to the Bayesian likelihood, such as the Bayesian Information Criterion, or BIC. The BIC contains two terms corresponding to the best fit that can be achieved by the theory and a penalty for complexity, which increases with the number of free parameters in the theory.

However, the existence of biologically altruistic behavioural traits presents a puzzle for the theory based only on a selection of individual organisms. Biologically altruistic traits are traits which increase the fitness of other organisms, whilst reducing the fitness of the altruistic organism itself. There are many examples of apparent altruism in the natural world. One is the 'wagon-training' of musk oxen.[3] This is a defensive formation that the oxen adopt when they are attacked by wolves. The larger stronger animals stand on the outside protecting the young and weaker members of the group. At first sight the theory of individual selection seems unable to explain such behaviour. Wouldn't a large musk ox who protects the weaker members of the herd do worse than a selfish ox who makes a run for it, or even stands behind the weaker members, making it more likely that they will be attacked instead of him? In general, why would an organism possessing altruistic traits have any selective advantage at the level of individual organisms, since possession of these traits is clearly detrimental to the organism's own ability to survive and reproduce?

Since Darwin, people have considered whether it might help to extend the theory of natural selection to allow not only selection between individuals, but also selection between groups. The idea, at least up to the 1960s, was that groups might themselves serve as units of selection. That is, they might reproduce, perhaps by some process of splitting, inheriting traits, and some groups might then out-compete other groups in a process of selection. So groups of altruistic musk oxen would be selected over groups of selfish oxen. Thus the reason for the existence of the trait in the male musk ox is its benefit to the group rather than to the ox himself.[4]

So we have two competing theories about the units of selection, each trying to explain the existence of biological altruism. One allows for individual organisms to serve as units of selection—call this theory IS. The other allows for selection to occur both at the individual and at the group level—call this theory GS. Prior to the 1960s, as Sober points out, the group selection theory was unpopular among population geneticists, but widely assumed to be true by field naturalists. The tide turned in 1966, when the biologist George C. Williams published a book criticizing the idea of group selection (Williams 1966).

Part of what Williams does in his book is to argue that, despite initial appearances, individual level selection can explain many cases of apparent biological altruism. Some cases, he says, can be accounted for as 'statistical effects' of individual level selection. For example, in the case of the musk oxen, what might really be going on is that the threat felt by an ox depends on its size relative to a predator. There is some threshold of predator size which determines whether the ox responds with 'counterthreat or with flight'. For predators of a certain size, larger oxen will be more inclined to stand their

[3] See Sober 1990, p. 80.
[4] Notice that this way of treating group selection should be taken in historical context. Since Williams' book, the idea of group selection has been revived in a different form, largely due to work of D. S. Wilson and Elliott Sober (Sober and Wilson 1998).

ground than to flee and the result is that they will end up in a more exposed position, seemingly protecting weaker members of the herd. Williams suggests that this case is one of a number of others which may also be explained in terms of different adaptive responses of individuals according to their own strengths.

In addition, Williams argues that whilst group selection is a possible evolutionary process, explanations which invoke group selection are relatively 'onerous'. Group selection hypotheses, he thinks, should only be invoked when 'simpler' (genic or individual) forms of natural selection failed to give an explanation. Williams can be taken to be making an argument that theories invoking only individual selection provide a better explanation of the biological phenomena than theories with group selection.

But what exactly is the reason that Williams sees group selection hypotheses as lacking in parsimony? Sober offers the following plausible interpretation. In a group selection explanation, the altruistic groups do better and so produce more offspring. This explains how we can have more altruists. The problem is that there isn't just group selection going on. There is also still selection between organisms in a group. And the effect of that will always be to favour the selfish. Groups will experience 'subversion from within', meaning they will tend to be overtaken by selfish organisms. What this means is that if group selection is to be effective it must occur at a fast enough rate to overcome the effects of individual selection. This imposes a number of constraints on acceptable rates of group reproduction and other parameters. Thus, as Sober says, 'The rational kernel of Williams' parsimony argument is that the evolution of altruism by group selection requires a number of restrictive assumptions about population structure' (Sober 1990, p. 83).

3.1 An explanationist Bayesian account

Sober argues that, in a Bayesian framework, the considerations involved in Williams' appeal to parsimony impact the prior and not the likelihood of the group selection theory.

Williams suggests that the hypothesis of group selection, if true, would explain the observations, and that the same is true for the hypothesis of individual selection that he invents. This means ... that the two hypotheses have identical likelihoods. If so, the hypotheses will differ in overall plausibility only if they have different priors.   (Sober 1990, p. 144)

Sober then provides an account of what is going on in Williams' argument which is supposed to justify the assignment of a lower prior to the group selection hypothesis, not in terms of a general argument, but for this particular case. Sober suggests that the reason that parsimony impacts on the plausibility of the hypotheses is because the foregoing argument about the restrictive assumptions on population structures amounts to a 'biological judgment about the relative frequency of certain population structures in nature' (Sober 1990, p. 146). The idea is that since the population structures supporting the group selection explanation are relatively rare, we may regard the group selection hypothesis as less plausible, and hence give it a lower prior.

However, making the connection between parsimony and plausibility via an appeal to relative frequency is not so convincing, since it is not clear how we should think of this relative frequency—in particular, it is not clear what the relevant population is for the population structures. In Section 3.2, I will suggest that in fact Williams' argument is much more naturally captured by the hierarchical Bayesian model, which does not require us to give even a local justification for a connection between parsimony and plausibility.

3.2 An emergent compatiblist account

The types of explanatory considerations that Sober has identified in Williams' argument are exactly those which can be easily modelled in the hierarchical terms described above. In this example, there are two competing frameworks: the individual selection framework in which natural selection is only allowed to operate on individual organisms, and the group selection framework in which both individual level and group level selection are allowed. In each case, filling in details of various parameters, such as selection rates, produces particular individual selection- or group selection-based fitted models.

When we ask whether the individual selection theory or the group selection theory provides the better explanation of biological phenomena, we are asking which of the two general frameworks IS or GS provides the better explanation. What Williams takes to be less explanatory about GS is that it needs to be combined with various very specific auxiliary hypotheses in order to account for the phenomena.

If it is true, as Williams alleges, that the group selection framework needs to be highly fine-tuned, this will be reflected in a lower Bayesian likelihood for that theory, since the theory will be automatically penalized for fine-tuning by the averaging process. The Bayesian then favours the more explanatory theory without assigning any higher prior probability to the individual selection theory. Indeed it is quite possible for the priors for the competing frameworks to be equal: $p(IS) = p(GS)$.

This case thus exemplifies the 'emergent compatibilist' position I argued for in Henderson (2014). According to emergent compatibilism, there is no need for explanatory considerations to play an autonomous role in the Bayesian machinery. Rather, given independently plausible choices of priors, the Bayesian favours the more explanatory hypotheses. We are then able to explain, using a Bayesian model, why explanatory considerations play a role in the inference, rather than stipulating that they do.

The explanatory considerations in Williams' argument are a feature of the relationship between framework and the auxiliaries required to produce a fully fitted theory. Therefore, the explanatory considerations are not naturally captured in Bayesian models which do not preserve the distinction between frameworks and fitted theories. In Sober's Bayesian treatment, the theories in question are fully fitted theories with parameters adjusted to account for the observations. In that account, $p(IS^*) > p(GS^*)$, and $p(D|IS^*) = p(D|GS^*)$ where $*$ denotes that the theory is fitted.

There is no conflict in formal terms between the hierarchical model I have advocated and Sober's explanationist Bayesian model. Sober's model works with fitted hypotheses and assigns a higher prior to the individual selection hypotheses. Indeed, we can always consider the support for specific hypotheses h by marginalizing out the frameworks:

$$p(h|D) = \sum_i p(h|F_i,D)p(F_i|D) \tag{1}$$

If we just consider the updating of $p(h)$ to $p(h|D)$ for the specific hypotheses h, it will be as if we did weight the prior in favour of the more explanatory hypotheses. However, although it is always possible to model episodes only in terms of specific hypotheses, by so doing, one misses important aspects of the structure of the reasoning which help to justify why the prior is higher for more explanatory specific hypotheses. Considering the role of the frameworks provides an independently motivated explanation of *why* explanatory considerations play a role in inference. Equation (1) makes it clear how the support for specific hypotheses h from D depends on two components:

i)   how well each framework F_i is supported by the data (as expressed in $p(F_i|D)$), and
ii) how well the specific h is supported by D, with respect to the framework F_i (as expressed in $p(h|F_i,D)$).

The first component, as we saw above, naturally reflects the explanationist concern with fine-tuning since it involves an averaged likelihood. By contrast, in Bayesian reconstructions which work with fitted models, explanatory considerations can be grafted onto the Bayesian machine by having them constrain the priors, but it remains rather opaque why they, rather than some other considerations, should be what determines the priors. On the alternative account that I am recommending, there is no need for explanatory considerations to constrain the priors explicitly. As we saw in the group selection example, the priors for the competing frameworks could even be equal.

Some assumptions about priors do play a role of course. In the calculation of the average likelihood, the likelihoods for specific hypotheses are weighted by a prior $p(h|F_i)$. Different assignments of these priors will result in different values for the average likelihood $p(D|F_i)$. However, a natural choice is a prior which is uniform over the different specific hypotheses, at least in situations where there is no background information favouring any particular hypothesis or range of hypotheses. For an objective Bayesian, this choice might be motivated by a consideration such as the principle of indifference. The important point is that, although the results do depend on assignments of priors to the specific hypotheses, there are generally reasons for these assignments, which can be motivated independently of the desire to ensure compatibility with IBE.

4. On the Generality of IBE

Part of the purpose of giving a Bayesian model of IBE is to elucidate the epistemic significance of explanatory considerations by connecting them to an established account of evidential support. We have discussed two different ways of modelling IBE in Bayesian terms: the explanationist Bayesian approach of grafting explanatory considerations onto the Bayesian machinery, and the hierarchical Bayesian approach in which explanatory considerations emerge. The first approach makes Bayesianism dependent upon explanationist guidance, whereas the second treats Bayesianism and IBE as independent, but compatible, characterizations of scientific inference.

The two approaches also have quite different implications for how 'local' we take IBE to be. The explanationist Bayesian model is compatible with, and indeed encourages the view that the epistemic significance of explanatory considerations such as parsimony is highly case-specific. For example, Sober has advocated the view that the epistemic significance of parsimony is highly 'local'.

When a scientist uses the idea [of parsimony], it has meaning only because it is embedded in a very specific context of inquiry. Only because of a set of background assumptions does parsimony connect with plausibility in a particular research problem. What makes parsimony reasonable in one context therefore may have nothing in common with why it matters in another. The philosopher's mistake is to think that there is a single global principle that spans diverse subject matters.   (Sober 1990, p. 140)

Sober argues that not only are the considerations which justify constraints on the prior specific to the case in cases like Williams', but there are also other cases where the considerations behind parsimony are quite different and are better seen as reflected in the Bayesian likelihood rather than the prior. He gives the example of the appeal to parsimony by cladists in phylogenetic inference. Overall, then, Sober's conclusion is that there is nothing much in common between the cases of parsimony, and no common representation in Bayesian terms.

However, Williams' argument only appears to be a very case-specific use of parsimony if we fail to identify and model the key structural feature that it has in common with other cases of IBE. In fact, the core explanatory considerations in Williams' arguments can be analysed in terms of the relationship between different levels of theory evaluation. Specifically, in Williams' argument, what is taken to be less explanatory about the theory involving group selection is that it needs to be combined with various very specific auxiliary hypotheses in order to account for the phenomena. This kind of concern is a very common feature of other cases of IBE.

In Henderson (2014), I discussed another example. This was the case of the explanation of retrograde motion by the Ptolemaic and Copernican theories. Planets generally traverse the sky in an eastward direction from night to night. However, from time to time, they have periods of 'retrograde motion' in which they go into reverse and move westward. The Ptolemaic theory, which has the Earth at the centre of the planetary orbits, explained retrograde motion by placing each planet on an epicycle, mounted on a deferent that carries the planet in its circular orbit around the Earth. When the planet moves backwards on its epicycle with respect to the motion of the deferent, it would appear from Earth to be in retrograde motion. By contrast, according to the

Copernican theory, the Sun is at the centre of the planetary orbits, including Earth's. Planets are then to be expected to be in retrograde motion whenever the Earth overtakes the planet in its orbit (or vice versa). Thus, the Copernican theory explains the phenomenon without the need to rely on the auxiliary hypothesis of an epicycle, or to account for the details of the motions by fine-tuning the rates of the epicycle orbits. This is the key reason that the Copernican theory provided a better explanation of retrograde motion than the Ptolemaic theory.

Other comparisons between theories can be understood along similar lines. As another example, consider the rivalry between the Newtonian corpuscular theory of light and the wave theory of light in the nineteenth century. The Newtonian theory of light treated light as a stream of corpuscles. It could explain diffraction phenomena, but it invoked different auxiliary hypotheses for different types of diffraction. The pattern of light observed on a screen as the result of shining light on a hair was explained in terms of an ether surrounding the hair whose variation in density produced the pattern on the screen. The observation of rings between two pieces of glass was explained by postulating that by passing through a medium such as glass, corpuscles of light acquire a disposition to either reflect or transmit through a subsequent surface. Newton called these dispositions 'fits of easy transmission and easy reflection'. The corpuscular explanation of thin slit diffraction relied on yet further auxiliary hypotheses. By contrast, the wave theory of light could use the same basic principles to explain all these different phenomena of diffraction. In that sense, wave theorists could reasonably claim that it provided a better explanation of these observations than the corpuscular theory.

The concern to avoid fine-tuning also explains why scientists often claim an explanatory advantage for their theory based on its ability to explain from its basic principles, rather than requiring a number of arrangements of auxiliary hypotheses. For example, Lavoisier points out that his new oxygen theory can explain various phenomena such as combustion and calcination of metals, saying

I have deduced all the explanations from a simple principle, that pure or vital air is composed of a principle particular to it, which forms its base, and which I have named the oxygen principle, combined with the matter of ore and heat. Once this principle was admitted, the main difficulties of chemistry appeared to dissipate and vanish, and all the phenomena were explained with an astonishing simplicity.[5]

Lavoisier claims that his theory provides simpler explanations of the different phenomena because they are all made in terms of his basic oxygen principle, and do not need to invoke particular assumptions on which the rival phlogiston theory required,

---

[5] Thagard's translation (Thagard 1978). French original is: Jai déduit toutes les explications dun principe simple, cest que l'air pur, l'air vital, est composé dun principe particulier qui lui est propre, qui en forme la base, et que jai nommé principe oxygine, combiné avec la matière du feu et de la chaleur. Ce principe une fois admis, les principales difficultés de la chimie ont paru sévanouir et se dissiper, et tous les phénomènes se sont expliqués avec une étonnante simplicité (Lavoisier 1783).

such as that the phlogiston given off in combustion has negative weight (see Thagard 1978, pp. 77–8).

In all these cases, scientists are taking theories to be less explanatory if they rely too heavily on fine-tuning auxiliary hypotheses. And in all these cases, a Bayesian account is available in which the frameworks which are overly fine-tuned receive lower Bayesian likelihoods, because of the way these framework likelihoods are computed as averages of specific theory likelihoods. The basic—and generally applicable—method of Bayesian model selection lies behind this approach. In Henderson (2014), I explain how the Copernican vs. Ptolemaic example may be handled in these terms. The phylo-genetic case that Sober discusses can also be modelled in a similar way. In that case, the competing frameworks are different tree structures that represent the branching structure of the evolutionary history. The auxiliary hypotheses are the particular assumptions about the probabilistic weighting of different branches. To compute the overall support for a particular tree, we compute the average likelihood over all the possible branch weightings. As in the case of Williams' argument, the main concern is to find the tree for which the explanation of the observed traits is due primarily to the structure of the tree itself, and not to particular choices of the branch probabilities.[6]

In cases such as these, it is not that there are no 'subject-matter-specific' assump-tions or judgements at work. In particular, the identification of the framework and its parametrization are important background assumptions in any given case. However, there is a structural commonality to the explanatory considerations involved, in terms of the relationship between different levels of theory evaluation, which makes possible a more unified connection to Bayesianism than Sober has suggested.

5. Conclusion

In this chapter, we revisited Williams' argument against group selection. I argued that, contrary to the treatment by Elliott Sober, this case is not best seen as one in which the scientist is recommending that the more explanatory (more parsimonious) theory gives a reason for assigning it a higher prior probability. Rather, the core consideration in Williams' argument is a concern that group selection hypotheses rely too heavily on fine-tuning of auxiliary hypotheses in providing their explanation of biological phe-nomena. Since the explanatory considerations are a feature of the relationship between levels of theory evaluation, their evidential significance is obscured in a Bayesian model like Sober's which works only with a hypothesis space of specific hypotheses. However, their significance can be illuminated in a Bayesian model which distin-guishes between the framework and specific theories. In such a model, we have a 'model-selection' process governing the comparison of frameworks, and a 'model-fitting' process determining the right parameter values for a given framework. Model

---

[6] This is discussed in detail in Sober 1988. There are a number of subtleties and controversies over the right way to do phylogenetic inference which need not detain us here.

selection, both in Bayesian and non-Bayesian settings, takes into account how fine-tuned the framework is, and hence accounts naturally for the explanatory considerations in IBE. In Bayesian model selection, this means that less explanatory hypotheses will typically (subject to reasonable, but independently motivated, assumptions about priors) have a lower likelihood than hypotheses which provide a better explanation.

Sober has argued against the idea that there is a 'single global principle' behind the explanatory virtue of parsimony which transcends different subject matters. Part of the reason is that he presents examples in which, he alleges, parsimony plays quite different roles with respect to evidential support. In some cases, he suggests, parsimony is reflected in Bayesian likelihoods. In others, such as Williams' argument against group selection, he argues that the prior probability is determined by the parsimony considerations. In my view, the kind of explanatory considerations found in Williams' argument are a feature of the relationship between levels of theory evaluation and can also be identified in various other cases of IBE. They can also be modelled in the Bayesian framework in a similar way. All this suggests that a more unified picture of IBE and its evidential significance may be available. But this conjecture does still need to be tested by going beyond the handful of cases so far considered. More work needs to be done to see if other cases of IBE can be analysed in a similar fashion, when they are examined in detail.

References

N. Copernicus. *De Revolutionibus Orbium Caelestium*. 1543.

C. Darwin. *The Origin of the Species by Means of Natural Selection*. Collier, New York, 6th edition, 1962 [1859].

L. Henderson. Bayesianism and inference to the best explanation. *British Journal for the Philosophy of Science*, 65: 687–715, 2014.

L. Henderson, N. D. Goodman, J. B. Tenenbaum, and J. F. Woodward. The structure and dynamics of scientific theories: A hierarchical Bayesian perspective. *Philosophy of Science* 77(2): 172–200, 2010.

M. Huemer. Explanationist aid for the theory of inductive logic. *Brit. J. Phil. Sci.*, 60: 345–75, 2009.

H. Jeffreys. *Theory of Probability*. Oxford University Press, Oxford, 3rd edition, 1998 [1939].

K. Kelly. Ockham's razor, empirical complexity and truth finding efficiency. *Theoretical Computer Science*: 270–89, 2007.

A. Lavoisier. Reflexions sur le phlogistique. *Memoires de l'Academie des Sciences*: 505–38, 1783.

P. Lipton. *Inference to the Best Explanation*. Routledge, London, 2nd edition, 2004.

D. J. C. MacKay. *Information Theory, Inference and Learning Algorithms*. Cambridge University Press, Cambridge, 2003.

T. Poston. *Reason and Explanation: A Defense of Explanatory Coherentism*. Palgrave Macmillan, Basingstoke, 2014.

E. Sober. *Reconstructing the Past: Parsimony, Evolution and Inference*. MIT Press, Cambridge, MA, 1988.

E. Sober. Let's razor Ockham's razor. In D. Knowles, ed., *Explanation and Its Limits*, 73–93. Cambridge University Press, Cambridge, 1990.

E. Sober and D. S. Wilson. *Unto Others: The Evolution and Psychology of Unselfish Behavior.* Harvard University Press, Cambridge, MA, 1998.

P. R. Thagard. The best explanation: Criteria for theory choice. *Journal of Philosophy*, 75: 76–92, 1978.

B. C. Van Fraassen. *Laws and Symmetry.* Clarendon Press, Oxford, 1989.

J. Weisberg. Locating IBE in the Bayesian framework. *Synthese*, 167: 125–44, 2009.

G. C. Williams. *Adaptation and Natural Selection: A Critique of Some Current Evolutionary Thought.* Princeton University Press, Princeton, NJ, 1966.

17

Inference to the Best Explanation and the Receipt of Testimony
Testimonial Reductionism Vindicated

Elizabeth Fricker

1. Explanationism and Testimonial Justification

Philosophers are perennially interested in the status of one's everyday beliefs about one's world, formed from perception and inference. Is one rationally justified in holding these beliefs? If so, what confers this justification? Explanationism offers a distinctive answer to this question: 'what justifies...the formation of any new belief...is that the doxastic move in question...increases the explanatory coherence of the subject's global set of beliefs. In particular, the explanationist holds that some beliefs are...justified by inference to the best explanation' (henceforth 'IBE') (Lycan 2002; see also Lipton 1991; Harman 1986).[1]

Explanationism so specified is a theory in epistemic dynamics, about what justifies one in forming a new belief. But it can also be formulated as a theory in epistemic statics, about the support relations that must hold between one's beliefs for them to be justified overall. Static and dynamic Explanationism are complementary theses, but one might plausibly hold dynamic but not static Explanationism.

Epistemically basic beliefs are justified, but not in virtue of support from other beliefs (see Audi 1998). Both Static and Dynamic Explanationism can be formulated to allow basic beliefs.

Static Explanationism: Every empirical belief is either basic or justified in virtue of relations of explanatory coherence in which inference to the best explanation plays a central role.

[1] This chapter develops a talk I gave at the May 2015 Orange Beach epistemology conference, and later in Bled in June 2015. My thanks to audiences at those events for valuable comments, especially to Susanna Rinard, Blake Roeber, Ernie Sosa, and Brad Weslake. Sandy Goldberg read my draft and made valuable comments. Thanks above all to Ted Poston and Kevin McCain, who organized a thoroughly enjoyable and stimulating conference, and gave me valuable comments as well as exercising great patience, as editors of this volume.

Dynamic Explanationism (DE):   All cases of justified formation of a new non-basic belief in response to an evidential input—either a new basic belief, or an experience or 'seeming' or some kind—are instances of IBE.

DE presupposes that Inference to the Best Explanation (IBE) is itself a legitimate form of non-demonstrative inference. I adopt this presupposition. My present project is a case study which, I shall argue, supports DE. DE is to be understood as a contingent thesis, about the status of matters of epistemic justification of belief for creatures broadly like us, and in a world sufficiently like our own in its main epistemic features.

DE implies a normative and a descriptive thesis. First,

Normative Dynamic Explanationism (NDE):   All legitimate forms of non-demonstrative inference to new beliefs from one's present belief base plus one's experiences or seemings are identical with, or reduce to, IBE.

Second, *Descriptive Dynamic Explanationism* (DDE) comprises two theses:

DDE 1:   A cognitively normal adult human, in worlds broadly similar to the actual world, is often placed to make inferences to new beliefs by means of IBE—her evidence base, together with new experiences and new basic beliefs, often provides epistemic basis for such new belief formation.

DDE 2:   When such individuals properly[2] form new non-basic beliefs by general cognitive capacities they have, it is on the basis of IBE.

NDE claims that other supposed methods of non-demonstrative inference, or epistemic principles for belief formation, that do not reduce to IBE, are not legitimate. So for NDE to be tenable, it must be shown how any apparently diverse methods of non-demonstrative inference that are legitimate reduce to IBE. This is the explanationist's tactic in relation to enumerative induction—where inductive inferences are well founded, it is claimed, they are so in virtue of a tacit IBE (Harman 1965; Lipton 1991). The explanationist may also maintain that certain other supposed epistemic principles or methods are not in fact legitimate. To do so without embracing scepticism, she will need to show how the domain of beliefs in question can be acquired justifiedly without recourse to any such special principle. Our enquiry below in our case study testimony will instance this strategy.

A constructively-minded explanationist will not want her theory to end up entailing scepticism about the justificatory status of one's empirical beliefs. Thus she will want more generally to show, regarding various domains in which humans form bodies of belief and act on them, that these beliefs can be exhibited as justified on the basis of the

---

[2] Of course people sometimes form beliefs via irrational procedures such as wishful thinking. For DDE 2 to be non-trivial requires that we have a positive account of how people's cognitive apparatus is designed by evolution to function; and an independent account of what are justified methods of forming beliefs. DDE maintains that these roughly coincide. It is familiar from various classic studies—e.g. of confirmation bias—that they do not do so perfectly. But this illustrates that DDE 2 can be filled out so as to be a plausible, while non-trivial, truth.

resources provided by explanationism. This leads her into the territory of descriptive dynamic explanationism. Rather than formulating the issues here in somewhat opaque abstract terms I will illustrate them by means of the case study that is my topic.

Testimony—the spoken or written word of another on some matter—is an indispensable source of knowledge and justified belief for each one of us in the conditions of division of epistemic skill and labour that characterize modern societies. For most of what each one of us believes, we do so only because we have trusted what we have been told—by our parents and carers, our school teachers, our friends and colleagues— or have read in a book or some other testimonial source—newspapers, radio and television, the internet. It is not possible for one now to separate out, in one's system of empirical beliefs, those which one acquired through reliance on trusted testimony versus those for which one had once some non-testimonial evidence. This is so because, as finitely cognitively resourced creatures, we generally operate so as to let beliefs in when the source is apparently reliable, and then keep the information thus acquired while throwing away the record of how one acquired it. (Cf. the 'clutter-avoidance principle' in Harman 1986.) Moreover, one's background beliefs, including many acquired via testimony, influence and inform the content of current perceptions. This being so it would be an epistemological disaster if investigation concluded that beliefs acquired from testimony and still epistemically dependent on that source are never justified. So it is imperative that one's general account of epistemic justification combines with one's account of how beliefs are acquired by a recipient of testimony (a T-recipient) to yield these results.

First, that epistemically justified belief from testimony[3] is humanly possible:

EJTBPoss: The nature and general epistemic circumstances of testimony, including the cognitive capacities possessed by recipients, allow for the acquisition of justified beliefs in what one is told.

Second, that in normal social conditions T-recipients frequently acquire justified belief through their receipt of testimony, hence that:

EJTBActual: In normal human societies, T-recipients frequently operate the suitable cognitive mechanisms they possess, which allow for justified acquisition of T-beliefs in response to the instances of testimony that they typically encounter.

EJTBPoss is necessary for a non-sceptical account of testimony. EJTBActual is a vaguer and less strictly binding constraint. While it is an imperative that one's account show how justified testimonial belief is within the reach of human capabilities it is, I suggest,

---

[3] More strictly, our concern is with beliefs acquired from taking the speaker's word for what she states, trusting her testimony. There are other ways in which one may justifiedly form beliefs as the upshot of witnessing testimony, but which do not instance the core process that testimony serves—which is its function in society to serve. These other ways are not cases of knowledge at second hand, epistemically dependent on the speaker's knowledge or other positive epistemic status with respect to what she states. See Fricker (2015a).

an open question at the start of theorizing to what extent people are sufficiently discriminating in their receipt of testimony, so as to always and only acquire justified beliefs from that source. A crucial point here is that what precisely is required of T-recipients depends on what precisely is provided by T-givers—recipients need to be adequately discriminating in their response to the types of testimony they in fact receive, but need not be to other types they might possibly but would not easily receive (see Fricker 2016b).

We can now see clearly the task of the proponent of non-sceptical Dynamic Explanationism, normative and descriptive (NDE plus DDE 1 and DDE 2) in relation to testimony: she must show how the sole resource for non-demonstrative inference that she allows—IBE—can, conjoined with a correct account of the nature of acts of testifying, and of the capacities of speakers and recipients, account for the truth of both EJTBPoss and EJTBActual. To do so her first task is to show how, on many occasions of aptly truthful testimony, the T-recipient has an evidential base available from which a well-supported inference to the truthfulness of the testimony, as part of the best explanation of why it was offered, can be made. This vindicates EJTBPoss. Her second task is to vindicate EJTBActual by making a convincing case that typical recipients of testimony are cognitively equipped to make the needed explanatory inferences, and that they in general manage their doxastic response to testimony by doing so—they construct an explanation of the speaker's utterance, why she offered her word on her topic to one, and take her word only if this explanation entails the accuracy of the testimony. My central Section 3 addresses this first task for the explanationist, Sections 4 and 5 then address the second task.

With this pair of tasks achieved Dynamic Explanationism accounts for how actual human testimonial beliefs are justified, when they are so: explanationism provides a convincing non-sceptical account of testimonial belief formation. This being so the case of testimony provides corroborative support for Dynamic Explanationism, and conversely, the antecedent plausibility of Dynamic Explanationism provides support for this 'reductive' account of testimonially justified belief. A happy coincidence of interest—coherence being a relation where justificatory support can run simultaneously in both directions—epistemic double-counting is OK!

This first project—providing support for Dynamic Explanationism by showing it provides a convincing account of how humans justifiedly form new testimonial beliefs—overlaps with another that has been one main focus of discussion in regard to testimony since the publication of Coady's classic innovatory treatment of what was up till then a neglected topic (Coady 1992). The issue is whether a 'reductionist' or 'fundamentalist'[4] account of justifiedness in testimonial beliefs ('testimonial justification') is correct. The dynamic explanationist account of testimonial justification I develop and defend below builds on the 'local reductionist'

---

[4] I follow Graham in using this aptly descriptive positive term rather than the negative mouthful 'anti-reductionism' (Graham 2006).

account of testimonial justification that I developed some years ago in 'Against Gullibility' (AG) (Fricker 1994).

One's *testimonial beliefs* (T-beliefs) are beliefs that were acquired via accepting as true what one was told, thereby incurring epistemic dependence on the speaker, and whose basis has not subsequently been changed or augmented by the acquisition of further confirming evidence. *Reductionism* about testimonial justification is the thesis that testimonial beliefs are justified in virtue of domain-general epistemic principles and methods of inference that confer epistemic justification, applied to beliefs from other sources such as perception and memory. A *global reductionist* account requires that one suspend belief in all one's previously acquired beliefs that depend epistemically on testimony, whether directly or indirectly, and reconstruct the entitlement to hold them using only such general justificatory principles and non-testimonial beliefs. A *local reductionist* account does not require this, but merely insists that, with regard to a fresh instance of testimony, the recipient should accept what she is told as true and form belief through trusting the speaker regarding her utterance only if she has a sufficient, and sufficiently independent, empirical basis to believe this speaker's testimony on this occasion about this topic to be trustworthy. This local reductionism is what Lipton calls a 'rule-reductionism': one's entitlement to accept a fresh instance of testimony is exhibited as holding in virtue of general principles of non-demonstrative empirical inference, applied to one's background of independently available evidence (Lipton 1998). It is a tricky matter to give a more specific account of what constitutes enough confirmatory independence to avoid epistemic circularity in this empirical basis to trust the speaker. Clearly if someone tells one: 'I am very trustworthy', and one accepts this as true, this past accepted testimony does not, intuitively, give one non-circular warrant to trust the next thing the speaker says. But I am confident there is in practice a range of rationally discriminating doxastic response to fresh instances of testimony to be found somewhere between the one extreme of unattainable global reductionism and the other extreme of uncritical acceptance of whatever one is told.[5] My present concern is with epistemic dynamics and so the question in virtue of what one's entire belief system, with its unredeemed global dependence on past trusted testimony, qualifies as justified is not urgent for my current dual-purpose project: first, to vindicate Dynamic Explanationism by showing it can give a non-sceptical account of the justified acquisition of new testimonial beliefs by ordinary recipients in their customary social circumstances; second, in doing so, to vindicate a local reductionist account of testimonial warrant, and see off fundamentalism.

One way in which testimonial justification would be fundamental is if T-beliefs were themselves epistemically basic beliefs—beliefs that are justified, and not in virtue of support from other beliefs. This option has not been seriously advocated in the

---

[5] Both the terms 'global' versus 'local' reduction, and the theoretical distinction they label, were introduced in Fricker (1994). Lipton (1998) makes a similar distinction between 'premise-reductive' versus 'rule-reductive' accounts of testimony.

literature, with the exception of Tyler Burge's writings, which can seem to suggest this (Burge 1993). Writers investigating the epistemology of testimony, myself included, have assumed that a T-recipient's position can be modelled thus: she knows that she has observed a certain audience-directed speech act, a telling;[6] the central epistemological question is what licenses her in inferring from this epistemic given to belief in the proposition thereby asserted. This model assumes first, that T-beliefs are not basic; second, that beliefs about the content and force of the understood speech act, call them U-beliefs, are typically epistemically basic.[7]

Sadly space does not permit me here to give a full defence explaining why T-beliefs are not basic, and so I content myself with two remarks. First, one must distinguish between a belief's being normatively inferential, versus it being formed by a real-time conscious psychological process of inference. The latter is not required for the former. Thus, the undeniable fact that on many occasions a T-recipient may just straightaway form belief in what she is told is consistent with holding T-beliefs to be normatively inferential. This epistemic status is a matter of what justification she needs to be able to offer for her belief if challenged, not of the phenomenology of its formation. Second, the deep reason why it is right to see everyday perceptual beliefs, but not T-beliefs, as basic, is to do with a contrast in the manner in which a certain state of affairs is present to a subject's consciousness when she on one hand sees that something is the case, and on the other enjoys the kind of representation of a state of affairs distinctive of grasping the truth condition of an assertoric utterance. It is right to regard U-beliefs as basic precisely because understanding an utterance—grasping its content and force –is a special sui generis type of representational state, a quasi-perception of this content and force. But this quasi-perception of content and force is not at all like a perceptual experience of the state of affairs which is its truth condition.[8]

Once treating T-beliefs as basic beliefs is ruled out, fundamentalism about testimonial justification assumes the form of this thesis: the denial of rule-reductionism. It amounts to the positing of an epistemic principle along these lines:

> *Fundamentalist Acceptance Principle* (FAP):   A recipient of testimony is entitled to accept as true what she is told, trusting the speaker's word on his topic, so long as this entitlement to trust is not overridden by other beliefs of hers that defeat the presumption of trustworthiness of the speaker with respect to her utterance.[9]

Hume famously observed that there is 'no a priori connection between testimony and reality' (Hume 1975). Lipton (1998) aptly interprets Hume as meaning by this simply that 'says P does not entail P'—there is no contradiction in supposing, of any instance

---

[6] Some writers hold it crucial whether she herself is the intended audience (see Hinchman 2005; McMyler 2011). My own view is that this makes no essential difference to her epistemic position (see Fricker 2006).

[7] For a defence of this view see Fricker (2003).   [8] See Fricker (2003, 2006).

[9] Burge's 'Acceptance Principle' is such a fundamentalist principle (Burge 1993). Other fundamentalists about testimony include Welbourne (1986) and Coady (1992).

of testimony, that what is said is false (barring certain peculiar propositions for which the fact of its assertion suffices to render true what is asserted).[10] More than this, everyday knowledge of human nature reveals how easily there can be, and often is, false testimony. Humans are entirely psychologically capable of lying, making assertions with deliberate deceptive intent, and there is very often motive for one person to mislead another (see Sperber et al. 2010). And deceptive intent is not the only risk for false testimony. Honest error, whether due to bad epistemic luck or simply to carelessness by a speaker in belief formation and in what he offers his word on, also poses an endemic risk to the accuracy of testimony.

This being so there is indeed no a priori link between the observed fact of someone's telling[11] one that P, and P. Instead there is a significant possibility of false assertion. So how is a T-recipient confronted with a telling that P to respond? Specifically, what epistemically justified route is available to her from the fact that she has been told that P to accepting as true P?[12] Fundamentalists bridge this evidential chasm with a dedicated epistemic principle, FAP. Their motivating intuition is that there is something special about testimony that makes it epistemically justified to respond to encountered testimony by accepting it as true, unless one has clear doxastic defeaters, despite the significant possibility of false testimony.

A local reductionist, in contrast, maintains that one is justified in accepting what one is told on some occasion only if one has sufficient empirical grounds, on that occasion, to believe the speaker to be telling the truth on that occasion—this comprising that she is neither lying, nor careless as to the truth of what she asserts, nor honestly deceived. (Antecedent grounds to believe the speaker trustworthy are not sufficient for justified trust since there must also be no contrary testimony, nor other strong counter-evidence to what is stated.)

One's commonsense appreciation of the nature of human acts of testifying makes local reductionism the prima facie correct account of its epistemology. Why ever would one be entitled to form belief in what one is told absent any evidence of the speaker's credentials when we all understand how easily it can happen, and all too often does happen, that a speaker lies or makes an honest error? The burden of proof is

[10] Lipton rejects Coady's stronger interpretation of Hume's dictum as entailing that it is a metaphysical possibility that there be a society that has the institution of testimony and all the testimony offered in that society is false. As Lipton observes, this clearly is not possible since there will be instances of contradictory testimony—A asserts P, while B asserts not-P.

[11] Ordinary language has this verb for the core case of testimony, face-to-face giving of one's word through linguistically mediated assertion to an intended audience. 'Telling' in ordinary language also has some other uses—'tell me that story about the little boy who found a magic ring, again'—but this is its core usage, and I adopt it to denote the speech act whose nature and epistemic force I am concerned to theorize.

[12] I am concerned with the normative structure of evidential support for her T-belief that a T-recipient needs to have for it to be justified. My somewhat metaphorical talk of an 'epistemic route' invokes such a normative structure. There is no implication that the real-time psychological process through which a T-recipient forms her accepting belief in what she is told is usually a sequential process of inference that starts with a belief that she has been told something and infers to the truth of what was asserted via ancillary premises about the speaker.

on fundamentalists to make a positive case why reductionism is wrong. Fundamentalists have risen to this task, and offered various distinct arguments. This is not the place to review and rebut all the arguments that have been or might be offered for Fundamentalism.

These arguments are mostly persuasive rather than compelling. But there is one argument that, if it went through, would be compelling for anyone committed to providing an anti-sceptical account of testimonial justification. This is the transcendental argument that is central to Coady's case for fundamentalism (Coady 1992). It goes thus: there is justified belief from testimony; on a reductionist account there could not be, since the independent empirical basis to trust testimony that reductionism requires is not available; hence reductionism is false. Coady maintains we must accept FAP as a correct epistemic principle on pain of espousing a disastrous scepticism about all of our beliefs that have epistemic dependence on testimony—which in modern conditions of division of epistemic labour is most of them. In AG I formulated and—in my view—rebutted this transcendental argument for fundamentalism about testimonial justification. I did so by showing how, on many occasions of testimony, a recipient has available to her an adequate independent empirical basis to trust the present speaker, on her present topic.

In AG I maintained that the epistemology of testimony is part of the epistemology of other minds. The epistemic means a recipient has available to assess a speaker for trustworthiness is to construct a mini-psychological interpretation of him. One should trust what he tells one just if one's best explanation of his utterance, as afforded by this mini-theory of his psychology, entails the fact that what he states is true. In my central Section 3 I elaborate the details of such a typical folk-psychological explanation of the speaker's act of telling revealing its truthfulness. The epistemology of testimony is indeed part of the epistemology of other minds, but it has as a further component: the nature of the socio-linguistic speech act of telling. In telling that P a speaker offers his word that P to his intended audience. Correlatively, telling is aptly governed by a social norm: tell that P only if you know that P. The social-norm-constituted force of the speech act of telling enables the distinctive content-type of an IBE entailing the truth of what is told that is often available to a T-recipient. This IBE entails that what was told is so as an inference from the confirmed explanatory hypothesis that the speaker, in telling as she did, was conscientiously conforming to the social norm for telling—she knew what she stated. I thus show that the means to attain justified testimonial beliefs is often available on a local reductionist account of what this requires. In Sections 4 and 5 I consider whether it is empirically plausible that actual T-recipients manage their response to received testimony via the IBE strategy described in Section 3. I argue this empirical hypothesis is plausible and well supported. The transcendental argument for Fundamentalism is thus refuted and room is made to espouse the plausible local reductionist account of testimonial justification. The local reductionist account of testimonial justification I provide is an explanationist account, corroborating the correctness of Dynamic Explanationism. So I trap two philosophical birds with one argument.

Fundamentalism about testimonial justification is inconsistent with Dynamic Explanationism: NDE entails there are no dedicated epistemic principles such as FAP. But the denial of fundamentalism does not entail the truth of explanationism. There could be a different, non-explanationist reductive account of testimonial justification. In Section 2 I examine and reject such an alternative reductionist account. The *brutely Humean account* posits that one's reliance on testimony is empirically grounded in enumerative induction from observed correlations of past instances of testimony with the obtaining of the asserted fact. I agree with Coady's verdict that this account is untenable. Coady (1992) makes several arguments against it. I give a further, crushing argument: there are no correlations to observe between pre-theoretically available utterance types and observed situations. To identify the speech act types made, as opposed to purely phonological types, a T-recipient must engage in some psychological interpretation of the speaker, recognizing his utterance as an intentional act aimed at communication made by another thinking, socially interacting agent. Even for these types the needed correlations will be scarce. And once one has got that far, why not go further and assess the likely truth of what is asserted with a bit more of the same psychological assessment of the speaker?

2. The Failure of Brutely Humean Reductionism

Hume states that our basis for accepting as true received testimony is the correlation we find between testimony and fact (Hume 1975). But there are very different possible choices when we specify what exactly the testimonial events are with which one is to seek factual correlations.

Imagine a Martian arriving on earth and seeking to gain information about the earthly world via inference going beyond what she can observe for herself. How must she go about things to glean information from human activity? It is helpful to start with a contrast. The Martian (assume she has perceptual capacities that enable her to detect and discriminate a range of states of affairs through observation similar to those that humans can detect) might observe the activities of a type of gregarious bird. She might discover through observing correlations that birds of the flock will make a certain distinctive kind of loud call when a predator is within perceptual range of the bird (but perhaps not of other birds); and that on hearing this call all the birds simultaneously take off and fly away in a flock. The Martian, having observed this correlation, could exploit what she has observed to herself infer that a bird-predator is present whenever she hears the distinctive bird call. She might also infer that this call is a simple communicative signal that has evolved within bird ecology to function as a warning call and serves to help the flock avoid being killed and eaten by predators. But she does not need to make this inference about the function of the call in order to exploit its informational value to her. She would not even need to realize that the birds are animals in order to exploit and benefit from the information carried by the bird call.

Human language is not at all like this. Consider the ways of typing human linguistic utterances that would be available to the Martian before she realizes that these utterances are communicative signals: these would be confined to phonological types. At this point we must make a large assumption about Martian perceptual capacities in relation to sounds: that these are sufficiently similar to human sound-perception 'similarity spaces' so that the Martian can come to perceptually discriminate the same phonological types that underlie the differences between distinct human syntactic utterance types. So: allow that the Martian comes to be able to discriminate the relevant phonological types. What informational use can she make of these? Not much. She will be unable to find any significant degree of correlation between the discriminated phonological types and types of state of affairs observable to her. This is due to two combining factors.

First, ex hypothesi, the Martian can only access pre-interpretively available phonological types—ones that can be discriminated before recognizing that what is going on is communication effected by means of a very sophisticated sign-system that also needs contextual resolution to access the intended message. But, as the scientific study of human language has made clear, phonological types have only a poor level of correlation with intentional speech act types (Sperber and Wilson 1986, 2002). So even if speech act types correlate with types of observable states of affairs, phonological types will not as a result do so.

Second, in any case, specific speech act types will mostly fail to correlate much with specific states of affairs observable by the Martian for several cumulative reasons. As Sperber and Wilson (2002) observe, a shared language enables users to successfully convey information about an enormous and finely discriminated range of states of affairs. Most of these will not be observable within the shared environment of speaker and listener—why tell someone about something she can see for herself already? So even if there were correlations between speech act types and states of affairs, the latter would not typically be epistemically accessible to the Martian. But there will not be much by way of such correlations. For there to be a correlation between a particular specific utterance type and a particular type of state of affairs there must be many occurrences of the utterance type. But the prior probability in human language use for any particular speech act type (by this I mean a specific force and a particular content, not broad classes of contents) of it occurring is typically very low (see Sperber and Wilson 2002).

It might seem that the claim just made, that phonological types do not correlate with states of affairs observable in the context of utterance, cannot possibly be right, since such correlations are what make it possible for a human child to learn her first language. This is indeed so, and of course there are certain settings in which such correlations are available to be picked up on, namely those in which child language learning is being enabled by adult teachers who aptly hone their utterances to this end. So a Martian who happens upon a group of carers and small children in a park will find some such correlations—e.g. between utterances of 'duck', and the presence of ducks. Will other conversational settings afford similar opportunities? As observed above, most adult conversation is about absent states of affairs, and is hopefully not too repetitive! But would a Martian who found herself co-present with, say, a group of enthusiasts at a football match commenting on the game progressing in front of them find some correlations? She would not find reliably projectible correlations without coming to grasp syntactic structure and its significance—'kick' may correlate with observable kicks, but not when prefixed with 'not', or featuring in a conditional construction. Nor will she find projectible correlations until she comes to appreciate the sophisticated social nature of the activity and appreciates that some utterances are jokes, not to be taken literally, that many involve hyperbole or sarcasm, and so forth. In short, in fully sophisticated human linguistic exchange, even when the talk is about the observable scene, there will not be many correlations of observable states of affairs with syntactic types, since these can effect many different types of speech act, identifiable only by a process of contextual resolution dependent on shared background information and social knowledge between speaker and intended audience.

These considerations show that there is no possibility to exploit the institution of human language use to extract fine-grained informational value from utterances without appreciating what is going on—that this is a broadly cooperative system of intentional communication by intelligent agents with beliefs and purposes. Phonological types, as occurring in language use generally, simply do not carry much information in the Dretskean sense. They are not at all like bird calls or tree rings. (Dretske 1981). Neither do syntactic types, since what speech act they effect is highly dependent on context.[13]

So the brutely Humean strategy for underwriting our basis to trust testimony does not work if the utterance types are phonological or syntactic types. Are its prospects any better if we take the types to be fully fledged speech act types: asserting that P, for some specific proposition P? They are not.

Human linguistic communication is pervaded by context-sensitivity in determination of the specific message whose communication is intended by an assertoric linguistic act.

An utterance is a linguistically-coded piece of evidence, so that verbal comprehension involves an element of decoding. However, the decoded linguistic meaning is merely the starting point for an inferential process that results in the attribution of a speaker's meaning... the linguistic meaning recovered by decoding vastly underdetermines the speaker's meaning. There may be ambiguities and referential ambivalences to resolve, ellipses to interpret, and other indeterminacies of explicit content to deal with. There may be implicatures to identify, illocutionary

---

[13]   But surely, for instance, utterances of the syntactic type 'It is raining' carry the information that it is raining? No: this sentence constitutes an assertion that it is raining only when not embedded in some longer construction, e.g. negation or a conditional; and when not being uttered for vowel practice, to rehearse a play, purely to annoy someone, etc. And even when it is being used literally to make an assertion, saying it is raining somewhere, contextual resolution is needed to identify where it is said to be raining.

indeterminacies to resolve, metaphors and ironies to interpret. All this requires an appropriate set of contextual assumptions, which the hearer must also supply.   (Sperber and Wilson 2002, 3)

The exact psychological mechanisms that enable T-recipients to successfully identify what exactly the speaker is trying to tell them by his assertion are a matter for psycholinguistic study (see Sperber and Wilson 1986, 2002). It may be that fast-track interpretation of utterances is effected by 'quick and dirty' heuristics that get the correct answer in usual conditions.[14] But for these to get it right the net effect must be that of arriving at an interpretation of the utterance that conforms with the contextual constraints on conversation regarding fixation of reference, etc., and is consistent with a plausible folk-psychological interpretation of what the speaker is up to: why she is acting to tell one just this fact on this occasion. So fast-track heuristics should be, and in normal comprehension they are, correctible by higher-level conscious deliberation sensitive to overall psychological plausibility when they yield a counter-intuitive and incorrect result (Sperber and Wilson 2002).

All this means that in order to have the capacity successfully to identify what speech acts are made in ongoing language use, a recipient needs a lot more than just knowledge of the 'code' that links lexical items with conventionally associated semantic features. On any particular occasion she must have the required contextual background information and be able to deploy it correctly, and she must have a tacit grasp of the various norms and rules that govern conversation—such as the Gricean maxims of relevance, etc. (Grice 1989). On top of all this, or as part of it, she must have sufficient grasp of 'theory of mind', folk psychology, to be able to construct a mini-psychological theory of the speaker sufficient to identify what she is likely to be saying in the context, and given the beliefs and purposes ascribed in this mini-theory.

This fact already suggests IBE-reductionism as preferable to Humean reductionism. If one has to construct a psychological interpretation of the speaker, or stand ready to do so, simply to find out what she is saying, then why not do a bit more of the same, and evaluate whether what she says is likely to be true by reference to her motives and competences as posited in one's psychological mini-theory of her?

In fact, evaluating the likelihood that what a speaker tells one is true by the method of constructing a psychological mini-theory of her is the only method widely available to T-recipients. We saw that phonological and syntactic utterance types do not carry much information. But, as our considerations showed, neither do specific assertoric speech act types. A research project setting out to discover whether there are significant correlations between various specific types of assertions and certain states of affairs would struggle to get started due simply to the lack of incidence of the speech act type for most assertible contents. As already noted, most speech act types have very low probability of occurrence. True, there may be a limited range of assertoric speech acts

---

[14] For instance it is well confirmed that direction of gaze of the speaker's eyes is tracked by infants to identify what the gazer is attending to; and also that this is used as an automatic heuristic to assign reference to demonstratives in a conversation (see Apperly 2011).

that are made quite often, and one could establish how strongly these correlate with what is asserted. Assertions that it is raining would likely be amongst these. So the situation is not as hopeless as it was with phonological types. But—here one can agree with Coady's pioneering critique of Hume—the evidential base is not going to be enough to establish for most possible speech act types that their occurrence correlates with the states of affairs therein asserted to obtain (Coady 1992). I conclude that our second kind of Humean reductionism, which seeks to ground acceptance of testimony on some topic in the fact that in the past testimony on that topic has correlated with what is asserted, i.e. has been true, founders for lack of an adequate evidential base.

Maybe we have not yet found the best way to explicate Humean reductionism. The problem, we just saw, with specific assertoric utterance types is that there are not enough instances of them to provide an empirical base to look for correlations. Maybe we need to consider a less brutal version of broadly Humean reductionism. Suppose instead we take broad classes of speech act, such that these do occur frequently, and see how these do for accuracy. For instance: everything Josh tells me. Coady has already argued effectively against there being an adequate evidential basis for establishing the reliability of testimony through such broader classes of speech act types: maybe one can gather enough instances for a trial, but there are not enough cases where one can independently confirm their truth or falsity to carry it through. Also note that, in this case, one is not confirming correlation with any one type of state of affairs, the correlandum is just the artificial one: whatever the truth-condition of said utterance is.

In AG I maintained that establishing the general reliability or otherwise of testimony, or even of various broad classes of testimony, is irrelevant to the question: should I trust this speaker on this topic on this occasion? What matters for this is not the statistical reliability or otherwise of testimony as a general category,[15] but whether this speaker is honest and competent on her topic. Now a speaker's track record on a topic can play an evidential role here. Suppose that all the things that Josh has told me on many past occasions about how cars work and how to repair them have been subsequently independently confirmed to me as true. This surely can justify me in expecting what he next tells me about cars also to be accurate. So is there not some truth to broadly Humean reductionism? Not really. Josh's track record of accuracy about all things mechanical is projectible to future cases only insofar as it is evidence of relevant characteristics of Josh—that he is both honest and knowledgeable about all matters mechanical. Evidence from track record can help to empirically ground trust in a specific person's testimony about some topic. But this method does not contrast with grounding trust in testimony via an assessment of the speaker's psychology—it is a part of it. It is only because, and to the extent that, I infer Josh's past flawless track record to be grounded in his character and expertise that I have a basis to extrapolate it to future occasions on which he tells me things about cars. If instead I knew that in the

---

[15] As I maintained in AG, testimony in general, with no restriction on type of subject matter or speaker as regards its likely truth, is not an epistemically unitary projectible category at all.

past he got everything right because his boss told him it all, and this boss has now left, then I would have no basis to be confident in the accuracy of his future assertions about what is needed to repair my car. For all I know he is totally incompetent left on his own to make a diagnosis.

The first moral of this section is that one cannot extract much information at all out of human communicative language use without recognizing it as what it is, intentional sharing of knowledge—or sometimes, intentional deception. And there is a second moral.

Distinguishing brutely Humean reductionism from my explanationist local reductionism is crucial to avoid reductionism about testimony getting bad press, and fundamentalism thereby gaining spurious currency. Reductionism is sometimes accused of failing to recognize speakers as what they are, agents acting intentionally to communicate, instead treating their testimony merely as a particular kind of natural sign or evidence. This is true of brutely Humean reductionism, which we have shown to fail precisely due to this feature. It is the very opposite of the truth about the explanationist local reductionism I set out in Section 3.

My explanationist reductionism does not neglect, but absolutely highlights and invokes, the fact that there is a difference of metaphysical kind between the intentional communicative acts of testifiers, persons aiming to share their knowledge with others (or to mislead them), versus mere instruments, such as fuel gauges, and natural signs, such as the rings of a tree indicating its age. My account shows how this metaphysical difference between mere non-intentional 'natural meaning' versus Gricean (Grice 1957) 'non-natural' or, as I prefer to call it, 'agential meaning', translates into an epistemic difference of kind in the means by which we may come to know things through accepting the offered word of others, as opposed to relying on the non-volitional correlations found in nature and in instruments. It also puts the nature of the norm-governed socio-linguistic transaction between a teller and her audience, two people cooperatively interacting, at the centre of the epistemology. Humean reductionism might aptly be accused of failing to see others as persons offering their word, instead treating them merely as bits of the natural world that happen to make noises carrying information.[16] My explanationist reductionism rests on the very opposite. It is through one's appreciation of acts of intentional communication by other cooperating agents, as being just that, that one is able to learn from them; but in a manner that involves apt empirically based discrimination between trustworthy and untrustworthy such acts.[17]

My account shows how one can obtain information from others' assertoric utterances aimed at informing an intended audience, in an empirically backed, suitably

---

[16] I think much of the discussion as to whether testimony is 'evidence' involves confusion over this issue. My explanationist local reductionism sees apt doxastic response to testimony as based in evidence as to the speaker's trustworthiness, a broadly psychological feature of an intelligent agent. This does not entail neglecting his status as an intelligent agent, or of his act as intentional norm-governed interpersonal communication. It affords empirical grounds regarding his motives and capacities, giving a basis in his intentional psychology to accept his word on the matter; not to ignore the fact that his act is an intentional offering of his word.

[17] Moran (2006) explores this contrast.

discriminating fashion. One does this not by ignoring the speaker's nature as an intelligent cooperating agent, but on the contrary by appreciating and exploiting that fact to gain the information she offers one. My account also shows that treating speakers with respect as what they are, intelligent agents offering their word on some topic, does not entail that one must take their word for what they tell one as an empirically ungrounded act of faith. A T-recipient often has available to her empirical evidence of the speaker's epistemic competence on her topic and good will in making her honest utterance. This gives empirical basis to trust her regarding her telling.[18] Where there is such evidence of a speaker's relevant moral and intellectual virtue there is ground to trust her. Where the evidence is of ill will or incompetence there is not. Being distrustful of a particular speaker's utterance does not issue from a failure to recognize her status as an agent acting intentionally, but on the contrary from a recognition of this and of the specific motives that the epistemically best available psychological interpretation of her ascribes.

3. Explanationist Reductionism: the Inference from Trustworthiness

How should a T-recipient manage her doxastic response to testimony? Well, where does she start from? When responding to testimony from a trusted friend one may immediately and effortlessly understand her speech act, and equally effortlessly and immediately add the information shared into one's stock of beliefs, perhaps at once adjusting one's course of action in light of it. In such a situation one may not, as it were, bother to pause to form an explicit belief that, say, Sara has told one that the lecture venue has changed. One apprehends the content and force of her utterance, and this transmits smoothly and directly into the laying down of a belief in the told fact. But this instant smooth phenomenology of immediate belief formation is consistent with the thesis that T-beliefs are normatively inferential. The latter requires not that T-beliefs are formed by a real-time process of conscious deliberation about the likely trustworthiness of the speaker's testimony with a positive conclusion, before it is accepted; but only that supporting justificatory beliefs about this must be available to support the acceptance of the testimony, if challenged. I will also argue in Section 4 that such beliefs are present in the recipient's cognitive background, switching her into acceptance mode, when this kind of smooth laying down of T-belief in response to testimony is ongoing.

Central in the needed support of any newly acquired T-belief is this explanation of how the recipient came to know the fact in question: I was told it. So belief about the

---

[18] In Fricker (2016a) I give an analytic account of trust. This shows that the idea that trust, to be such, must be empirically ungrounded, resting only on epistemic faith, rests on a conceptual confusion. Rational trust is based in an empirically grounded estimation of the trustee as worthy of trust.

content and force of the speech act must always be able to be formed, even if, when there is no challenge, it is not actualized. I think it is very plausible that any normal adult forming belief in response to testimony is epistemically and cognitively placed to form such a belief. If she could not, then her own new belief would be absurd to her—she would have no explanation of how she came to believe it, cogent in the context of her general theory of the world and her own means of epistemic access to it; and it would be hard for the belief to survive this realization.[19]

Beliefs about what speech act was made—U-beliefs—are basic beliefs. When, as is usual, the hearer achieves effortless, phenomenally immediate comprehension of the content and force of the heard utterance, she enjoys a distinctive kind of phenomenal state: she hears its meaning, its truth-conditional content, in the utterance itself. She enjoys a quasi-perception of its meaning. Her U-belief, if formed, is grounded in this quasi-perception, and is justified in virtue of it (see Fricker 2003).

There is a tension to be resolved between the thesis that U-beliefs are basic beliefs, and the point noted earlier, that to arrive at correct understanding the hearer must deploy processes that utilize information about context in accordance with principles of relevance, etc., and that make sense of the utterance in relation to the speaker's postulated psychology. The tension is resolved by noting that, when understanding is achieved effortlessly, these processes are deployed automatically at a pre-attentional level, and the conscious result is a phenomenally 'given' quasi-perception of the computed content and force. When understanding cannot be so achieved, conscious puzzling out of, say, which item the speaker was referring to, deploying principles of conversational relevance, is carried out via conscious attention. This is consistent with U-beliefs being, in the usual case, basic beliefs; but defeasible by contrary information. Language use would not be possible without the effortless phenomenology that goes with the basicness of U-beliefs (see Fricker 2003).

In considering how a T-recipient should manage her doxastic response to received testimony we may thus assume she starts from the epistemic given: 'I have been told that P.' Our question is what normative route is available to her from this U-belief to justified acceptance of what she is told in that speech act. Or to put it non-metaphorically: what other beliefs are needed for justified acceptance of what she has been told?

The fundamentalist reply to this question is: nothing—the belief that she has been told is sufficient. The fundamentalist must acknowledge that a T-recipient who trusts what she is told is thereby rationally committed to holding also that the speaker was honest and competent on her topic.[20] But on the fundamentalist view she is entitled to

---

[19] It is important to remember that we are considering how new beliefs are formed and made sense of, epistemically supported, at that time. Of course we humans are built to then discard information about how we came by the belief once it is established.

[20] There are important details here which I gloss over due to space constraints in this discussion. All the recipient is rationally committed to by her acceptance of the testimony is that the speaker spoke truth. But the usual basis for this being so is that she spoke honestly and out of competence. See Fricker (2016b).

assume these facts about the speaker on no evidence, in the absence of defeaters. While these are commitments of her acceptance of the testimony, evidenced belief in them does not play a positive role in supporting her belief. The burden of this chapter is to show that we do not need to espouse fundamentalism to explain how justified T-beliefs are often available to speakers, and to offer an alternative explanationist account of how this is so. This is offered as preferable to fundamentalism on various grounds, including the fact that it coheres with Dynamic Explanationism, an independently plausible account of how human empirical beliefs are justified.[21]

In Section 2 we examined and rejected the view that an empirical basis to accept what she is told is often provided to a T-recipient by means of enumerative induction from observed past regularities: correlation of past instances of certain types of testimony with the obtaining of what is asserted to be so in that testimony. We acknowledged that a good track record on a certain topic by a particular testifier is an empirical basis sometimes available to trust fresh tellings by her on that topic. But, we noted, this regularity in accuracy of past testimony is only projectible to future instances when it serves as evidence of the speaker's relevant psychological and character traits: evidence that she is knowledgeable about that kind of matter and also honest. This being so, the fact that a good track record can justify reliance on someone's testimony does not count against but supports the general theory set out in the rest of this chapter.

I shall argue first that ordinary human T-recipients both can and do manage their receipt of testimony by empirically based assessment of speakers for honesty and competence about their topic; second, that this is effected by means of the construction of a fragment of a psychological interpretation of the speaker; and third, that this exercise in 'mind-reading' is an instance of IBE. In this section I argue first, that ordinary T-recipients possess the general cognitive capacities needed to aptly infer this sort of psychological interpretation of the speaker; second, that they often have access to enough suitable evidence to deploy this method to achieve a correct and well-evidenced explanation of why the speaker told one what she did. In Section 4 I review the broadly empirical case to think that T-recipients usually do in this manner monitor speakers for trustworthiness, and conclude there are good reasons to think they do. All this being so, the case of testimony supports Dynamic Explanationism, while Dynamic Explanationism in turn supports my explanationist reductionism about testimony.

My present account builds on ideas I first put forward in AG. The new feature is my emphasis on the nature of the speech act of telling, in particular the fact that it is governed by a social norm: Tell that P only if you know that P. This social fact, I argue, affords the primary basis to expect a speaker to tell one thing only if she is honest and competent—only if she knows what she is talking about. I now spell out how this works in detail.

---

[21] I do not attempt to offer a full refutation of fundamentalism in this chapter. This would involve expounding and countering various other persuasive arguments for fundamentalism.

It is useful to define a property of testifiers:

Trust (T, P, O, R):   A testifier T is trustworthy with respect to a proposition P on occasion O as regards an intended audience R just if she has this property: not easily would T tell R that P on O unless she knew that P.

A T-recipient has an epistemic route to justified acceptance of what she is told if she has a basis to know these two premises: 'T told me that P', and 'Not easily would T tell me that P on this occasion unless he knew P'. These two premises allow a highly probable inference to: 'T knows that P', from which it is an obvious entailment that: 'P'. Our definition of trustworthiness captures the property of the testifier invoked in the second key premise in this basis for trusting a speaker. Let us call this inference to the truth of what is told the *inference from trustworthiness*. The inference from trustworthiness is not the only possible basis on which one can infer to the truth of what is asserted by a speaker on an occasion. But it is the usual way, and the way such that testimony as a social institution for the transfer of knowledge has evolved, and persists, because it serves it (see Fricker 2015).[22]

In what follows I will show how deployment of everyday folk-psychological understanding and knowledge of others, including everyday social knowledge about the nature of speech acts and communicative practices, often renders a T-recipient placed to know the relevant instance of the key second premise, that the person telling her something is trustworthy with respect to her current telling. So she is evidentially equipped to make the inference from trustworthiness. The inference from trustworthiness to the truth of what is told is a case of IBE employing folk psychology. Trustworthiness is a character property of the testifier (albeit not always a stable one), and establishing that he has this property enables one to infer that the correct explanation of why he said what he did includes the fact that he knows what he asserted. In favourable cases this explanation is both correct and well evidenced for the recipient by her relevant social and personal knowledge. Thus we see how explanationism can supply a positive account of how justified belief can be and often is obtained from testimony.

A T-recipient *accepts* a speaker's testimony when she forms belief in what the speaker tells on his say-so—she takes his word for it that P. It is apt to describe this as trusting the speaker with respect to his utterance. Why so? Because, when I take your word for it that P, I rely on you to have told me what you know (not deceived me or erred through epistemic carelessness), in virtue of relevant epistemic and character virtues of yours. But trust is to be distinguished from what I call epistemic faith—belief in or reliance on some fact in the absence of evidence that it obtains. One should

---

[22] Trustworthiness as defined is not an esoteric concept. It features in ordinary T-recipients' evaluations of whom to trust—'T told me, and she would not do so unless she knew' is exactly the kind of justification someone might offer for having trusted T's testimony.

not trust a speaker with respect to his assertion unless one has good evidence of his trustworthiness regarding it.[23]

The thesis that T-recipients have the general cognitive capacities required to construct an explanation of a testifier's utterance invoking her motives and beliefs is not controversial. This is an exercise in 'mind-reading', ascribing mental states to others that explain their observed actions, and all cognitively normal humans can do this to some extent. I next illustrate one way in which we can get knowledge about the world from observing others' non-communicative actions via constructing explanations of them in terms of motives and beliefs; and will then contrast this with how one is able to learn from another's acts of testifying, telling one things.

We all—cognitively normal adult humans—are to some extent masters of 'mind-reading', the deployment of folk psychology (FP) to ascribe beliefs and motives to others. We have at least a tacit appreciation of how beliefs and desires produce action: intentional action is the pursuit of one's most pressing desire or goal, the apt action being selected in the light of one's beliefs. Schematically:

Rational Action Schema: If A wants that P, and believes by φ-ing she can bring about P, then A has a (defeasible) reason R to φ; and if R is undefeated, she will do so if she is able.

Tacitly appreciating this schema of rational action, by deploying it,[24] we often impute beliefs and desires to others. Consider this example:

COAT: I see you reach for your warm coat before leaving, and infer that you believe it is cold outside and wish to stay warm on your walk. In ascribing to you the belief that it is cold, and the desire to stay warm, to explain your action, I somehow infer to the correct and best-evidenced explanation of your action.[25]

Clearly an extensive background of facts about what are normal human desires and beliefs, and other folk-psychological, social, and environmental knowledge is required, to select this particular rationalization as the best explanation out of an enormous number of possibilities consistent with the Rational Action Schema. That schema itself is a deeply embedded platitude of folk psychology.

Having come to know by means of a cognitive process equivalent in its outcome to IBE that you believe that it is cold outside, I may further infer, again via an IBE regarding your mental states, that your belief is very probably knowledge. What best explains why you believe that it is cold outside? Answer: that you have by some

---

[23] See Fricker (2014, 2016a) for my analytic account of trust that explains how it relates to reliance and contrasts it with epistemic faith. See Fricker (2015b) for my full account of the speech act of telling and different possible responses to it.

[24] More cautiously: we deploy cognitive mechanisms by some means aptly sensitive to this schema in what results they produce. See Section 4.

[25] So it is a datum that our actual methods of 'mind-reading' approximate genuine IBE in their outcomes. In Section 4 I defend the thesis that these mechanisms are in fact tantamount to the deployment of IBE, and can be described as such.

means come to know this. A non-knowledgeable belief would need a very different, less likely kind of explanation—for instance that you had been tricked by a practical joker who wanted you to be overdressed for the weather.[26] But once I know that you know that P, I am epistemically equipped to come to know P myself, via an obvious entailment.[27] COAT instances a way in which one can very frequently acquire knowledge oneself, through ascribing mental states, including knowledge to others, via IBE from their observed actions to the explaining cause of this, their knowledgeable beliefs and desires.

The inference from trustworthiness to acceptance of what one is told goes via ascribing knowledge of what she tells to the speaker. So is the way one gets knowledge from accepting testimony just another instance of the mechanism described in COAT—inference to the correct explanation of another's action by reference to established lawlike generalizations of folk psychology? If so, then there would be nothing special about learning from what others tell us, as distinct from this broader pattern. The epistemology of testimony really would be just part of the epistemology of other minds with possibilities for learning about the non-mental world oneself no more than a beneficial side-effect. No: there is something special and distinctive about the act of telling, and correlatively of how we are able to learn from others' testimony, as I will now show. What is special about how one can come to know things about the non-mental world from observing others' actions of telling, addressed to oneself or another, is the detail of why precisely it is that the best explanation of S telling one that P is that she knows what she states.

We saw that brutely Humean reductionism does not work. Looking for correlations between utterances typed in ways available to an observer who does not realize these are intentional communicative actions is hopeless. Unlike our hypothesized mechanism that signals presence of a predator among a species of birds, there are no such correlations to be found.[28]

Gaining information from what others tell us requires appreciating that language use is intentional action and inferring to the best explanation of this. But, you object, this is already true of COAT! Indeed it is. But, as will be revealed below, there is a further key distinction between COAT and this contrasting case:

COAT-TELLS: I am about to leave our warm apartment and go out into the night to walk to a friend's house to deliver something she needs. You say to me: 'It's cold out tonight, you had better wear your coat.'

---

[26] Of course there are other less outlandish explanations of a non-knowledgeable belief: you misheard the weather forecast. But knowledge, for such everyday matters, is the default explanation of belief.

[27] Actually one may move directly to ascribing knowledge, not going there via the route of ascribing belief. There is a lot of evidence that knowledge is a more easily dealt with concept than belief, and the concept is acquired earlier (see Nagel 2013). This is helpful, not harmful, to my overall argument.

[28] Of course there will be some—for instance between raised volume and pitch, and emotional arousal of utterer. My thesis is that this method has no chance at all of tapping into the rich informational content that language as intentional communication in fact offers up to the fully cognitively equipped receiver, who understands what is being offered.

Putting on one's coat is an action. But it is not, in itself, an action made with communicative intent.[29] Telling is an action of a very special type. It is, qua type, an intentional act aimed at communication. Moreover, unlike one-off non-conventionally mediated acts of Gricean communication, as an act of communication it is a folk-socio-linguistic type regulated by a social norm.[30] This last feature in particular gives rise to and allows this phenomenon:

TELLING: Someone who tells that P to an intended recipient R offers to R her word that P; she offers to R the right to believe P on her say-so.

One can marry someone by uttering suitable words only when the needed surrounding institutions are in place. Similarly, one can effect a performance with the upshot that one has offered one's word that P to an intended recipient, thereby enabling her rightfully to rely on one's word as to P, only when the needed institutions are in place. In the case of telling, the main institution enabling such performances is the existence of the socio-linguistic speech act type of telling in the community of the shared language of teller and recipient. Telling is governed by this social norm:

T-Norm: One must: tell someone that P only if one is epistemically so placed as to properly guarantee to them the truth of P.

In telling that P one gives one's word that P to one's intended audience, enabling them properly to rely on it. It is coeval with this fact that telling, in a community that contains this socio-linguistic type, is governed by the T-norm. That the T-norm is a social norm in a community C does not require explicit knowledge of the norm by participants in C. What matters is that it is tacitly appreciated, so that sensitivity to it influences their responses to tellings by others and controls their own tellings. For tellings in all but some unusual contexts[31] this T-norm equates with the norm:

K-Norm: One must: tell someone that P only if one oneself knows that P.[32]

Exploiting this institution of telling, a teller is able intentionally to undertake a performance by means of which she renders herself responsible, to her intended audience, for the truth of what she asserts in her utterance to be so.[33] Telling is a performative act-type which, like promising, creates an 'ought' from an 'is'. The occurrence of an act of telling adds to the normative landscape. It has the upshot that the teller binds herself,

---

[29] Of course putting one's coat on could be the pre-arranged signal between myself and my co-conspirator to set off the fire alarm. Or it could be the mechanism of an entirely one-off non-conventionally mediated act of Gricean communication.

[30] Fricker (2012) argues for an important contrast between these non-linguistically mediated acts, versus tellings. On social norms see (Bicchieri 2006, Graham 2015).

[31] See Williamson (2000), Goldberg (2015). Fricker (2015a) argues that the K-Norm holds because it provides one fix on our standards for knowing.

[32] Contextualism about standards for knowing messes this up. This is a powerful argument against it. Our T-practices are inconsistent with it, which shows it to be false as a descriptive account of our ordinary language concept 'knows'.

[33] See Fricker (2015a) for a fuller account.

giving her word that P to her intended audience. By her act of telling she renders herself responsible for the truth of P to her audience. The 'bottom line' of this responsibility is that she can be criticized and complained to if what she tells is not something she can properly vouch for the truth of—that is, she does not know it. In some circumstances more serious sanctions result from norm-violating tellings, the offence of perjury being the strongest.

We have articulated this dual fact about telling: it is intentional communication; and it is governed by the K-Norm, so that the teller is answerable to her audience for the truth of what she tells. This dual fact is essential to the specifics of a type of explanation, that is very often correct, of why someone told something to someone—an explanation that is, moreover, often epistemically accessible from the recipient's standpoint. Consider again our example COAT. Suppose it is not cold tonight, and you never thought it was, but you needed to take your coat with you because it was likely to be cold tomorrow, and so you just wore it as the easiest way to carry it even though you knew you would be too hot. So I made a mistake in inferring from your putting on your coat to its being cold, via the hypothesis that you knew it was cold. This was bad luck for me. But I have no basis to complain to you. You did not mislead me by putting your coat on in my presence. You had no communicative intentions at all in doing so. I made an over-ambitious inference, and have only myself to take to task for it.

In contrast if, as in COAT-TELLS, you say to me: 'It is going to be cold tonight, you had better wear your coat', and it is not cold at all—then I have a basis for complaint. Certainly I can complain to you, and about you, if you lied. But I can also complain if you were sincere, but had jumped to a conclusion about the temperature on inadequate evidence. Part of conscientious conforming to the K-Norm for telling is that one must be epistemically conscientious—at least, one must be so as regards the things one tells to others. One must do one's best to keep an accurate check on what one does and does not know, and only tell people things one knows. It is not just lying, but epistemic carelessness, that the social K-Norm sanctions.

In both COAT and COAT-TELLS folk-psychological knowledge is deployed to ascribe motives and knowledge that explain action. But a further feature sets COAT-TELLS apart from COAT and other cases where one infers from someone's observed action to an item of their knowledge, but where this action was not aimed at communication. The further feature is our dual fact about telling: it is an intentional communicative act governed by the K-Norm, and hence with the normative upshot noted—the speaker renders herself responsible for the truth of what she asserted to her intended audience. What difference does this dual fact make to the IBE that is sometimes available from the fact that A has told that P to the truth of P?

As we have seen, a T-recipient will typically justify her belief in what she is told via the inference from trustworthiness: 'S told me that P, and not easily would he do so unless he knew that P; so P.' How is a T-recipient often able to know, have adequate grounds to believe, that a speaker is trustworthy with respect to his utterance? When someone tells one something one can deploy one's psychological knowledge—both general and about that particular person—to form an explanation of why he was motivated to do so. There will be specifics in particular cases. There are also general precepts, such as the Gricean norms of conversation (Grice 1989). And even if one does not have a positive theory about precisely why someone was motivated to tell one something one can, even so, often know that he is very probably trustworthy as regards his utterance. One can often know this due to the holding of the K-Norm.

Tellings are regulated by the social norm 'tell that P only if you know that P', and so it features as the default stance in one's explanation of a speaker's telling that he satisfies this norm. One needs only to rule out the various ways in which the norm may be violated to have grounds to expect the speaker to be trustworthy as regards his utterance. This itself requires doing some psychology. The epistemically vigilant T-recipient must construct a fragment of an explanation of why the speaker made his utterance, enough to be sure that whatever in fact led him to say what he did, the explanation rules out the speaker's violating the K-Norm. That is to say it rules out epistemic-cum-assertoric carelessness,[34] bad epistemic luck, or a deceptive intent. Whatever the full positive explanation is, so long as these three ways in which the K-Norm may be violated are ruled out by one's theory of the speaker and his circumstances, then it is safe to infer trustworthiness and to accept what one is told.

In short: because the K-Norm holds as a social norm governing tellings, one can safely infer that a particular telling is trustworthy, so long as the several particular manners of violation of it have been ruled out. Thus the holding of the K-Norm as a social norm sanctioning non-knowledgeable tellings plays an essential role in the empirical basis often available to a T-recipient to expect the speaker to be trustworthy as regards his utterance.

This point is reinforced by a contrast. Suppose someone utters the words: 'I wonder whether P.' Does this type of utterance, as a rule, give its audience any basis to accept as true P? Of course it does not. Expressing one's wondering whether P is not subject to the K-Norm, nor any similar epistemic norm. So the fact that someone has expressed their wondering about P has no epistemic significance. There is normally no reason at all to expect someone S to have the property: not easily would S wonder whether P, unless P. But when someone tells that P, this does have epistemic significance because the sanction of the K-Norm means that, special circumstances such as deceptive intent or chronic epistemic carelessness aside, one does not easily tell that P unless one knows that P.

A certain regularity in behaviour that is ongoing in a community C reflects a social norm obtaining in C just if violations of this regularity are sanctioned by participants, and these sanctions maintain the regularity. Sanctions include the external sanctions of disapproval and sometimes further punishments, plus the internal sanction of

---

[34] This covers both carelessness in formation of beliefs, leading one to fail to know that one does not know something, and carelessness in speaking, failing to regulate one's tellings to state only what one believes oneself to know. This includes what is now sometimes unattractively described as 'bullshitting'.

conscience. A social norm does not exist in C unless the sanctions are sufficiently motivating to ensure conformity to the norm, except in situations where there is an even stronger motivation to violate it.[35]

This is why conformity to the K-Norm is the default stance in explaining a teller's utterance. This default stance is overridden, in inferring the best explanation of the speaker's utterance, when circumstances indicate he has a motive to lie, or indicate he may have been misinformed or misled, or when his reputation based on track record is for epistemic carelessness. There are many likely human motivations to lie, on many occasions, and all of us are subject to bad epistemic luck, or lack of vigilance, from time to time. So the default stance of trustworthiness needs empirical work to vindicate it, and will often be overridden. But the fact that in telling one that P a speaker assumes to one responsibility for the truth of P, and that there is a social norm sanctioning doing this when one is not properly placed for it, gives prima facie reason to expect a speaker—so long as he has no motive in the circumstances to deceive—to be trustworthy, allowing the inference from trustworthiness to the truth of what he tells.

We have set out how our dual fact about the speech act of telling underwrites the possibility of a distinctive kind of inference to the best explanation from the fact of being told something to the truth of what one is told, via the ascription of trustworthiness to the speaker. Moreover this special kind of explanation of the speaker's action that exists, when it is a telling, is often epistemically accessible to a recipient. Thus she is placed to know, and to supply, this justification for accepting what she was told: 'S told me that P, and he would not have done so unless he knew.'

The reader I hope has been convinced of the first part of my thesis here: that there is a distinct style of explanation for tellings due to the dual fact. But more needs to be said to vindicate my claim that such an inference to the truth of what was told via the hypothesis of trustworthiness is often epistemically accessible to T-recipients. I have explained above why the default stance is for trustworthiness—it can be ascribed, in the absence of evidence indicating a source of norm violation. But given the many human motives and risks for violation, there needs to be an empirical basis for detecting motives and tendencies for norm violation, and for checking that these trust-defeating circumstances are absent, when they are. I have written elsewhere about the empirical grounds often available to a T-recipient to assess a speaker for trustworthiness, and space prevents more than briefly mentioning some resources available. I suggested in AG that there is a general default in favour of sincerity, as part of the principles that constrain psychological ascription generally; and, for reports of everyday facts that the speaker has witnessed, a presumption of competence. These presumptions are not a priori, but are deeply embedded principles of folk psychology. (Of course these would not be so, except in the context of our dual fact about telling. There is no default in favour of competence for wonderings.) Often one does not need personal knowledge

---

[35] Compare: it is against the law in the UK to drive a car at above 70mph. But this law is widely disregarded by the vast majority of drivers with no disapproval from other drivers. It is the law, but it is not a social norm.

of the individual speaker, general social knowledge about social roles and associated competences suffices—the receptionist in one's hotel can be expected to know about what the latest checkout time is, but not about how to make the best soufflé au Grand Marnier (see Fricker 2002).

I finish this section by dealing with two likely objections. First, has the theory just given not proved too much? Surely our dual fact about telling gives the fundamentalist exactly what she wants—an argument showing it is epistemically permissible to trust what one is told without any evidence of trustworthiness, so long as this presumption is not defeated by other beliefs one has?

This worry misunderstands the sense in which our dual fact licenses a default stance in favour of trustworthiness. This is not an entitlement to believe what one is told without engaging one's cognitive resources in assessment of the speaker for trustworthiness, but instead a principle that constrains this project. Moreover the default stance for trustworthiness is empirically grounded in the nature of telling as socio-linguistic act. My thesis is that appreciation of the nature of telling informs a recipient's aptly discriminating assessment. It is not that the nature of telling underwrites a special epistemic principle dispensing a T-recipient from any requirement to assess the speaker as part of the process of accepting what she tells or from any need to appreciate the nature of telling.

The second objection pushes in the other direction: just because there is a norm in force in a community, this does not entail that one can presume it is conformed to as one's default interpretative stance. The remarks made above concerning the contrast between a social norm and a law provide the basis to rebut this objection. A social norm does not exist unless its sanctions are largely effective—unless they have psychological motivating force influencing what participants do, and how they react to violators. This does not entirely see off the objection, since a social norm will be violated when there are strong enough motives for violation, these overcoming the force of the internal and external sanctions against violation. So the default stance in favour of trustworthiness is an empirically backed presumption in interpretation that flows not just from the existence of the norm, but from that together with contingencies in a particular community about the effectiveness of sanctions and the fact that violaters are likely to be found out. This theme is pursued in Section 4.

I have shown how the epistemic circumstances of a recipient of knowledgeable testimony often equip her to infer to the correct explanation of the testifier's utterance, namely that he spoke from knowledge out of a desire to share what he knows. The recipient is able to conclude this because she has evidence indicating the trustworthiness of the speaker as regards his utterance. Such an inference can be made from the available evidential base by IBE deploying general folk-psychological and socio-linguistic knowledge. So I have shown that my explanationist local reductionist account satisfies our constraint EJTBPoss. In Section 4 I consider whether it also satisfies EJTBActual: do ordinary T-recipients usually invest the cognitive resources to make an assessment of testifier trustworthiness and, if they do so, is the means they employ IBE?

4. Defending EJTBActual

I have shown how a T-recipient often has available to her the evidential resources to estimate the trustworthiness as regards her telling of a speaker, through inference to the best explanation of the speaker's utterance by reference to her motives and capacities. So justified testimonial belief is often available to a T-recipient within the constraints of DE-style local reductionism. But to show that Dynamic Explanationism explains how actual human T-recipients in fact often acquire justified T-beliefs, it must also be shown that actual T-recipients manage their receipt of testimony using something like the local reductionist strategy outlined. This breaks down into two questions. First, do ordinary T-recipients usually invest the cognitive resources to make an assessment of testifier trustworthiness? Second, if indeed they do so, is the means they employ IBE?

This is an empirical matter. Here I merely make some remarks in light of which EJTBActual is plausible given the requirements of explanationist local reductionism. Remember that, unlike EJTBPoss, EJTBActual is not a tightly binding constraint. It is not a datum that actual human practices of testimony reception, however empirical research reveals them to be, are epistemically faultless. This is extremely unlikely. Causal empiricism suggests there are endemic human tendencies to gullibility. Epistemology should show this up as bad policy in belief formation needing improvement, not rubber-stamp it as epistemically fine. But my optimism inclines to the view that ordinary T-recipients apply an effective filter to the testimony they receive in many circumstances. I first sketch a benign picture of human T-reception practices, and then offer an argument indicating things are approximately like that.

There is a huge contrast in how human T-recipients respond to testimony, according to the circumstances. At one extreme one just 'drinks in' what is being told to one and lays down belief in it. One is switched into this receptive mode when, for instance, one is listening to a BBC news broadcast on a factual topic, or listening to an account from a trusted friend of how she has just spent her day. At the other extreme one responds to testimony by thinking explicitly about whether the source is to be trusted. In these conscious reflections one will articulate and bring to bear considerations of motive and competence of the kind discussed in Section 3 to answer the questions: Why is he telling me this? What is he up to? Is he sincere? Does he 'know what he is talking about'? Deliberations of this kind are exercises in IBE deploying folk psychology, everyday social knowledge, and ad personam knowledge about the particular speaker.

But what about when one is unreflectively drinking in what one is told? Such phenomena need not and typically do not instance pure gullibility—that is, a recipient who would believe absolutely anything she was told, whomsoever it was that told her it. There are two ways in which drinking-in can depart from this. First, the recipient can be monitoring the T-inputs, so that if the speaker were to say something implausible, this would trigger a critical reaction. This monitoring is switched on all the time, although it produces a double-take only if something unlikely interrupts the smooth acceptance of testimonial input to lay down belief (see AG). Second, the recipient may be in drinking-in mode in the first place only because this is a testifier she has good empirical basis to trust. Background beliefs about the speaker—often not about him personally, but about his social position and hence stereotypical competences and motives—function as enabling or inhibiting switches that determine the flowpath that the T-inputs take into the recipient's cognitive organization.[36]

Fine, it may be objected, allow that this sort of process occurs; it does not vindicate Dynamic Explanationism because this pre-attentional T-filter process, while broadly adaptive, does not instance IBE. Reinforcing this worry are some results from cognitive psychology about our human capacities for mind-reading. It seems we are built sometimes to use 'quick and dirty' heuristics for ascribing mental states to others. These, in the situations in which they are used, approximate what genuine IBE mind-reading would result in, but via a cognitively much less costly shortcut. One example: ascription of what is seen by another, and hence what they know, is ascribed via a simple heuristic of extrapolation from direction of gaze.[37] If these sorts of heuristics are how humans actually assess testifiers for whether they know what they are talking about, this is surely a problem for the IBE account of Section 3—how does it fit with EJTBActual? I think the tension here can be eased. I emphasized earlier that portraying a belief as normatively inferential is a matter of what justifying premises a believer must be able to offer in defence of her belief, not a claim about real-time processes of belief formation. This is why it is not a problem for my explanationist local reductionism if we often ascribe mental states to others—including, on occasion, the supposed fact that a speaker knows what she is talking about—via a simple quasi-behaviourist heuristic: a rule that takes one from a certain type of observable situation directly to a particular mental state ascription. What matters for the IBE account of testimony management is, first, that T-recipients are epistemically sensitive in their belief formation to the considerations that would feature as evidence in an inference to the best explanation of the speaker's telling. But the active-monitoring proposal above achieves this. Second, that T-recipients are able, if the matter is raised, to construct a fully fledged folk-psychological inference to the best explanation of the speaker's utterance. And this they clearly can do, since they actually do it on some occasions.

That is the benign picture. Why should we think it actual? Here I argue by an appeal to authority. Sperber et al. (2010) articulate a strongly persuasive case that human T-recipients generally engage in effective monitoring of speakers for trustworthiness. There is good reason to think they must do so, they argue, since first, while it is almost always in recipients' interest for speakers to speak knowledgeably, speakers often have

---

[36] These sub-attentional filtering processes are approximations to the evidentially most justified conclusion, and are not always accurate. They can lead to systematic biases, which may have socially negative side-effects, including inducing systemic testimonial injustice. This phenomenon is powerfully highlighted and theorized in M. Fricker (2007).

[37] See Apperly (2011).

motives to deceive. But testimony would not persist as a social institution unless it was useful. It would not be useful if recipients were deceived a lot. So recipients must have, and employ, effective methods to detect untrustworthy testimony. Second, since recipients are often able to detect the falsity of false testimony, this successful 'epistemic vigilance' provides a motive for speakers to be trustworthy—they are likely to be caught, and sanctioned, if they are not. This postage-stamp resume does not do justice to the full case offered by Sperber et al., who provide a sort of transcendental argument from social evolution for the effectiveness of testimonial monitoring by human recipients. There I must leave the matter on this occasion.[38]

5. Two Objections Defused

Two objections are salient regarding the account of human justified T-beliefs that I have offered, which is both normative and descriptive. First, that it is descriptively too sanguine about actual humans' management of received testimony. Second, that it is too restrictive as to what is required for someone to gain knowledge from testimony. I finish by briefly explaining and then doing something to defuse these objections.

The account of Section 4, which claims humans generally operate an aptly discriminating[39] response to received testimony, is at odds with this fact: on what we may call theoretical matters, as opposed to easily established observational facts, there is within human culture the wide circulation of different and mutually inconsistent views. At most one of such contrary opinions can be true, yet more than one is widely circulating, each believed by large groups of people. Almost all of these people hold these theoretical beliefs—whether about economics, about religion, about climate change, about politics, etc.—on the basis of testimony. This entails that, on theoretical topics, some testimony management is not aptly discriminating. Here I can only note a couple of pointers in the direction an explanation of this phenomenon must take.

The account of the previous sections certainly needs to be nuanced to deal with this recalcitrant datum. Either the idea that T-recipients engage in effective monitoring at all must go; or alternatively—the more plausible option I suggest—we must explain why monitoring is more effective in some subject domains than in others. As already noted, T-recipients' ability to screen out false testimony is relative to a domain of likely T-inputs. So we must ask: is there more dishonest, or more incompetent testimony in theoretical domains? Motives for deception may be different, but seem unlikely to be more prevalent in theoretical domains: there are many likely motives for deception about personal and everyday matters. But it may be easier to effect a deception

---

[38] See however Michaelian (2013) and Shieber (2015) for a more pessimistic view. Fricker (2016b) examines how what is required for effective filtering by a T-recipient depends on her testimonial environment, which can also be tuned and constrained to be more or less benevolent.

[39] Though see note 37.

when the testimony is received through, say, the internet rather than in a face-to-face interaction. And deception on difficult-to-establish theoretical topics is less likely to be subsequently found out through independent discovery of the falsity of what is testified to. Insofar as the fear of discovery is a motive for honesty, it may operate less powerfully on theoretical topics. The threat of sanction is less, and hence deception may be more endemic.

But I think the main contrast is more likely to be regarding competence: both its possession by testifiers, and the resources effectively to assess for it. As with deceptive intent, honest but frequently mistaken testifiers on everyday topics will tend to get found out, often very quickly. ('You told me we had enough petrol left to get to Scotland without filling up the tank!') They will lose their credibility and not be trusted in the future; and/or learn to be more epistemically conscientious before they give their word to others on some practically important fact. But it is much harder to conclusively establish that, say, a socially recognized expert on the topic of global warming has empirically unjustified false beliefs. On most important large theoretical issues like this, the layperson is confronted with a bewildering array of diverse opinions from various self-styled or socially certified experts. There are two cumulative problems that obstruct the lay T-recipient in establishing who is right in these sorts of matters: the impossibility of independent checking on which proffered theory is correct, plus the difficulty of effective monitoring for competence—defined as likely correctness of opinion on the topic.

Necessarily one does not check directly for competence, if this means truth of the proffered testimony. If one knew, independently of the testimony, whether what is told is true, one would not need to rely on the testimony regarding it. I have defined speaker trustworthiness, which incorporates competence, as the speaker's possession of this modal property: 'Not easily would she testify that P, unless she knew that P.' This property of a speaker is also not one that one can check for directly; one must assess for it via what are hopefully reliable proxies for its presence. Very often the best available indicators of competence are social certifications of an individual as qualified and so competent in a certain subject. For instance, one will trust a qualified doctor to diagnose the cause of one's symptoms and recommend a suitable treatment. This suggests a general explanation for why false testimony from incompetent sources is frequently offered, and its content circulates accepted and unchallenged, in certain kinds of areas, such as political and religious belief. First, in these areas there are not reliable observable proxies for genuine competence; second, the falsity of testimony in these areas is unlikely to be discovered. (This unlikelihood of discovery is time-relative of course: future generations will find out, to their cost, whether human activities are causing global warming with disastrous consequences for the planet!) I tentatively conclude that the datum that contrary testimony on many theoretical topics circulates widely does not put in doubt my account of effective monitoring by T-recipients of testimony on everyday topics, where competence is more able to be assessed, and lying is likely to be quickly discovered.

Our first objection was that our account is too optimistic about the effectiveness of human filtering out of non-knowledgeable testimony. The second objection pushes in the other direction: that it is too stringent in what it requires of T-recipients, in order for them to get knowledge from accepting what they are told. An ideal T-filtering process would be tuned to 'let in' all and only tellings made by knowledgeable testifiers. This ideal is clearly unattainable. How nearly it can be approximated is relative to the domain of likely testifyings a recipient must be equipped to respond aptly to—that is, with apt sensitivity to whether the offered testimony is knowledgeable. What, as a recipient, one would aspire to is to tune one's filtering so as to let in only a few false testimonies while not sacrificing too many opportunities to gain knowledge. We can think of an account of conditions for justified acceptance of offered testimony as articulating what makes for such an apt policy of testimony reception, the nearest approximation achievable to the ideal T-filter. In this discussion I have approached the issue within an internalist framework, looking at the type of grounds a recipient should have when she accepts what she is told. But the idea of an aptly-tuned policy of T-reception can be spelled out also in externalist terms.

So: one has a justified T-belief just if that belief is formed via an apt general policy for T-reception. How does this relate to gaining knowledge from testimony? On the classic view justification is necessary for knowledge. But we can divorce the two distinct ideas of having, and exercising, an apt general policy, and gaining knowledge. Maria Lasonen-Aarnio has made this theoretically liberating suggestion in relation to understanding whether knowledge can survive the acquisition of misleading defeaters (Lasonen-Aarnio 2009). And I have elsewhere suggested we may need to do something like this to explain knowledge from testimony (Fricker 2016b). There is a strong intuition that if someone tells one something, expressing her knowledge, and one believes her, one gets knowledge oneself thereby. This intuition is one driver of fundamentalism. In Fricker (2016b) I suggest one gets knowledge if someone tells one something, expressing his knowledge, and one believes him with epistemic propriety, even if one's belief fails a modal reliability test—a non-knowledgeable speaker might easily have told one something similar but false, and one would have believed him also. One gains knowledge, one gets the goods, even though one might easily have been fooled instead. So I there suggested that modal reliability may be too strong a condition on testimonial knowledge. Our review here suggests the further thought: perhaps justification, theorized as an apt general policy for T-reception, is not required to obtain knowledge from testimony on a particular occasion. Even if you should have been more wary in whom you believed about some topic, if you through luck hit upon a knowledgeable testifier, then you gained knowledge from him. If this is right, then while I have argued against fundamentalism about testimonial justification—the conditions under which one should accept what one is told as part of an apt general policy—maybe there is room for fundamentalism about how knowledge can be acquired from testimony. One should seek to avoid gullibility; but even the gullible may get knowledge when they hit a lucky streak. The fact that one would wrongly accept non-knowledgeable testimony does not mean one can never acquire knowledge from testimony, even when the speaker one believes with undiscriminating trust is a benevolent and highly competent expert expressing her knowledge.

References

Apperly, I. (2011). *Mindreaders: The Cognitive Basis of Theory of Mind*. Hove, Psychology Press.

Audi, R. (1998). *Epistemology: A Contemporary Introduction to the Theory of Knowledge*. London, Routledge.

Bicchieri, C. (2006). *The Grammar of Society: The Nature and Dynamics of Social Norms*. Cambridge, Cambridge University Press.

Burge, T. (1993). 'Content Preservation'. *Philosophical Review* 102(4): 457–88.

Coady, C. A. J. (1992). *Testimony: A Philosophical Study*. Oxford, Clarendon Press.

Dretske, F. (1981). *Knowledge and the Flow of Information*. Oxford, Blackwell.

Fricker, E. (1994). 'Against Gullibility'. In *Knowing from Words: Western and Indian Philosophical Analysis of Understanding and Testimony*, eds. B. K. Matilal and A. Chakrabarti. Dordrecht, Kluwer: 125–61.

Fricker, E. (2002). 'Trusting Others in the Sciences: A Priori or Empirical Warrant?' *Studies in History and Philosophy of Science* 33(2): 373–83.

Fricker, E. (2003). 'Understanding and Knowledge of What Is Said'. In *Epistemology of Language*, ed. A. Barber. Oxford, Oxford University Press: 325–66.

Fricker, E. (2006). 'Second-Hand Knowledge'. *Philosophy and Phenomenological Research* 73(3): 592–681.

Fricker, E. (2012). 'Stating and Insinuating'. *Aristotelian Society Supplementary Volume* 86(1): 61–94.

Fricker, E. (2014). 'Epistemic Trust in Oneself and Others: An Argument from Analogy'. In *Religious Faith and Intellectual Virtue*, ed. T. C. O'Connor. Oxford, Oxford University Press: 174–203.

Fricker, E. (2015a). 'How to Make Invidious Distinctions amongst Reliable Testifiers'. *Episteme* 12: 173–202.

Fricker, E. (2015b). 'Know First, Tell Later: The Truth about Craig on Knowledge'. In *Epistemic Evaluation*, ed. J. H. Greco. Oxford, Oxford University Press: 46–86.

Fricker, E. (2016a). 'Doing (Better) What Comes Naturally: Zagzebski on Rationality and Epistemic Self-Trust'. *Episteme* 13(2): 151–66.

Fricker, E. (2016b). 'Unreliable Testimony'. In *Alvin Goldman and His Critics*, ed. H. M. Kornblith. Oxford, Blackwell.

Fricker, M. (2007). *Epistemic Injustice: Power and the Ethics of Knowing*. Oxford, Oxford University Press.

Goldberg, S. (2015). *Assertion: On the Philosophical Significance of Assertoric Speech*. Oxford, Oxford University Press.

Graham, P. (2006). 'Liberal Fundamentalism and Its Rivals'. In *The Epistemology of Testimony*, eds. J. Lackey and E. Sosa. Oxford, Clarendon Press: 93–115.

Graham, P. (2015). 'Epistemic Normativity and Social Norms'. In *Epistemic Evaluation: Purposeful Epistemology*, ed. J. H. Greco. Oxford, Oxford University Press: 247–73.

Grice, P. (1957). 'Meaning'. *Philosophical Review* 66: 337–88.

Grice, P. (1989). 'Logic and Conversation'. In *Studies in the Way of Words*. Cambridge, MA, Harvard University Press.

Harman, G. (1965). 'The Inference to the Best Explanation'. *Philosophical Review* 74: 88–95.

Harman, G. (1986). *Change in View: Principles of Reasoning*. Cambridge, MA, MIT Press.

Hinchman, E. (2005). 'Telling as Inviting to Trust'. *Philosophy and Phenomenological Research* 70(3): 562–87.

Hume, D. (1975). 'An Enquiry Concerning Human Understanding'. In *Enquiries Concerning Human Understanding and Concerning the Principles of Morals*, eds. P. H. Niddich and L. A. Selby-Bigge. Oxford, Clarendon Press.

Lasonen-Aarnio, M. (2009). *Indefeasible Knowledge*. Oxford, Oxford University Press.

Lipton, P. (1991). *Inference to the Best Explanation*. London, Routledge.

Lipton, P. (1998). 'The Epistemology of Testimony'. *Studies in History and Philosophy of Science* 29(1): 1–31.

Lycan, W. (2002). 'Explanation and Epistemology'. In *The Oxford Handbook of Epistemology*, ed. P. K. Moser. Oxford, Oxford University Press.

McMyler, B. (2011). *Testimony, Trust, and Authority*. Oxford, Oxford University Press.

Michaelian, K. (2013). 'The Evolution of Testimony: Receiver Vigilance, Speaker Honesty and the Reliability of Communication'. *Episteme* 10(1): 37–59.

Moran, R. (2006). 'Getting Told and Being Believed'. In *The Epistemology of Testimony*, eds. J. Lackey and E. Sosa. Oxford, Clarendon Press.

Nagel, J. (2013). 'Knowledge as a Mental State'. In *Oxford Studies in Epistemology*, Volume 4, eds. T. Gendler and J. Hawthorne. Oxford, Oxford University Press.

Shieber, J. (2015). *Testimony: A Philosophical Introduction*. New York, Routledge.

Sperber, D. and D. Wilson (1986). *Relevance: Communication and Cognition*. Oxford, Blackwell.

Sperber, D. and D. Wilson (2002). 'Pragmatics, Modularity and Mind-Reading'. *Mind and Language* 17: 3–23.

Sperber, D., C. Heintz, O. Mascaro, H. Mercier, G. Origgi, and D. Wilson (2010). 'Epistemic Vigilance'. *Mind and Language* 25(4): 359–93.

Welbourne, M. (1986). *The Community of Knowledge*. Aberdeen, Aberdeen University Press.

Williamson, T. (2000). *Knowledge and Its Limits*. Oxford, Oxford University Press.

Index

Index of Names